Extracellular Fine Particles

Yoshinobu Baba · Rikinari Hanayama ·
Hidetaka Akita · Takao Yasui
Editors

Extracellular Fine Particles

Springer

Editors
Yoshinobu Baba
Nagoya University
Nagoya, Japan

Rikinari Hanayama
Kanazawa University
Kanazawa, Japan

Hidetaka Akita
Tohoku University
Sendai, Japan

Takao Yasui
Institute of Science Tokyo
Yokohama, Japan

ISBN 978-981-97-7069-4 ISBN 978-981-97-7067-0 (eBook)
https://doi.org/10.1007/978-981-97-7067-0

This work was supported by Japan Science and Technology Agency.

This Springer imprint is published by the registered company Springer Nature Singapore Pte Ltd.
The registered company address is: 152 Beach Road, #21-01/04 Gateway East, Singapore 189721, Singapore

Contents

Extracellular Fine Fiber-Induced Carcinogenesis and Its Prevention

Shinya Toyokuni, Yuki Maeda, Qinying Lyu, Danyang Mi, and Yingyi Kong

1 Cancer as a Major Cause of Human Mortality

Cancer, dysregulated proliferation of basically monoclonal autologous cells causing destruction of the preexistent normal structure, has been present from the ancient days (David and Zimmerman 2010). At present, cancer is a major cause of human mortality worldwide, and cancer has been the top cause of death in Japan since 1981 (https://ganjoho.jp/reg_stat/statistics/stat/short_pred_en.html). It is established that alteration of the genomic information is the primary cause of cancer though epigenetic changes are also important. Carcinogenesis is therefore a process to obtain mutation(s), which is regulated by environmental, somatic and hereditary factors (Vogelstein and Kinzler 1998) (Fig. 1). Fe(II) is a cofactor for ribonucleotide reductase, which catalyzes the first reaction for the de novo synthesis of DNA (Petersson et al. 1980). No life on the earth can survive without iron. On the other hand, excess iron may cause oxidative DNA damage, resulting in a variety of mutations (Toyokuni 1996).

S. Toyokuni (✉) · Y. Maeda · Q. Lyu · D. Mi · Y. Kong
Department of Pathology and Biological Responses, Nagoya University Graduate School of Medicine, 65 Tsurumai-Cho, Showa-Ku, Nagoya 466-8550, Japan
e-mail: toyokuni@med.nagoya-u.ac.jp

S. Toyokuni
Center for Low-Temperature Plasma Sciences, Nagoya University, Furo-Cho, Chikusa-Ku, Nagoya 464-8603, Japan

Center for Integrated Sciences of Low-Temperature Plasma Core Research (iPlasma Core), Tokai National Higher Education and Research System, Furo-Cho, Chikusa-Ku, Nagoya 464-8603, Japan

Y. Baba et al. (eds.), *Extracellular Fine Particles*,
https://doi.org/10.1007/978-981-97-7067-0_1

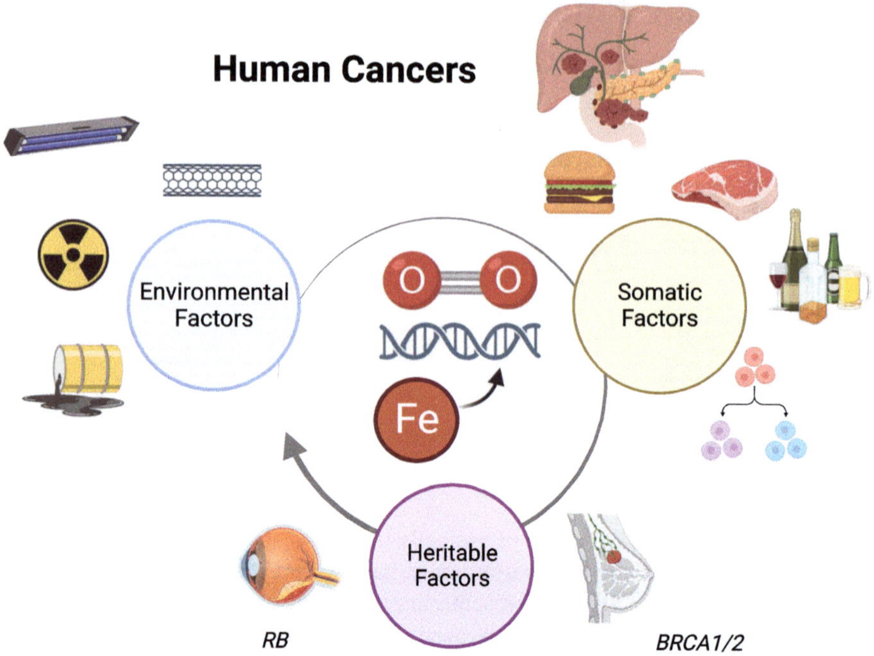

Fig. 1 Carcinogenesis and modifying factors. Genomic alterations are in the center of carcinogenesis, where iron plays a dual role. Whereas iron is essential for de novo DNA synthesis, excess iron in the presence of oxygen is a cause of mutation(s), a driving force for carcinogenesis. There are three major modifying factors for carcinogenesis; environmental, somatic as well as heritable factors

2 Development and Utilization of Nanomaterials in the Society

The development and utilization of nanomaterials have brought about a paradigm shift in various aspects of society. Nanotechnology, which involves the manipulation and engineering of materials at the nanoscale, typically defined as <100 nm in one or more dimension(s), has opened up a world of possibilities that were once considered science fiction. From healthcare to electronics, energy to environmental conservation, nanomaterials have become indispensable in advancing our technological prowess and addressing some of the most pressing challenges of our time (Riley and Narayan 2021).

One of the most profound impacts of nanomaterials has been in the field of healthcare. Nanoparticles, such as liposomes and quantum dots, have revolutionized drug delivery systems (Jacob et al. 2018). These tiny carriers can transport drugs directly to target cells, reducing side effects and improving treatment efficacy. Additionally, nanomaterials enable the development of highly sensitive diagnostic tools (Sharifi et al. 2019). Quantum dots, for instance, have been used in imaging techniques to detect diseases like cancer at an early stage, allowing for more effective treatment (Iannazzo et al. 2021).

The use of nanomaterials has also given rise to innovative therapies, like nanoparticle-based cancer treatments that can specifically target tumor cells while sparing healthy ones (Dang and Guan 2020). Moreover, researchers are exploring the potential of nanorobots for tasks such as targeted drug delivery within the body, showcasing the limitless possibilities for nanotechnology in the healthcare sector (Suhail et al. 2022).

Nanomaterials have significantly impacted the electronics industry. The miniaturization of electronic components, driven by advancements in nanotechnology, has led to the development of smaller and more efficient devices. Transistors made from nanomaterials have increased processing power while reducing energy consumption, making electronics faster, more powerful, and energy-efficient (Syedmoradi et al. 2019). This has paved the way for the proliferation of portable devices, the Internet of Things (IoT), and the advancement of artificial intelligence.

Furthermore, nanomaterials play a vital role in energy production and storage. Nanoscale materials have been used in the development of high-capacity batteries and supercapacitors, leading to longer-lasting and faster-charging devices. Solar cells employing nanomaterials have demonstrated improved efficiency, making energy sources more viable and sustainable (Pomerantseva et al. 2019).

Addressing environmental challenges is another area where nanomaterials have proven invaluable. Nanotechnology-based solutions are being deployed for water purification, air filtration, and waste management (Mauter and Elimelech 2008). For example, nanoparticles like titanium dioxide are used in photocatalytic coatings that can break down pollutants when exposed to light (Nasr et al. 2018). This technology can be applied to reduce air pollution in urban environments and purify water sources contaminated with organic compounds and heavy metals.

Nanomaterials are also aiding in the development of lightweight, high-strength materials that enhance fuel efficiency in transportation, reducing greenhouse gas emissions. In addition, nanotechnology is contributing to the creation of more efficient and cost-effective methods for desalination, crucial for addressing global water scarcity issues (Teow and Mohammad 2019).

While the development and utilization of nanomaterials hold immense promise, they also raise ethical and safety concerns. The potential environmental impact of engineered nanoparticles, as well as their effects on human health, must be thoroughly researched and addressed. Striking a balance between technological advancement and responsible stewardship of these materials is essential.

3 Biological Significance of Iron and Iron Metabolism

There is a consensus that no life on the earth can survive without iron from bacteria to humans (Weber et al. 2006). This is associated with the availability of iron on the earth during evolution. Adult humans possess as much as 2.5–4 g of iron inside the whole body, which is the highest amount among the metals (Toyokuni 2009a). Iron is a transition metal, thus revealing redox activity between Fe(II) and Fe(III) and facilitating the electron transfer (Koppenol and Hider 2019). Iron is a major cofactor for a variety of enzymes in the form either of Fe(II), *Fe-S* cluster and heme (Harigae 2018). Fe(II) is necessary for the catalytic activity of ribonucleotide reductase, the first step of de novo DNA synthesis (Petersson et al. 1980). Furthermore, approximately 60% of iron is present as heme in hemoglobin to transport oxygen. In principle, only Fe(II) is soluble at physiological neutral pH whereas Fe(III) is insoluble (Fig. 2). Iron is always in the form of Fe(II) when it goes through the membrane. On the other hand, iron is stored in ferritin and is transported through transferrin in the safe insoluble form of Fe(III) (Toyokuni et al. 2020a). There are specific mechanisms to take out of these Fe(III), both of which use lysosomes where the pH is acidic. The mechanism for taking out Fe(III) from ferritin is called ferritinophagy, which was recently discovered (Mancias et al. 2014).

Iron is absorbed through duodenal villous epithelial cells in higher species via two distinct iron transporters, DMT1 (SLC11A2) and ferroportin-1 (FPN1; SLC40A1) sequentially (Yanatori and Kishi 2019). Each transporter is accompanied by specific reductase (Dcytb; Fe[III] to Fe[II]) and oxidase (hephestin; Fe[II] to Fe[III]), finally to be captured by transferrin in the portal venous system. Iron is transported to every

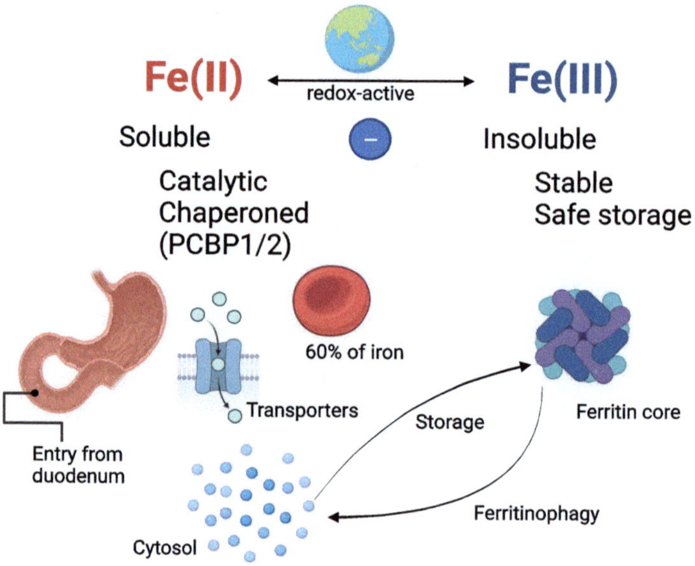

Fig. 2 Characteristics of iron metabolism. Iron is essential for all the lives on the earth. Iron is abundant on the earth and redox-active, thus allowing electron transfer. Only Fe(II) is soluble at neutral pH. Iron is absorbed through duodenal epithelial cells in higher species, and 60% of iron is present in erythrocytes. Insoluble Fe(III) is used for safe storage of iron in ferritin. Various fine molecular mechanisms, including ferritinophagy and cytosolic iron chaperones, have been discovered recently. PCBP1/2, poly rC binding protein 1/2. Refer to text for details

cell in the body through the transferrin and transferrin receptor system (Kawabata 2019). Transferrin Fe(III) is taken out in the endosomes/lysosomes, reduced through STEAP3 to Fe(II) and transported to cytosol again by DMT1 (SLC11A2) to be used for various cellular functions (Toyokuni et al. 2023).

As already described, Fe(II) is a risk for the Fenton reaction and Fe(III) is insoluble. How iron exists in the cytosol has been a huge challenge in this research area. The concept of labile iron exists from the 1970's (Kakhlon and Cabantchik 2002). Since 10 years ago, it has become possible to measure and locate catalytic Fe(II) (catalytic fraction of labile iron) with specific fluorescent probes (Hirayama 2019). Recently, poly rC-binding protein 1 and 2 (PCBP1 and PCBP2) were identified as iron chaperones in the cytosol and these chaperoned Fe(II) is not catalytic (Philpott et al. 2020). Both PCBP1 and PCBP2 can bind three Fe(II), and PCBP1 comes from a pseudogene of PCBP2. PCBP2 can receive Fe(II) from DMT1 and heme oxygenase and load Fe(II) to ferroportin-1 (Yanatori et al. 2020) whereas PCBP1 can load Fe(II) to ferritin core (Shi et al. 2008). Indeed, PCBP1 and PCBP2 work competitively in the cytosol. Knockout mice of PCBP1/2 are both embryonic lethal.

4 Novel Role of Extracellular Vesicles in Iron Metabolism

Serum ferritin is a marker of iron storage in humans. It is thought that serum ferritin is mostly empty inside the core (Torti and Torti 2021). However, how it is secreted outside of the cell has not been elucidated. It is now established that cells secrete small vesicular structures with cargo molecules to communicate and support the other cells of their own. The cargo molecules are quite various, including proteins, miRNA, mRNA and DNA. These extracellular vesicles are classified according with the diameter size, defined by 30~120 nm, 100~1,000 nm and 800~5,000 nm, as exosomes, microvesicles and apoptotic bodies, respectively (Niel et al. 2018).

We recently found that exosomes are closely associated with iron metabolism. Major surface markers for exosome include CD9, CD63 and CD81. Out of these, expression of only CD63 is regulated by iron-regulatory protein/iron-responsive element system at the 5' end of the mRNA (Yanatori et al. 2021). This means that iron deficient status in cells blocks the translation of CD63 whereas iron sufficient status deblocks its translation (Fig. 3). Furthermore, CD63-positve exosomes secreted from iron-sufficient cells include ferritin core as a cargo, and the ferritin core include insoluble Fe(III). This pathway involved also NCOA4 for loading holo-ferrintin to exosomes. The whole process means that excess iron is safely shared only among cells within the identical individual (Toyokuni et al. 2023). No life on the earth can live without iron, which is the same for prokaryotes and parasites. Infection by bacteria or parasites to higher species needs a strategy to take away iron from the host (Toyokuni et al. 2020b).

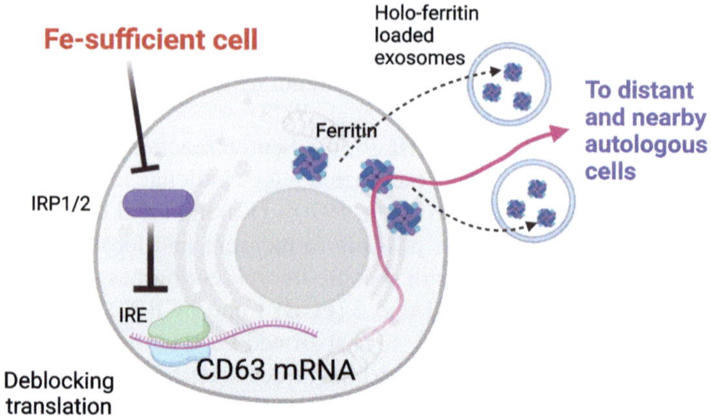

Fig. 3 Association of iron metabolism and extracellular vesicles. CD63, a surface marker of exosomes, is under the regulation of iron-responsive element (IRE)/iron-regulatory protein (IRP) system. In iron sufficient cells, surplus iron is stored as holo (iron-loaded)-ferritin, which is secreted as holo-ferritin loaded exosomes for distant or nearby autologous cells. Refer to text for details

5 Iron and Carcinogenicity

Despite its necessity, excess iron has been associated with carcinogenesis (Toyokuni 1996, 2009a; Torti et al. 2018). There are three distinct levels of evidence for this. The first one is a human epidemiological data on the mass population. The second one is also a human epidemiological data but on specific diseases. The last one is an accumulated data on animal experiments.

Regarding the first one, one study on the US population is well cited. This study showed that regular reduction of iron with phlebotomy for 5 years significantly reduced the incidence and mortality of cancer in the middle aged to old population with peripheral arterial disease (Zacharski et al. 2008). Other findings are summarized elsewhere (Toyokuni 1996; Torti et al. 2018).

The second category includes hereditary hemochromatosis, chronic viral hepatitis, ovarian endometriosis and asbestos exposure-induced mesothelioma in humans (Toyokuni 2009a). Hereditary hemochromatosis consists of genetic diseases of 5 different genes (https://www.ncbi.nlm.nih.gov/omim/?term=hemochromatosis), where continued iron absorption causes excess iron in the liver and other organs. Hepatocellular carcinoma is the most prevalent cancer in that liver is a major iron storage iron (Elmberg et al. 2003). Other findings are summarized in Table 1.

Regarding the third category of animal experiments, there are many reports on this and the readers are invited to refer to those review articles (Toyokuni 1996, 2009a; Torti et al. 2018). Here we describe only ferric nitrilotriacetate (Fe-NTA)-induced renal carcinogenesis briefly. This model was discovered with serendipity in rats by Shigeru Okada and Osamu Midorikawa in Kyoto University in 1982 after working on a hemochromatosis-mimic model by Fe-NTA administration (Toyokuni et al. 2022). Intraperitoneal injection of Fe-NTA to rats or mice causes ferroptosis specifically

Table 1 Human diseases with excess iron-associated cancer risk

Human disease	Target organs/ cells	Cancer risk	Molecular mechanism(s)
Hereditary hemochromatosis	Hepatocytes (liver)	Hepatocellular carcinoma	Hereditary; pathogenic variants either of five distinct iron metabolism-associated genes *(HFE1, HJV, HAMP, TFR2, SLC40A1, FTH1)*, which all increase total iron uptake)
Viral hepatitis B & C	Hepatocytes (liver)	Hepatocellular carcinoma	Infection; persistent hepatocyte loss causes hepcidin depletion, increasing iron uptake; vaccination for hepatitis B and directly acting antiviral drugs for hepatitis C dramatically decreased these diseases
Endometriosis	Ovarian surface epithelial cells	Endometrioid adenocarcinoma; serous adenocarcinoma	Ectopia of endometrial tissue in ovary causes iron excess due to monthly hemorrhage in situ
Exposure to asbestos	Mesothelial cells in the somatic cavities	Mesothelioma	Thin and biopersistent nanofibers reach pulmonary alveoli and kills macrophages with ferroptosis; iron coated asbestos finally reach mesothelial cells in the somatic cavities and cause oxidative DNA damage (refer to text)

in the proximal renal tubular cells via Fenton-like reaction (Zheng et al. 2021). Repeated intraperitoneal injections of Fe-NTA to rodents induce proliferation of ferroptosis-resistant cells, which finally leads to a high incidence of aggressive renal cell carcinoma, a half of which are accompanied by pulmonary metastasis. In rats the cancer incidence is ~90% after 12 weeks-injection and the 1.5-year observation thereafter. Most common genomic alteration is the homozygous deletion of *Cdkn2a/2b (p16/p15)* tumor suppressor gene(s) (Tanaka et al. 1999; Akatsuka et al. 2012). This was significantly promoted in Brca1 haploinsufficient rats (Kong et al. 2022). Regarding mice, there is a distinct species difference, which is explained partially by the difference in iron metabolism and ferroptosis-resistance (Cheng et al. 2022). Accordingly, excess iron is an inevitable cause of carcinogenesis.

6 Foreign Body Exposure Dysregulates Iron Metabolism and is Associated with Carcinogenesis

We have been always exposed to exogenous nanomaterials through respiratory and gastrointestinal mucosa or skin. Even in the ancient days, live volcanos or fire should have produced various nanomaterials. It was after the industrial revolution that people recognized the association between foreign materials and carcinogenesis, which started by the findings of Sir Percival Pott that scrotal skin cancer is frequent in chimney sweepers (Brown and Thornton 1957). Then, in the last century, asbestos-induced various diseases have become the social problem (Pathology of Asbestos-Associated Diseases 2014).

Asbestos is the natural fibrous mineral which has been mined in mountainous areas. There are three major types of asbestos fibers which have been used in industry; blue asbestos (crocidolite), brown asbestos (amosite) and white asbestos (chrysotile). Crocidolite and amosite contained 30% of iron in its component, but chrysotile contained no iron (IARC 2012). It is well established that the fiber size is important for the pathogenicity, namely diameter <250 nm and length <8 μm in each fiber, which is usually in normal distribution (Stanton and Wrench 1972). These physical parameters are important for the fibers to go through the respiratory system to reach alveoli in the lung. Exposure to asbestos fibers have been associated with a rare cancer mesothelioma, which originates from mesothelial cells covering somatic cavities but not exposed to the outside. ~80% of mesothelioma is of pleural origin and ~20% of peritoneal origin (IARC 2012).

It has been a long-time mystery how asbestos fibers causes mesothelioma, but it is now established that these fibers go through the pulmonary parenchyma using decades of time (30~40 y), depending on the negative pressure of pleural cavity (Toyokuni 2019). This is consistent with the very long incubation period of mesothelioma after asbestos exposure. It is now thought that ~80% of mesothelioma is associated with past asbestos exposure.

What is happening as biological responses when asbestos fibers have reached the alveoli? Alveolar macrophages take up the fibers, but they die because they cannot manage the fibers due to the rigidity and the length. Here only the fibers remain and they go peripherally, which is phagocytosed by another macrophage. This is repeated permanently and the fibers finally reach the pleural space by disrupting the visceral pleural tissue. Then, the fibers would be stuck in the parietal mesothelia, which are in line with the lymphatic vessels. In the clinical observation, most of the mesothelioma arises in the parietal mesothelia (Toyokuni 2019, 2009b).

The important point is that during these decades of process asbestos fibers collect iron on the surface (Fig. 4). We have done a comprehensive study on the specific adsoption of asbestos fibers and found that hemoglobin and histones are in common for all the major asbestos fibers (Nagai et al. 2011a). Importantly, chrysotile, white asbestos, causes hemolysis and causes more prominent iron excess at the local site of exposure than crocidolite and amosite. This is confirmed by an animal experiment, where chrysotile exposure was the earliest to cause mesothelioma among the three

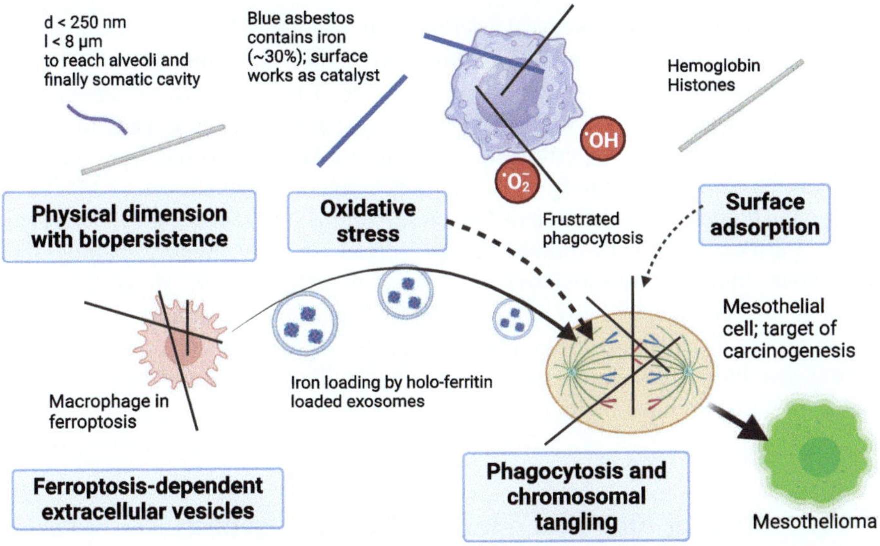

Fig. 4 Carcinogenic mechanisms of fibrous nanomaterials. The target cells of carcinogenesis for fibrous nanomaterials is mesothelial cells, covering the somatic cavity. All the important points are included in the figure. Iron-coated fibrous nanomaterials, including asbestos fibers, are phagocytosed by mesothelial cells, where those fibers directly damage genomic DNA with double-strand breaks and modifications. Most of the fibrous nanomaterials are phagocytosed by macrophages, which indirectly burden oxidative stress to mesothelial cells via frustrated phagocytosis and ferroptosis-dependent holo-ferritin loaded extracellular vesicles (mostly exosomes). Refer to text for details

major asbestos after intraperitoneal injection of 10 mg in rats. Of note, the target gene of this carcinogenesis of the homozygous deletion of *Cdkn2a/2b* tumor suppressor gene, which is the flag ship mutation of excess iron-induced carcinogenesis (Jiang et al. 2012). This is true for human mesothelioma and is used as a diagnostic marker for mesothelioma (Nabeshima et al. 2016; Husain et al. 2017).

Looking into the pathology involved at the cellular level, macrophages exposed to asbestos fibers collects iron by altering the iron metabolism to take up more iron and not to release iron (Ito et al. 2020). We believe that this is a protective reaction for infection not to provide microorganisms with iron. In the case of macrophage death, the accumulated iron is carried by the remaining biopersistent asbestos fibers, which generates asbestos bodies.

Recently, we found that this macrophage death is classified into ferroptosis and these dying macrophages secrete holo-ferritin loaded exosomes (Ito et al. 2021). Furthermore, this is received by the mesothelial cells nearby in the parietal pleura, the target cells of asbestos-induced carcinogenesis (Fig. 4). Of note, mesothelial cells that received these holo-ferritin loaded exosomes reveals oxidative DNA damage, including double-strand breaks and base modifications. Thus, exogenous and endogenous fine particles are closely linked (Toyokuni et al. 2023).

Some of the nanomaterials reveal a similar dimension to asbestos fibers and biopersistence, which includes a subset of multiwalled carbon nanotubes (MWCNT). We studied a variety of carbon nanotubes to evaluate their toxicity and carcinogenicity. Our conclusion is that MWCNT of ~50 nm diameter reveals the highest mesothelial carcinogenicity by intraperitoneal injection to rats and thus has to be handled with extreme care. Interestingly, most of the generated mesothelioma by MWCNT revealed the homozygous deletion of *Cdkn2a/2b* tumor suppressor gene (Nagai et al. 2011b). Further, MWCNTs showed the specific adsorptive activity to hemoglobin and histone, but also holo-transferrin (Wang et al. 2016). These suggest that the carcinogenic mechanism of asbestos and MWCNT of ~50 nm is similar and depends on iron excess. Finally, we stress that the diameter of MWCNT is critical because ~50 nm diameter with rigidness penetrates mesothelial cells, which appears to be associated with carcinogenicity (Nagai et al. 2011b).

7 Prevention of Fine Fiber-Induced Carcinogenesis

It was unexpected that natural fibrous stones and nanomaterials of similar physical dimension are carcinogenic to humans. Furthermore, the carcinogenic mechanism was associated with the fundamental mechanism of living cells, namely dysregulation of iron metabolism apparently to protect themselves. However, this iron sequestration caused intracellular excess iron, which finally led to the carcinogenic pathway through Fenton-like reaction and a variety of indirect mechanisms including holo-ferritin loaded exosomes (Ito et al. 2021).

Considering these mechanisms, appropriate iron removal may be a good strategy to prevent extracellular fine fiber-induced carcinogenesis. Of note, there is no active pathway to excrete iron out of the body in higher species once iron is taken up from duodenum, and 60% of body iron is in red blood cells as heme in hemoglobin. The only available methods would be phlebotomy or the use of iron chelator drugs. We have performed both of them in asbestos-induced mesothelial carcinogenesis in rats. Both the use of oral iron chelator, deferasirox (Nagai et al. 2013), and phlebotomy (Ohara et al. 2018) worked to prevent asbestos-induced mesothelial carcinogenesis, either in the longer survival, less tumor mass burden or histological-subtype fractions for the better prognosis. Here sarcomatoid subtype is a more aggressive histotype than epithelioid subtype. In practice, taking an oral iron chelator, deferasirox, for a long period shows a high possibility to obtain some side effects, such as renal toxicity, and thus presents some ethical problem. We believe that regular phlebotomy is much safer, also established as blood donation and thus is easier to be accepted to the people already exposed to asbestos.

Alternatively, we recently found that Brca1 haploinsufficiency promotes chrysotile-induced mesothelial carcinogenesis in rats, suggesting a higher risk for chrysotile-induced mesothelioma in BRCA1 pathogenic variants (Luo et al. 2023). Asbestos exposure is also associated with a higher risk in lung cancer, which we did not have enough space to describe in this chapter.

8 Conclusion

Natural or synthetic fibrous nanofibers can be carcinogenic, depending on the physical dimensions (diameter and length) and biopersistence after exposure. The major target cells are mesothelial cells covering somatic cavities, and the major genetic alteration is the homozygous deletion of *Cdkn2a/2b* tumor suppressor genes. Local iron excess is the pathogenesis, which includes specific adsorptive function of asbestos/multiwalled carbon nanotube of hemoglobin, iron sequestration as defense mechanism as well as iron transport via ferroptosis-derived holo-ferritin loaded exosomes to mesothelial cells.

9 Competing Interest

The authors declare that they have no competing interests.

Acknowledgements BioRender.com was used for the Figures.

Author Contribution ST designed the review work and wrote the manuscript. YM, QL, DM and YK contributed to the discussion and revision of the manuscript.

Funding This work was supported in part by JST CREST (JPMJCR19H4) and JSPS Kakenhi (JP19H05462, JP20H05502) to ST.

References

Akatsuka S, Yamashita Y, Ohara H, Liu YT, Izumiya M, Abe K *et al.* Fenton reaction induced cancer in wild type rats recapitulates genomic alterations observed in human cancer. PLoS ONE 2012;7(8):e43403.

Brown JR, Thornton JL. Percivall Pott (1714-1788) and chimney sweepers' cancer of the scrotum. Br J Ind Med 1957;14(1):68–70.

Cheng Z, Akatsuka S, Li GH, Mori K, Takahashi T, Toyokuni S. Ferroptosis resistance determines high susceptibility of murine A/J strain to iron-induced renal carcinogenesis. Cancer Sci 2022;113(1):65–78.

Dang Y, Guan J. Nanoparticle-based drug delivery systems for cancer therapy. Smart Mater Med 2020;1:10–19.

David AR, Zimmerman MR. Cancer: an old disease, a new disease or something in between? Nat Rev Cancer 2010;10(10):728–733.

Elmberg M, Hultcrantz R, Ekbom A, Brandt L, Olsson S, Olsson R *et al.* Cancer risk in patients with hereditary hemochromatosis and in their first-degree relatives. Gastroenterology 2003;125(6):1733–1741.

Harigae H. Iron metabolism and related diseases: an overview. Int J Hematol 2018;107(1):5–6.

Hirayama T. Fluorescent probes for the detection of catalytic Fe(II) ion. Free Radic Biol Med 2019;133:38–45.

Husain AN, Colby TV, Ordóñez NG, Allen TC, Attanoos RL, Beasley MB *et al.* Guidelines for Pathologic Diagnosis of Malignant Mesothelioma 2017 Update of the Consensus Statement From the International Mesothelioma Interest Group. Archives of pathology & laboratory medicine 2017;142(1):89–108.

Iannazzo D, Espro C, Celesti C, Ferlazzo A, Neri G. Smart Biosensors for Cancer Diagnosis Based on Graphene Quantum Dots. Cancers 2021;13(13).

IARC W. Asbestos (chrysotile, amosite, crocidolite, tremolite, actinolite, and anthophyllite). IARC Monographs on the Evaluation of Carcinogenic Risks to Humans. A Review of Human Carcinogens; Part C: Arsenic, Metals, Fibres, and Dusts. Volume 100C. Lyon, France2012. p 219–309.

Ito F, Yanatori I, Maeda Y, Nimura K, Ito S, Hirayama T *et al.* Asbestos conceives Fe(II)-dependent mutagenic stromal milieu through ceaseless macrophage ferroptosis and beta-catenin induction in mesothelium. Redox Biol 2020;36:101616.

Ito F, Kato K, Yanatori I, Murohara T, Toyokuni S. Ferroptosis-dependent extracellular vesicles from macrophage contribute to asbestos-induced mesothelial carcinogenesis through loading ferritin. Redox Biol 2021;47:102174.

Jacob J, Haponiuk JT, Thomas S, Gopi S. Biopolymer based nanomaterials in drug delivery systems: A review. Materials Today Chemistry 2018;9:43–55.

Jiang L, Akatsuka S, Nagai H, Chew SH, Ohara H, Okazaki Y *et al.* Iron overload signature in chrysotile-induced malignant mesothelioma. J Pathol 2012;228:366–377.

Kakhlon O, Cabantchik ZI. The labile iron pool: characterization, measurement, and participation in cellular processes(1). Free Radic Biol Med 2002;33(8):1037–1046.

Kawabata H. Transferrin and transferrin receptors update. Free Radic Biol Med 2019;133:46–54.

Kong Y, Akatsuka S, Motooka Y, Zheng H, Cheng Z, Shiraki Y *et al.* BRCA1 haploinsufficiency promotes chromosomal amplification under Fenton reaction-based carcinogenesis through ferroptosis-resistance. Redox Biol 2022;54:102356.

Koppenol WH, Hider RH. Iron and redox cycling. Do's and don'ts. Free Radic Biol Med 2019;133:3–10.

Luo Y, Akatsuka S, Motooka Y, Kong Y, Zheng H, Mashimo T *et al.* BRCA1 haploinsufficiency impairs iron metabolism to promote chrysotile-induced mesothelioma via ferroptosis resistance. Cancer Sci 2023;114(4):1423–1436.

Mancias JD, Wang XX, Gygi SP, Harper JW, Kimmelman AC. Quantitative proteomics identifies NCOA4 as the cargo receptor mediating ferritinophagy. Nature 2014;509(7498):105–109.

Mauter MS, Elimelech M. Environmental applications of carbon-based nanomaterials. Environ Sci Technol 2008;42(16):5843–5859.

Nabeshima K, Matsumoto S, Hamasaki M, Hida T, Kamei T, Hiroshima K *et al.* Use of p16 FISH for differential diagnosis of mesothelioma in smear preparations. Diagn Cytopathol 2016;44(9):774-780.

Nagai H, Ishihara T, Lee WH, Ohara H, Okazaki Y, Okawa K *et al.* Asbestos surface provides a niche for oxidative modification. Cancer Sci 2011a;102:2118–2125.

Nagai H, Okazaki Y, Chew S, Misawa N, Yamashita Y, Akatsuka S *et al.* Diameter of multi-walled carbon nanotubes is a critical factor in mesothelial injury and subsequent carcinogenesis. Proc Natl Acad Sci U S A 2011b;108(49):E1330–1338.

Nagai H, Okazaki Y, Chew SH, Misawa N, Yasui H, Toyokuni S. Deferasirox induces mesenchymal-epithelial transition in crocidolite-induced mesothelial carcinogenesis in rats. Cancer Prev Res (Phila) 2013;6:1222–1230.

Nasr M, Eid C, Habchi R, Miele P, Bechelany M. Recent Progress on Titanium Dioxide Nanomaterials for Photocatalytic Applications. ChemSusChem 2018;11(18):3023–3047.

van Niel G, D'Angelo G, Raposo G. Shedding light on the cell biology of extracellular vesicles. Nat Rev Mol Cell Biol 2018;19(4):213–228.

Ohara Y, Chew SH, Shibata T, Okazaki Y, Yamashita K, Toyokuni S. Phlebotomy as a preventive measure for crocidolite-induced mesothelioma in male rats. Cancer Sci 2018;109(2):330–339.

Pathology of Asbestos-Associated Diseases, Third edition. Oury TD, Sporn TA, Roggli VL, editors. Berlin/Heidelberg: Springer; 2014.

Petersson L, Graslund A, Ehrenberg A, Sjoberg BM, Reichard P. The iron center in ribonucleotide reductase from *Escherichia coli*. J Biol Chem 1980;255(14):6706–6712.

Philpott CC, Patel SJ, Protchenko O. Management versus miscues in the cytosolic labile iron pool: The varied functions of iron chaperones. Biochimica Et Biophysica Acta-Molecular Cell Research 2020;1867(11).

Pomerantseva E, Bonaccorso F, Feng X, Cui Y, Gogotsi Y. Energy storage: The future enabled by nanomaterials. Science 2019;366(6468).

Riley PR, Narayan RJ. Recent advances in carbon nanomaterials for biomedical applications: A review. Current Opinion in Biomedical Engineering 2021;17.

Sharifi M, Avadi MR, Attar F, Dashtestani F, Ghorchian H, Rezayat SM *et al.* Cancer diagnosis using nanomaterials based electrochemical nanobiosensors. Biosensors & Bioelectronics 2019;126:773–784.

Shi HF, Bencze KZ, Stemmler TL, Philpott CC. A cytosolic iron chaperone that delivers iron to ferritin. Science 2008;320(5880):1207–1210.

Stanton MF, Wrench C. Mechanisms of mesothelioma induction with asbestos and fibrous glass. J Natl Cancer Inst 1972;48:797–821.

Suhail M, Khan A, Rahim MA, Naeem A, Fahad M, Badshah SF *et al.* Micro and nanorobot-based drug delivery: an overview. J Drug Target 2022;30(4):349–358.

Syedmoradi L, Ahmadi A, Norton ML, Omidfar K. A review on nanomaterial-based field effect transistor technology for biomarker detection. Mikrochim Acta 2019;186(11):739.

Tanaka T, Iwasa Y, Kondo S, Hiai H, Toyokuni S. High incidence of allelic loss on chromosome 5 and inactivation of p15 INK4B and p16 INK4A tumor suppressor genes in oxystress-induced renal cell carcinoma of rats. Oncogene 1999;18:3793–3797.

Teow YH, Mohammad AW. New generation nanomaterials for water desalination: A review. Desalination 2019;451:2–17.

Torti SV, Torti FM. CD63 orchestrates ferritin export. Blood 2021;138(16):1387–1389.

Torti SV, Manz DH, Paul BT, Blanchette-Farra N, Torti FM. Iron and Cancer. Annu Rev Nutr 2018;38:97–125.

Toyokuni S. Iron-induced carcinogenesis: the role of redox regulation. Free Radic Biol Med 1996;20:553–566.

Toyokuni S. Role of iron in carcinogenesis: Cancer as a ferrotoxic disease. Cancer Sci 2009a;100(1):9–16.

Toyokuni S. Mechanisms of asbestos-induced carcinogenesis. Nagoya J Med Sci 2009b;71(1-2):1–10.

Toyokuni S. Iron addiction with ferroptosis-resistance in asbestos-induced mesothelial carcinogenesis: Toward the era of mesothelioma prevention. Free Radic Biol Med 2019;133:206–215.

Toyokuni S, Kong Y, Cheng Z, Sato K, Hayashi S, Ito F *et al.* Carcinogenesis as Side Effects of Iron and Oxygen Utilization: From the Unveiled Truth toward Ultimate Bioengineering. Cancers (Basel) 2020;12(11):3320.

Toyokuni S, Yanatori I, Kong Y, Zheng H, Motooka Y, Jiang L. Ferroptosis at the crossroads of infection, aging and cancer. Cancer Sci 2020b;111:2665–2671.

Toyokuni S, Kong Y, Zheng H, Maeda Y, Motooka Y, Akatsuka S. Iron as spirit of life to share under monopoly. J Clin Biochem Nutr 2022;71(2):78–88.

Toyokuni S, Kong Y, Katabuchi M, Maeda Y, Motooka Y, Ito F *et al.* Iron links endogenous and exogenous nanoparticles. Arch Biochem Biophys 2023;745:109718.

Vogelstein B, Kinzler KW. The genetic basis of human cancer. New York: McGraw-Hill; 1998.

Wang Y, Okazaki Y, Shi L, Kohda H, Tanaka M, Taki K *et al.* Role of hemoglobin and transferrin in multi-wall carbon nanotube-induced mesothelial injury and carcinogenesis. Cancer Sci 2016;107:250–257.

Weber KA, Achenbach LA, Coates JD. Microorganisms pumping iron: anaerobic microbial iron oxidation and reduction. Nat Rev Microbiol 2006;4(10):752–764.

Yanatori I, Kishi F. DMT1 and iron transport. Free Radic Biol Med 2019;133:55–63.
Yanatori I, Richardson DR, Toyokuni S, Kishi F. The new role of poly (rC)-binding proteins as iron transport chaperones: Proteins that could couple with inter-organelle interactions to safely traffic iron. Biochim Biophys Acta Gen Subj 2020;1864(11):129685.
Yanatori I, Richardson DR, Dhekne HS, Toyokuni S, Kishi F. CD63 is regulated by iron via the IRE-IRP system and is important for ferritin secretion by extracellular vesicles. Blood 2021;138(16):1490–1503.
Zacharski L, Chow B, Howes P, Shamayeva G, Baron J, Dalman R *et al.* Decreased cancer risk after iron reduction in patients with peripheral arterial disease: Results from a randomized trial. J Natl Cancer Inst 2008;100:996–1002.
Zheng H, Jiang L, Tsuduki T, Conrad M, Toyokuni S. Embryonal erythropoiesis and aging exploit ferroptosis. Redox Biol 2021;48:102175.

Toward Digital Bioanalysis of Extracellular Vesicles

Hajime Shinoda and Rikiya Watanabe

Abstract A wide array of extracellular vesicles within living organisms plays a crucial role in mediating physiological functions and is involved in significantly to various diseases and viral infections. Understanding the composition of these vesicles is essential to identify disease origins and develop effective diagnostics and therapeutics. However, current technologies typically analyze groups of vesicles together, leading to averaged information that lacks insight into individual vesicle components, such as the number of molecular species and their correlation with diseases. To enhance our understanding of extracellular vesicles, we have developed SATORI, a cutting-edge technology capable of identifying and rapidly detecting RNA from viruses, which are examples of extracellular vesicles, at the single-molecule level. Additionally, we created an automated platform (opn-SATORI) and a portable platform (COWFISH) based on SATORI, which serve as the initial step in clinical applications for the high-throughput diagnosis of viral infections, such as COVID-19. In this chapter, we introduce our latest technology, SATORI, and discuss its future potential for the comprehensive analysis of extracellular vesicles and use in the diagnosis of viral infections.

Keywords Digital bioanalysis · CRISPR-Cas · bioMEMS · Single-molecule imaging

1 Introduction

In recent years, extensive basic and diagnostic research has focused on a wide variety of extracellular vesicles in vivo, including microsomes and exosomes. Nucleic acids, proteins, and lipids in such vesicles contain information about the state of originated cell and the state of underlying disease; therefore, they have the potential to serve as emerging liquid biopsies (Hoshino et al. 2020). For example, several proteins in vesicles can be detected in plasma to diagnose various cancers including breast,

H. Shinoda · R. Watanabe (✉)
Molecular Physiology Laboratory, Cluster for Pioneering Research, RIKEN, Saitama, Japan
e-mail: rikiya.watanabe@riken.jp

Y. Baba et al. (eds.), *Extracellular Fine Particles*,
https://doi.org/10.1007/978-981-97-7067-0_2

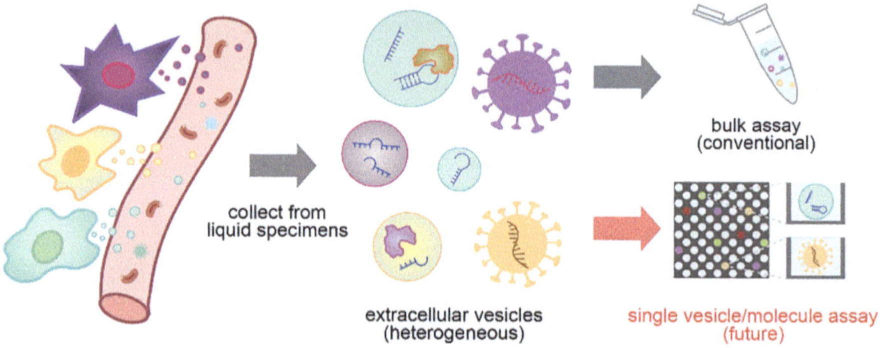

Fig. 1 Analysis of extracellular vesicles

lung, and pancreatic cancers. Extracellular vesicles in vitro, such as viruses, have received great attention owing to the recent pandemics. In particular, the emergence of COVID-19 has triggered rapid research on its mechanisms of infection and evolution, as well as the development of new vaccines, drugs, and diagnostics (Xu et al. 2020; Valera et al. 2021; Liu and Rusling 2021; Bruijns et al. 2022).

The components of extracellular vesicles have been extensively analyzed using various bioanalytical methods, such as RNA-Seq for genetic material and ELISA and mass spectrometry for proteins (Hoshino et al. 2020; Shao et al. 2018). These methods are powerful for the quantification of biomolecules; however, they generally analyze a large number of vesicles together because of their low sensitivity, making it difficult to accurately characterize individual extracellular vesicles, which vary in chemophysical parameters such as size, stiffness, and molecular composition (Pegtel and Gould 2019). To address this issue, extensive studies have sought to develop new technologies based on single molecule analysis to quantify the components at the level of individual extracellular vesicles (Fig. 1) (Wang et al. 2019; Liu et al. 2018). Recently, we developed the new method, SATORI, which can identify and detect RNA at the single molecule level (Shinoda et al. 2021; Iida et al. 2023a), and successfully applied it to the highly sensitive and rapid detection of genetic material derived from virus as an example of an extracellular vesicle. Here, we introduce the latest technologies of SATORI and present a future perspective for their practical application in the diagnosis of viral infections.

2 Diagnostic Methods for Viral Infections

Antigen and genetic testing are the most commonly used diagnostic methods for viral infections. Antigen testing is a simple and rapid diagnostic method targeting antigenic proteins (Liu and Rusling 2021). Although it is widely used for the initial screening of viral infections, its relatively low sensitivity and accuracy pose technical challenges. In contrast, genetic testing such as PCR targets viral genes and generally amplifies them for detection with high sensitivity, allowing for a definitive diagnosis; however, it requires approximately one hour (Petralia and Conoci 2017). Recently, genetic tests have been developed that can detect viruses in approximately 30 min (Gosert et al. 2019); nonetheless, further reduction in detection time remains a challenge. Thus, the existing diagnostic methods have technical trade-offs in sensitivity, accuracy, and throughput, and the development of a new technology that meets all these requirements is needed to prepare for future viral pandemics.

Recently, CRISPR-Cas, an immune protein well conserved in bacteria, has attracted attention as a key component of new genetic tests for viral infection diagnosis (Kaminski et al. 2021). Typical examples of genetic tests using CRISPR-Cas include SHERLOCK by Zhang et al. (Gootenberg et al. 2017) and DETECTR by Doudna et al. (Chen et al. 2018). Cas13a, which is a CRISPR-Cas widely used in genetic testing, is an RNA-guided RNase that forms a binary complex with CRISPR RNA (crRNA) (Fig. 2). The Cas13a-crRNA complex can bind to a target RNA with a sequence complementary to the spacer region in the crRNA, inducing its own conformational change and leading to its activation as an RNase. Activated Cas13a randomly cleaves the bound target RNA and surrounding single-stranded RNA (ssRNA). When fluorescent reporters, ~5 nt of ssRNAs with fluorescent and quenching groups attached to both ends, are added to the reaction solution, activated Cas13a cleaves them, producing a fluorescence signal. The fluorescence signal increases over time as the Cas13a reaction proceeds, and the presence of the target RNA in the sample is determined by detecting the signal.

CRISPR-based nucleic acid detection itself is not sufficiently sensitive to detect low copies of viral RNA genes in a sample and are mainly used in combination with nucleic acid amplification processes such as PCR. Without amplification, the detection sensitivity of the target RNA is low at ~50 pM (3×10^7 copies/μL) (East-Seletsky et al. 2016; Li et al. 2019), which is difficult for diagnosis of viral infections such as COVID-19 (infected patient samples: ~10^3–10^6 copies/μL (Wölfel et al. 2020)).

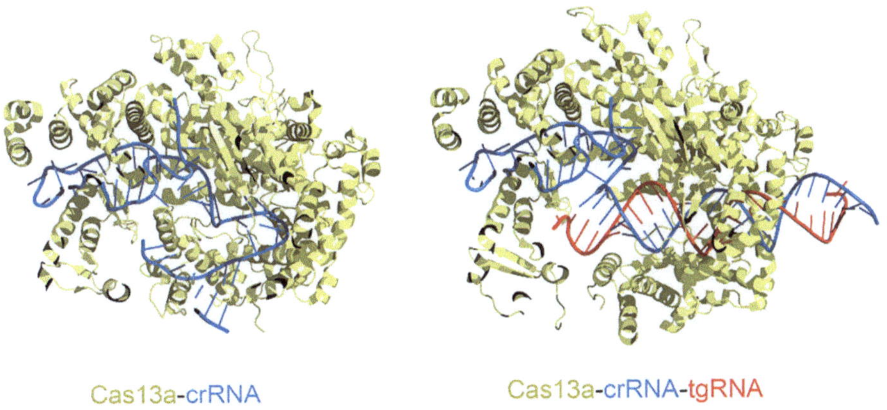

Fig. 2 Structure of CRISPR-Cas13a. PDB: 5XWY (binary), 5XWP (ternary)

In contrast, incorporating the amplification process improves detection sensitivity to ~2–20 aM (~1–10 copies/μL) (Gootenberg et al. 2017; Li et al. 2019; Broughton et al. 2020); however, it requires at least several tens of minutes for the amplification, which still represents a trade-off between sensitivity and throughput even using CRISPR-Cas.

3 SATORI

We have developed "SATORI" (CRISPR-based amplification-free digital RNA detection) (Shinoda et al. 2021; Iida et al. 2023a), which enables highly sensitive, accurate and rapid "digital bioanalysis" of ssRNA, thereby satisfying almost all requirements of viral infection diagnostics. SATORI was achieved by introducing the Cas13a-based RNA detection technology into microchip-based digital bioanalysis, which has recently attracted considerable attention as an exceptionally sensitive approach. The principles of microchip-based digital bioanalysis are as follows (Noji et al. 2022):

(i) Target biomolecules of interest are fractionated into microchambers on a chip at the single-molecule level for subsequent detection, which is achieved by coupling them to fluorogenic enzymatic reactions.
(ii) The presence of biomolecules fractionated in the microchamber is detected as a binary fluorescent signal ("1" representing presence and "0" absence) based on a defined threshold, giving rise to the term "digital" bioanalysis.

Detection modalities for digital bioanalysis can be broadly classified into three categories: direct detection of target enzymes, indirect detection of target by one-to-one binding with fluorogenic enzymes via antibodies, and indirect detection of target by activation of fluorogenic enzymes. Some cutting-edge technologies categorized as direct detection of target molecules have been developed using multicolor fluorogenic substrates, e.g., SEAP (Sakamoto et al. 2020). These technologies enable enzyme identification at the isozyme level, leading to early diagnosis of diseases such as pancreatic cancer. Digital ELISA (Rissin et al. 2010), categorized as indirect detection of target by one-to-one binding with fluorogenic enzymes via antibodies, is a commercially available digital bioanalysis, which has been used for early disease diagnosis and continuous disease monitoring. SATORI is categorized as indirect detection of target by activation of fluorogenic enzymes, i.e., Cas13a, and its RNA detection procedures are as follows (Fig. 3):

(i) Target RNA, such as viral RNA, is mixed with the reaction solution. The reaction solution contains the pre-prepared Cas13a-crRNA complex and fluorescent reporters.
(ii) The mixture is encapsulated in each chamber and sealed with oil by sequential introduction of the above mixture and oil on the microchip.

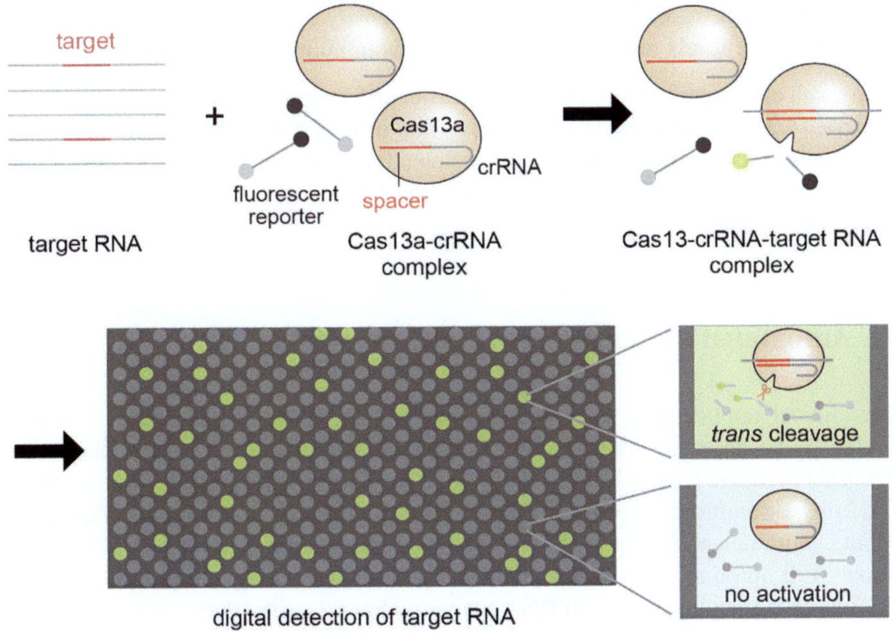

Fig. 3 Digital RNA detection by SATORI

(iii) In the chamber, the Cas13a-crRNA complex is activated by specific binding to the target RNA. When the fluorescent reporter is cleaved by the activated Cas13a-crRNA-target RNA complex, the fluorescent group physically dissociates from the quenching group, leading to the emission of a fluorescence signal.

(iv) Fluorescence images of hundreds of thousands of microchambers are acquired within 2 min by using a fluorescence microscope. The presence or absence of the target RNA in the chamber is determined by binarizing the fluorescence signals. In addition, the copy number of the target RNA in each sample is quantified by counting the number of chambers that emit a fluorescent signal.

In SATORI, the most important point is that the series of processes (i)–(iv) can be completed within 5 min. To achieve rapid detection, a detection system is constructed based on the following two points:

(i) Each chamber contains several molecules of the Cas13a-crRNA complex to enhance its binding efficiency to the target RNA. The presence of a single molecule of the target RNA results in the rapid formation of Cas13a-crRNA-target RNA, leading to the rapid activation of RNase activity.

(ii) The latest SATORI uses LtrCas13a (from *Leptotrichia trevisanii*). Among the Cas13a orthologs, LtrCas13a has a particularly high RNase activity, which was first identified in our previous study (Shinoda et al. 2022). Owing to the high activity of LtrCas13a, the fluorescence intensity in the microchamber reached detectable levels more rapidly within a few minutes after oil sealing.

As a result of the optimization, SATORI can quantify target RNAs with a sensitivity of 720 aM (4.3×10^2 copies/μL) and a detection time of 5 min (Shinoda et al. 2022). This represents a ten-fold reduction in detection time compared to conventional genetic tests such as RT-qPCR, making it the fastest digital bioanalysis for RNA developed to date.

4 Automated Platform of SATORI (opn-SATORI)

For clinical applications, it is necessary to automate the whole process so that SATORI can be easily used in clinical settings. We developed a fully automated SATORI platform (opn-SATORI: aut**o**mated **p**latform o**n SATORI**) (Shinoda et al. 2022) by assembling a confocal fluorescence microscope and a liquid-dispensing robot, communicating via TTL or serial communication (Fig. 4). opn-SATORI allows full automation of entire SATORI process, including sample preparation, fluorescence imaging, and viral RNA copy number quantification.

In the first generation SATORI (described above), the target molecules are stochastically encapsulated in a microchamber by free diffusion. The total volume of the microchamber arrays is very small (~15 nL) compared to that of the reaction solution applied to the chip (~100 μL). Therefore, most target RNAs are not encapsulated in the chamber and are discarded during the SATORI assay, which reduces the detection sensitivity owing to the low capture efficiency of target RNAs. To solve this problem, we developed a novel system to capture the Cas13a-crRNA-target RNA complex on

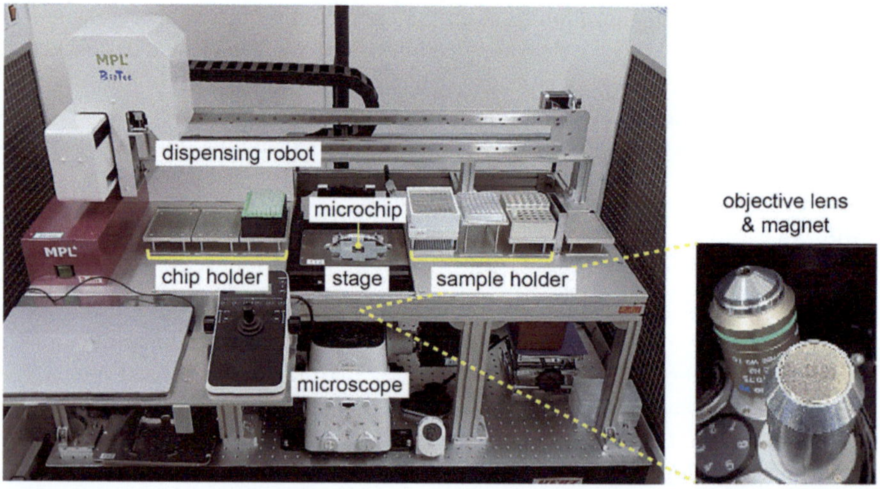

Fig. 4 Photograph of opn-SATORI

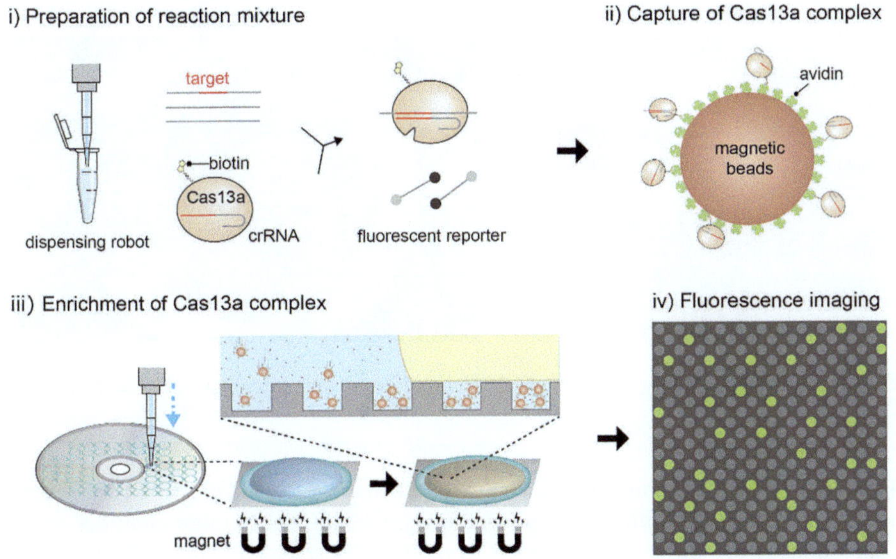

Fig. 5 Schematics of digital RNA detection by opn-SATORI

magnetic beads via biotin-avidin interactions and to enrich them in the microchambers by magnetic force, resulting in a high capture efficiency of the target RNA in the chamber (Fig. 5). This system is fully automated through the development of a custom-made magnetic holder mounted on an electric objective lens turret (Fig. 4), which helps complete the entire process within 9 min. This strategy dramatically improved detection sensitivity by ~46-fold to 1.4 copies/μL, almost equivalent to conventional RT-qPCR (Ct: 39), a level sufficient to diagnose viral infections such as COVID-19.

5 Small Portable Platform (COWFISH)

The opn-SATORI is large, measuring 160 cm wide, 70 cm deep, and 160 cm high. The fluorescence imaging system consists of a confocal microscope, which is extremely expensive. Although it could be installed in large hospitals and testing facilities, installing it in city clinics and quarantine stations was expected to be challenging, limiting the versatile use of SATORI for viral infection diagnosis. To address this issue, we developed a compact and inexpensive alternative fluorescence imaging system for digital bioanalysis, named COWFISH (**Co**mpact **W**ide-Field **F**emtoliter-Chamber **I**maging **S**ystem for **H**igh-Speed Digital Bioanalysis) (Iida et al. 2023b) as an alternative to the confocal microscope (Fig. 6).

Assembled with commercially available LEDs, optical filters, telecentric lenses, and an SLR camera, COWFISH is small and inexpensive, measuring 35 cm wide,

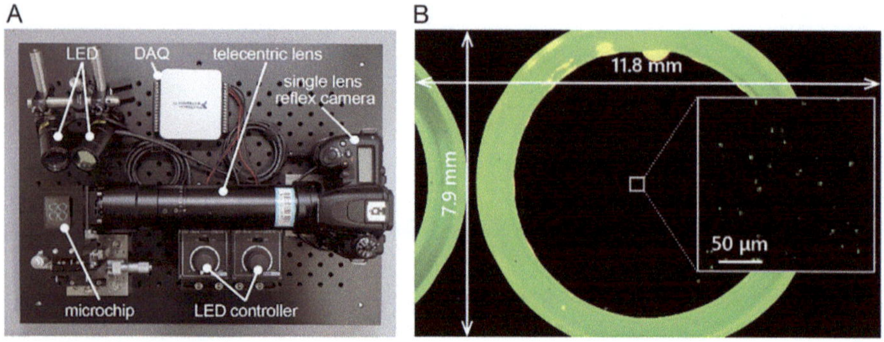

Fig. 6 COWFISH. **a** Photograph of COWFISH. **b** Fluorescence image obtained by SATORI with COWFISH

45 cm deep, and 30 cm high, one-fifth the footprint of opn-SATORI. COWFISH uses a telecentric lens to capture fluorescence images of the entire microchamber array. These images are directly transmitted to the CMOS sensor of the SLR camera at nearly the same magnification and without distortion. Thus, it is possible to acquire approximately 600,000 microchamber images in a single exposure. Owing to the fine pixel size of latest CMOS sensors, the spatial resolution is around 3 μm for green (~532 nm) imaging and 4 μm for red (~640 nm) imaging. Although this resolution is not as high as the sub-micron level achieved by confocal microscopes, fluorescence imaging can be easily conducted by COWFISH using microchambers with a diameter of 3.5 μm, spaced 8 μm apart.

The performance of COWFISH was evaluated using SATORI without magnetic bead enrichment. The detection sensitivity of COWFISH was 480 aM (290 copies/μL) for viral RNA of SARS-CoV-2, which was comparable to that obtained with a confocal microscope. In addition, the large field of view of COWFISH (11.8 mm wide and 7.9 mm high) allows fluorescence images of approximately 600,000 microchambers to be acquired simultaneously, reducing image acquisition time to less than one-tenth that of a confocal microscope. Notably, the small and portable nature of COWFISH allowed it to be installed in a hospital and successfully detected patient-derived SARS-CoV-2 in clinical settings (Lida et al. 2024), demonstrating the feasibility of point-of-care testing for viral infections and its potential application in the digital bioanalysis of extracellular vesicles.

6 Future Perspective

We developed SATORI, a rapid RNA detection technology at the single-molecule level, along with its application in the point-of-care testing of viral infections, including COVID-19 (Valera et al. 2021). SATORI fulfills almost all the requirements for diagnosing viral infections; therefore, it can be adapted for use not only

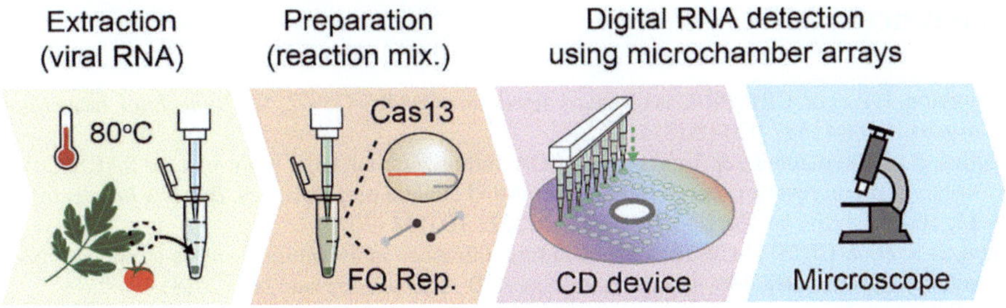

Fig. 7 Schematics of direct-SATORI

for humans but also for other living organisms, such as plants. For example, Direct-SATORI, a derivative of SATORI, has been recently used to detect viral RNA directly from tomato leaves without a purification step (Fig. 7) (Ueda et al. 2023). This capability extends the versatility of SATORI to diagnose viral infections in various living organisms and further shortens the testing time, paving the way for practical applications not only in medicine but also in agriculture, fisheries, and animal husbandry.

SATORI is sufficiently versatile to be applied for the detection of a variety of RNAs, not only from viruses but also from extracellular vesicles in vivo, such as exosomes/microsomes, in liquid specimens. However, further technological developments are necessary to comprehensively analyze the components of individual vesicles. Particularly, it is necessary to:

(i) Increase the variety of molecular species that can be measured.
(ii) Develop multiplex detection techniques to analyze the correlation between multiple molecules in individual vesicles.

Regarding the first point, one possible approach is the simultaneous use of Cas13a and Cas12a targeting DNA (Chen et al. 2018) in SATORI, which allows the simultaneous detection of multiple genetic materials. In addition, using digital bioanalysis techniques categorized into other modalities, such as SEAP and digital ELISA, will significantly expand the variety of molecular species that can be detected, including proteins. Regarding the second point, one possible approach is to develop a novel platform to consistently perform the following processes: fractionation of individual vesicles in a microchamber, extraction of vesicle components, and multiplexed digital bioanalysis developed in the first point. Accordingly, SATORI represents the initial step toward the digital bioanalysis of extracellular vesicles, and ongoing efforts to actively develop and advance digital bioanalysis are highly anticipated in the future for a deeper understanding of extracellular vesicles.

References

Broughton, J. P. *et al.* CRISPR-Cas12-based detection of SARS-CoV-2. *Nat Biotechnol*, https://doi.org/10.1038/s41587-020-0513-4 (2020).

Bruijns, B., Folkertsma, L. & Tiggelaar, R. FDA authorized molecular point-of-care SARS-CoV-2 tests: A critical review on principles, systems and clinical performances. *Biosens Bioelectron X* **11**, 100158, https://doi.org/10.1016/j.biosx.2022.100158 (2022).

Chen, J. S. *et al.* CRISPR-Cas12a target binding unleashes indiscriminate single-stranded DNase activity. *Science* **360**, 436–439, https://doi.org/10.1126/science.aar6245 (2018).

East-Seletsky, A. *et al.* Two distinct RNase activities of CRISPR-C2c2 enable guide-RNA processing and RNA detection. *Nature* **538**, 270–273, https://doi.org/10.1038/nature19802 (2016).

Gootenberg, J. S. *et al.* Nucleic acid detection with CRISPR-Cas13a/C2c2. *Science* **356**, 438–442, https://doi.org/10.1126/science.aam9321 (2017).

Gosert, R., Naegele, K. & Hirsch, H. H. Comparing the Cobas Liat Influenza A/B and respiratory syncytial virus assay with multiplex nucleic acid testing. *J Med Virol* **91**, 582–587, https://doi.org/10.1002/jmv.25344 (2019).

Hoshino, A. *et al.* Extracellular Vesicle and Particle Biomarkers Define Multiple Human Cancers. *Cell* **182**, 1044–1061 e1018, https://doi.org/10.1016/j.cell.2020.07.009 (2020).

Iida, T., Shinoda, H. & Watanabe, R. SATORI: Amplification-free digital RNA detection method for the diagnosis of viral infections. *Biophys Physicobiol* **20**, e200031, https://doi.org/10.2142/biophysico.bppb-v20.0031 (2023a).

Iida, T. *et al.* Compact wide-field femtoliter-chamber imaging system for high-speed and accurate digital bioanalysis. *Lab Chip* **23**, 684–691, https://doi.org/10.1039/d2lc00741j (2023).

Kaminski, M. M., Abudayyeh, O. O., Gootenberg, J. S., Zhang, F. & Collins, J. J. CRISPR-based diagnostics. *Nat Biomed Eng* **5**, 643–656, https://doi.org/10.1038/s41551-021-00760-7 (2021).

Li, Y., Li, S., Wang, J. & Liu, G. CRISPR/Cas Systems towards Next-Generation Biosensing. *Trends Biotechnol* **37**, 730–743, https://doi.org/10.1016/j.tibtech.2018.12.005 (2019).

Iida, T. *et al.* Portable wide-field femtoliter-chamber imaging system for point-of-care digital bioanalysis. *iScience.* https://www.sciencedirect.com/science/article/pii/S2589004224020935 (2024).

Liu, G. & Rusling, J. F. COVID-19 Antibody Tests and Their Limitations. *ACS Sens* **6**, 593–612, https://doi.org/10.1021/acssensors.0c02621 (2021).

Liu, C. *et al.* Single-Exosome-Counting Immunoassays for Cancer Diagnostics. *Nano Lett* **18**, 4226–4232, https://doi.org/10.1021/acs.nanolett.8b01184 (2018).

Noji, H., Minagawa, Y. & Ueno, H. Enzyme-based digital bioassay technology - key strategies and future perspectives. *Lab Chip* **22**, 3092–3109, https://doi.org/10.1039/d2lc00223j (2022).

Pegtel, D. M. & Gould, S. J. Exosomes. *Annu Rev Biochem* **88**, 487–514, https://doi.org/10.1146/annurev-biochem-013118-111902 (2019).

Petralia, S. & Conoci, S. PCR Technologies for Point of Care Testing: Progress and Perspectives. *ACS Sens* **2**, 876–891, https://doi.org/10.1021/acssensors.7b00299 (2017).

Rissin, D. M. *et al.* Single-molecule enzyme-linked immunosorbent assay detects serum proteins at subfemtomolar concentrations. *Nat Biotechnol* **28**, 595–599, https://doi.org/10.1038/nbt.1641 (2010).

Sakamoto, S. *et al.* Multiplexed single-molecule enzyme activity analysis for counting disease-related proteins in biological samples. *Sci Adv* **6**, eaay0888, https://doi.org/10.1126/sciadv.aay0888 (2020).

Shao, H. *et al.* New Technologies for Analysis of Extracellular Vesicles. *Chem Rev* **118**, 1917–1950, https://doi.org/10.1021/acs.chemrev.7b00534 (2018).

Shinoda, H. *et al.* Amplification-free RNA detection with CRISPR-Cas13. *Commun Biol* **4**, 476, https://doi.org/10.1038/s42003-021-02001-8 (2021).

Shinoda, H. *et al.* Automated amplification-free digital RNA detection platform for rapid and sensitive SARS-CoV-2 diagnosis. *Commun Biol* **5**, 473, https://doi.org/10.1038/s42003-022-03433-6 (2022).
Ueda, T. *et al.* Purification/Amplification-Free RNA Detection Platform for Rapid and Multiplex Diagnosis of Plant Viral Infections. *Anal Chem* **95**, 9680–9686, https://doi.org/10.1021/acs.analchem.3c01691 (2023).
Valera, E. *et al.* COVID-19 Point-of-Care Diagnostics: Present and Future. *ACS Nano* **15**, 7899–7906, https://doi.org/10.1021/acsnano.1c02981 (2021).
Wang, C. *et al.* Droplet digital PCR improves urinary exosomal miRNA detection compared to real-time PCR. *Clin Biochem* **67**, 54–59, https://doi.org/10.1016/j.clinbiochem.2019.03.008 (2019).
Wölfel, R. *et al.* Virological assessment of hospitalized patients with COVID-2019. *Nature* **581**, 465–469, https://doi.org/10.1038/s41586-020-2196-x (2020).
Xu, M. *et al.* COVID-19 diagnostic testing: Technology perspective. *Clin Transl Med* **10**, e158, https://doi.org/10.1002/ctm2.158 (2020).

Mechanisms of Asymmetrical Exosome Release from Polarized Epithelial Cells: Implications for the Molecular Basis of Exosomal Heterogeneity

Mitsunori Fukuda

Abstract Exosomes are small extracellular vesicles of endocytic origin and are secreted into the extracellular space by fusion between intraluminal vesicle (ILV)-containing multivesicular bodies (MVBs) and the plasma membrane. It has recently been shown that a single cell such as an epithelial cell secretes distinct types of exosomes having different protein compositions (referred to as "exosomal heterogeneity"). Polarized epithelial cells have two distinct plasma membrane domains (an apical plasma membrane and a basolateral plasma membrane), which are separated by tight junctions. Two distinct exosome subtypes, i.e., apical exosomes and basolateral exosomes, are asymmetrically secreted from the apical side and basolateral side, respectively, of the cell. These two subtypes are independently regulated by distinct mechanisms in terms of ILV (exosome precursor) biogenesis and MVB transport to the plasma membrane. On the apical side, ILV formation is regulated by the ALIX–syntenin-1–syndecan-1 complex and the MVB formed is transported to the apical plasma membrane by small GTPases Rab27 and Rab37. On the basolateral side, ILV formation is regulated by the neutral sphingomyelinase 2 (nSMase2)-mediated ceramide production machinery, and the MVB formed is transported to the basolateral plasma membrane by Rab39–UACA–BORC. These mechanisms are also retained in nonpolarized cells and are likely to contribute to the genesis of exosomal heterogeneity in nonpolarized cells.

Keywords Ceramide · ESCRT · Exosome · Multivesicular body (MVB) · Rab small GTPase · SNARE

M. Fukuda (✉)
Laboratory of Membrane Trafficking Mechanisms, Department of Integrative Life Sciences, Graduate School of Life Sciences, Tohoku University, Aobayama, Aoba-Ku, Sendai, Miyagi 980-8578, Japan
e-mail: nori@tohoku.ac.jp

Y. Baba et al. (eds.), *Extracellular Fine Particles*,
https://doi.org/10.1007/978-981-97-7067-0_3

1 Introduction

Release of extracellular vesicles (EVs) is a novel means of cell-to-cell communication, and they are generally classified into two types based on their origins: plasma membrane-derived EVs and endomembrane-derived EVs (D'Angelo et al. 2023; Dixson et al. 2023). Plasma membrane-derived EVs include microvesicles, protrusion-derived vesicles, migrasomes, and apoptotic bodies, and their size is generally large, ranging from 100 nm to 5 μm in diameter (thus collectively referred to as large EVs). By contrast, exosomes, which are representative of endomembrane-derived EVs, are small EVs (50–150 nm in diameter) of late endocytic or multivesicular body (MVB) origin. Although the biogenesis of large EVs apparently differs with the large EV species, exosome biogenesis and release appears to be achieved by a common pathway as follows. First, intraluminal vesicles (ILVs), exosome precursors, are formed in late endosomes ((i) "ILV formation step"). The resulting ILV-containing late endosomes (named "multivesicular bodies (MVBs)") are then transported to the plasma membrane ((ii) "MVB transport step"). Finally, the MVBs fuse with the plasma membrane and release their content in the form of exosomes ((iii) "MVB fusion step").

It is now widely recognized that a single cell secretes distinct types of exosomes in terms of their size and content (Fig. 1, left) (Colombo et al. 2013; Kowal et al. 2016; Zhang et al. 2018), and considerable attention is now being paid to exosomal heterogeneity (Kalluri and LeBleu 2020; Mathieu et al. 2019). However, the mechanism of production and the physiological significance of the individual exosome subtypes are poorly understood. To understand the significance of exosome subtypes, we have recently focused our attention on polarized epithelial cells, which have two distinct plasma membrane domains, an apical plasma membrane facing the lumen and a basolateral plasma membrane facing the extracellular matrix or adjacent cells (Fig. 1, right). These two membrane domains are physically and spatially separated by tight junctions, and thereby preventing the exosomes released from the apical plasma membrane (named "apical exosomes") and the exosomes released from the basolateral plasma membrane (named "basolateral exosomes") from intermingling either in vivo or in in vitro 2-dimensional (2D) monolayer cultures.

Recent proteomic analyses of apical exosomes and basolateral exosomes have revealed that their protein compositions differ, and that most of the known exosome markers, including CD63 and CD9, are enriched in only one or the other of the two types of exosomes (Davies et al. 2020; Matsui et al. 2021). In addition, the molecular mechanisms responsible for the biogenesis and transport of the exosome-precursor-containing MVBs of the apical exosomes and basolateral exosomes have been shown to be completely different. In this chapter, I provide an overview of the mechanisms of asymmetrical exosome release from polarized epithelial cells and discuss the possible involvement of these mechanisms in producing exosomal heterogeneity in nonpolarized cells and in neurons exhibiting axon–dendrite polarity.

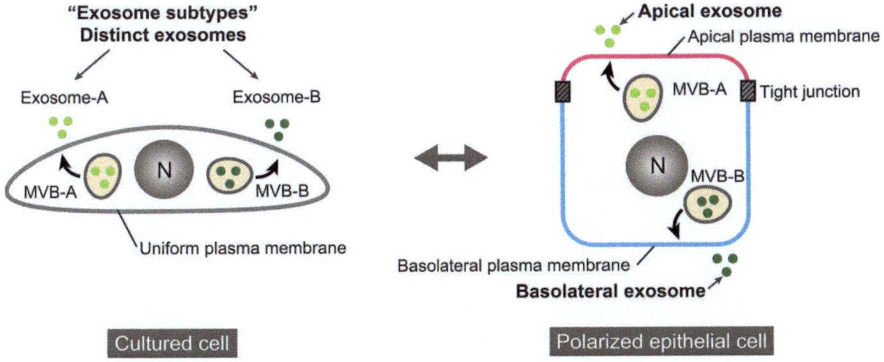

Fig. 1 Exosome subtypes in cultured cells and polarized epithelial cells. (Left) Most cultured mammalian cells have a uniform plasma membrane. However, they secrete distinct types of exosomes (referred to as "exosome subtypes" or "exosomal heterogeneity") (Colombo et al. 2013; Kowal et al. 2016; Zhang et al. 2018), suggesting the presence of distinct exosome biogenesis machineries, but their physiological significance has been poorly understood. (Right) Asymmetrical exosome rerelease from polarized epithelial cells. The cells have two distinct plasma membrane domains, ie, an apical plasma membrane (magenta line) and basolateral plasma membrane (cyan line), which are spatially and physically separated by tight junctions. Thus, the exosomes secreted from the apical plasma membrane (named "apical exosomes"; light green) and the exosomes secreted from the basolateral plasma membrane (named "basolateral exosomes"; dark green) do not intermingle in our body (Matsui et al. 2021)

2 Asymmetrical Exosome Rerelease from Polarized Epithelial Cells

Polarized MDCK cells in 2D monolayer culture secrete small extracellular vesicles (sEVs), including exosomes, differently from their apical side and basolateral side. The results of electron-microscopic examinations together with nanoparticle tracking analysis (NTA) have indicated that there is no clear difference in the size and morphology of the sEVs on the two sides, e.g., they are approximately 50–100 nm in diameter. Proteomic analyses have indicated that most of the known exosome markers are enriched in the exosomes on only one side or the other, e.g., the tetraspanin CD63 and flotillin 1 (FLOT1) are enriched in apical exosomes, whereas the tetraspanins CD9 and CD81 are enriched in basolateral exosomes (Fig. 2), and that GPRC5C, a G protein-coupled receptor (GPCR), is the only apical-exosome-specific marker (Matsui et al. 2021). However, the fact that CD63 is a marker for apical exosomes does not mean that it is exclusively present in apical exosomes, because a smaller amount of CD63 than in apical exosomes is always present in basolateral exosomes, and the same is true in regard to CD9 and CD81 in apical exosomes. Although CD9 is also known to be secreted in the form of microvesicles, which are plasma membrane-derived large EV (lEVs) (Mathieu et al. 2021), the purified apical and basolateral exosome fractions used for mechanistic analyses are not contaminated

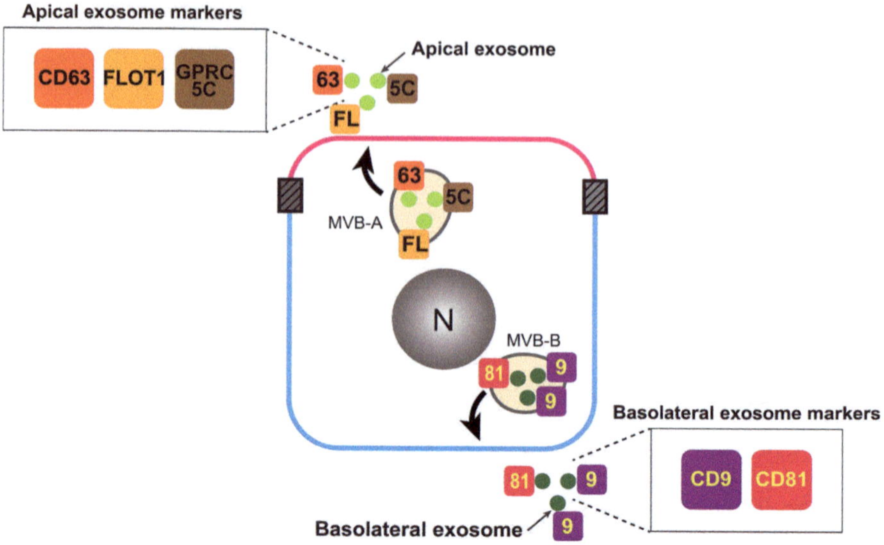

Fig. 2 Apical exosomes and basolateral exosomes have distinct protein compositions. Proteomic analyses of apical and basolateral exosomes from polarized epithelial cells have shown that they consist of different proteins (or exosomal markers) (Davies et al. 2020; Matsui et al. 2021). In polarized MDCK cells, CD63 and flotillin-1 (FLOT1) are enriched in apical exosomes, and GPRC5C, a kind of G protein-coupled receptors, is the only known apical-exosome-specific marker (Matsui et al. 2021). By contrast, CD9 and CD81 are enriched in basolateral exosomes

by these CD9-containing lEVs, because lEVs can be easily removed by centrifugation and/or filtration through a 0.22-μm pore filter (Matsui et al. 2021). In addition, no annexin I, a known microvesicle-specific marker (Jeppesen et al. 2019) is contained in immuno-affinity-purified exosome samples, strongly indicating that the exosomes used in the study by Matsui et al. (2021) were bona fide exosomes.

3 Mechanisms of the Biogenesis of Apical Exosomes and Basolateral Exosomes

Exosome precursors are formed in late endosomes by inward invagination of their limiting membrane to form ILVs ("ILV formation step"), and ILV-containing late endosomes are called MVBs. The best characterized regulator of ILV formation in MVBs is the ESCRT (endosomal sorting complexes required for transport) machinery (Hurley 2015; Vietri et al. 2020), although involvement of the ESCRT machinery in exosome biogenesis has been a matter of controversy, and certain MVBs can be formed in an ESCRT-independent manner (Colombo et al. 2013; Trajkovic et al. 2008). Inhibition of the ESCRT machinery in polarized MDCK cells, however, has been shown to dramatically promote both apical and basolateral exosome release, presumably by inhibiting lysosomal functions (Matsui et al. 2021). Similar phenotypes have also been observed in bafilomycin A1 (a V-ATPase inhibitor)-treated MDCK cells. Consistent with these findings, inhibition of fusion between MVBs and lysosomes has recently been reported to also promote exosome release (Shelke et al. 2023).

Although the inhibition of the ESCRT machinery in MDCK cells did not inhibit exosome release, depletion of the ESCRT-associated protein ALIX has been found to specifically inhibit apical exosome release without affecting basolateral exosome release (Matsui et al. 2021). On the apical side, ALIX, together with syntenin-1 and syndecan-1, promotes ILV formation independently of the ESCRT machinery (Baietti et al. 2012) ("MVB-A" in Fig. 3). Basolateral exosome release is also ESCRT-independent, and neutral sphingomyelinase 2 (nSMase2)-mediated ceramide production machinery specifically promotes ILV formation on the basolateral side (Trajkovic et al. 2008) ("MVB-B" in Fig. 3). Actually, pharmacological inhibition of nSMase2 by GW4869 inhibits basolateral exosome release without affecting apical exosome release. Importantly, the ALIX–syntenin-1–syndecan-1 complex and the nSMase2-mediated ceramide production machinery independently regulate apical exosome release and basolateral exosome release, respectively (Matsui et al. 2021).

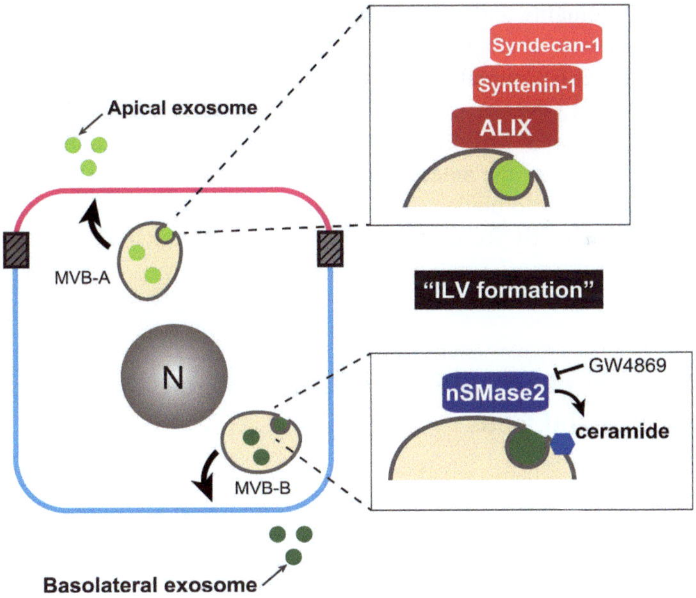

Fig. 3 Distinct exosome biogenesis machineries in polarized epithelial cells. Precursors of apical exosomes (ILVs in MVB-A) are formed by the ALIX–syntenin-1–syndecan-1 complex (Baietti et al. 2012). Precursors of basolateral exosomes (ILVs in MVB-B) are formed by nSMase2-mediated ceramide production machinery independently of the ESCRT function (Trajkovic et al. 2008). GW4869 is a specific inhibitor of nSMase2. These two machineries are regulated independently in polarized MDCK cells (Matsui et al. 2021)

4 Mechanisms of Rab-Mediated MVB Transport in Epithelial Cells

The two distinct MVBs formed by different ILV formation mechanisms described above must be specifically transported to either the apical plasma membrane or the basolateral plasma membrane ("MVB transport step"). Rab small GTPases are conserved regulators of membrane traffic in all eukaryotes, and approximately 60 different Rab isoforms are present in mammals (Homma et al. 2019, 2021; Stenmark 2009). Distinct sets of Rab isoforms separately mediate apical exosome release and basolateral exosome release (Matsui et al. 2022). Rab27 (specifically Rab27A and Rab27B in vertebrates) and Rab37 are independently involved in the apical transport of MVBs ("MVB-Aa" and "MVB-Ab", respectively, in Fig. 4), whereas Rab39 (specifically Rab39A and Rab39B in vertebrates) are involved in the basolateral transport of MVBs ("MVB-B" in Fig. 4). Rab27 mediates apical exosome release through interaction with its specific effector(s), e.g., Slac2-b, Slp4-a, and Munc13-4 (Fukuda 2013; Messenger et al. 2018; Ostrowski et al. 2010). By contrast, the mechanism of Rab37-mediated exosome release (i.e., identification of the Rab37 effector) is completely unknown. Rab39 mediates basolateral exosome release through interaction with its specific effector uveal autoantigen with coiled-coil domains and ankyrin repeats (UACA) (Mori et al. 2013), which then recruits lyspersin, a subunit of the BLOC1-related complex (BORC) (Fig. 4). BORC is an MVB/lysosomal membrane-resident complex consisting of eight subunits (BLOS1, BLOS2, snapin, KXD1, myrlysin/BORCS5, lyspersin/BORCS6, diaskedin/BORCS7, and MEF2BNB/BORCS8), and through interaction with a kinesin motor it regulates the anterograde transport of MVBs and lysosomes to the cell periphery along microtubules (Bonifacino and Neefjes 2017). Thus, it is likely that the Rab39–UACA–BORC complex promotes MVB transport to the basolateral plasma membrane along microtubules.

5 Mechanisms of SNARE-Mediated Exosome Release from Epithelial Cells

MVBs transported to just beneath the plasma membrane eventually fuse with the plasma membrane and release ILVs into the extracellular space in the form of exosomes ("MVB fusion step" in Fig. 5). In general, membrane fusion is promoted by a specific soluble *N*-ethylmaleimide-sensitive factor attachment protein receptor (SNARE) complex, that consists of one R-SNARE (formerly called vesicle-SNARE) on the transport vesicles/organelles and two or three Q-SNAREs (Qa-, Qb-, Qc, or Qbc-SNARE; formerly called target-SNAREs) on the target membrane (Hong 2005; Jahn and Scheller 2006). The SNARE complex forms a parallel four-helix bundle, with each SNARE contributing one helix (or two helices from Qbc-SNARE) to the bundle, that drives membrane fusion. Several individual SNARE proteins have been

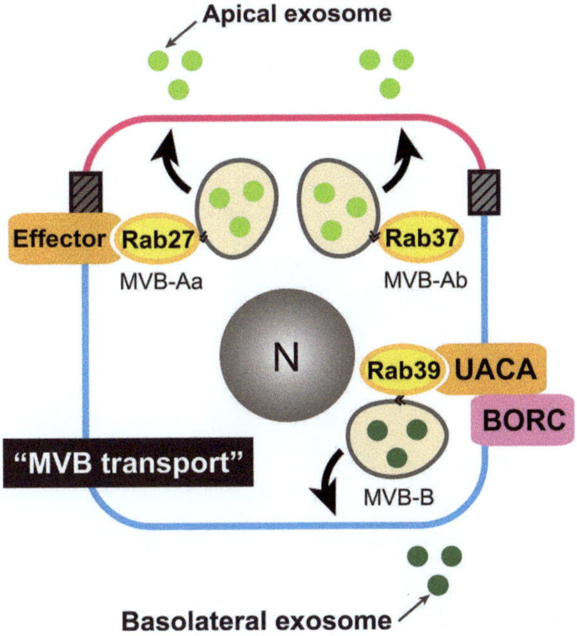

Fig. 4 Distinct MVB transport machineries in polarized epithelial cells. MVB-Aa and MVB-Ab contain apical exosome precursors, and they are independently transported to the apical plasma membrane by the small GTPases Rab27 and Rab37, respectively. Rab27 effectors, Slp4-a, Slac2-b, and/or Munc13-4, are presumably involved in this process (Ostrowski et al. 2010; Messenger et al. 2018). MVB-B, which contains basolateral exosome precursors, is transported to the basolateral plasma membrane by the Rab39–UACA–BORC complex (Matsui et al. 2022)

reported to be involved in exosome release (Fader et al. 2009; Hessvik et al. 2023; Hyenne et al. 2015; Mitani et al. 2022; Peak et al. 2020; Verweij et al. 2018; Wang et al. 2021b; Wei et al. 2017), but the specific SNARE complex that regulates the MVB–plasma membrane fusion has only been identified recently (Liu et al. 2023; Matsui et al. 2023). Among the mammalian SNARE proteins, R-SNARE vesicle-associated membrane protein 5 (VAMP5) is well localized at CD63-positive MVBs in MDCK cells, and its depletion causes a dramatic reduction in both apical and basolateral exosome release (Matsui et al. 2023). At least two distinct SNARE complexes, i.e., VAMP5–synaptosome-associated protein 47 kDa (SNAP47)–syntaxin 1 (STX1) and VAMP5–SNAP47–syntaxin 4 (STX4), mediate exosome release from polarized MDCK cells (Fig. 5). The VAMP5–SNAP47–STX1 complex mediates both apical and basolateral exosome release, whereas the VAMP5–SNAP47–STX4 specifically mediates basolateral exosome release alone. Consistent with these findings, Qa-SNARE STX1 is localized both at the apical plasma membrane and at the basolateral plasma membrane, whereas another Qa-SNARE, STX4, is exclusively localized at the basolateral plasma membrane (Fig. 5). Thus, exosomal heterogeneity is produced even at the MVB–plasma membrane fusion level.

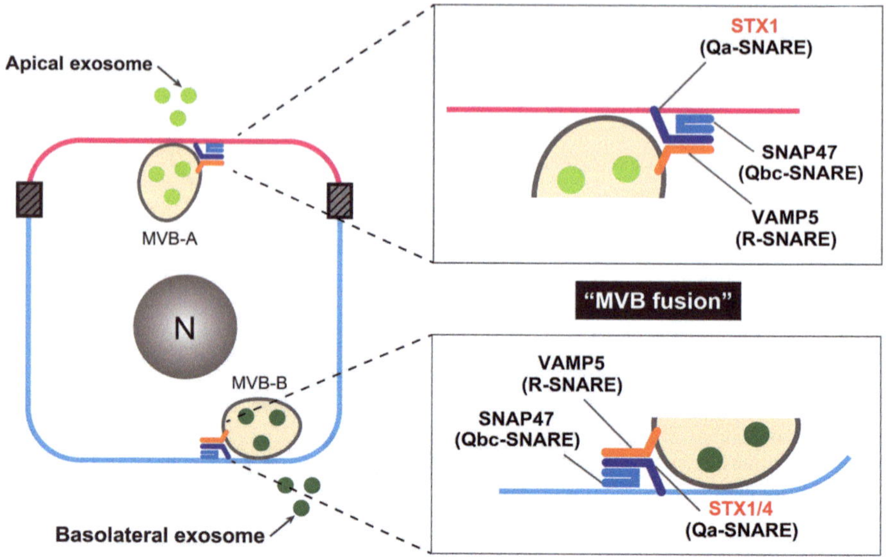

Fig. 5 Role of distinct sets of SNAREs in MVB–plasma membrane fusion in polarized epithelial cells. On the apical side, Qa-SNARE STX1, Qbc-SNARE SNAP47, and R-SNARE VAMP5 mediate fusion between MVB-A and the apical plasma membrane, and thereby enabling the release of apical exosomes. On the basolateral side, a STX1–SNAP47–VAMP5 complex or STX4–SNAP47–VAMP5 complex mediates fusion between MVB-B and the basolateral plasma membrane, thereby enabling the release of basolateral exosomes (Matsui et al. 2023). STX4 is specifically localized at the basolateral plasma membrane, whereas STX1 is localized at both membranes

6 Molecular Basis of Exosomal Heterogeneity in Nonpolarized Cells

Almost all of the exosome regulators (e.g., Rabs and SNAREs) identified in polarized epithelial cells are retained in nonpolarized cells, including HeLa cells and MCF10A cells, and are involved in exosome release (Matsui et al. 2022, 2023), suggesting that the release of heterogeneous exosomes from nonpolarized cells is also regulated by distinct mechanisms at the ILV formation step, MVB transport step, and/or MVB fusion step. For example, depletion of either STX1 or STX4 in HeLa cells suppresses exosome release, although residual exosome release activity is still observed. Intriguingly, simultaneous depletion of STX1 and STX4 in HeLa cells strongly inhibited exosome release, and an additive inhibitory effect was observed (Matsui et al. 2023), indicating that STX1 and STX4 regulate exosome release independently. Since VAMP5 is expressed in a variety of human and mouse tissues (NCBI gene IDs: 10,791 and 53,620, respectively), VAMP5–SNAP47–STX1/4 complexes are presumably the general and major fusion regulators between MVBs and the plasma membrane. However, since several R-SNAREs other than VAMP5 are localized at CD63-positive MVBs (Matsui et al. 2023), an alternate SNARE complex or tissue- or cell-type-specific SNARE complex also contributes to the fusion between MVBs and the plasma membrane. Moreover, since VAMP5 is only conserved in vertebrate species, another R-SNARE must mediate the MVB fusion step in invertebrates. Actually, it has recently been reported that VAMP7–SNAP23–STX4 also regulates the fusion process for exosome release (Liu et al. 2023).

7 Perspectives

Recent accumulating evidence has indicated that heterogeneous exosomes are released from a single cell, but their physiological significance is poorly understood, even in polarized epithelial cells. Exosomes that can activate immune cells (i.e., chemokine and cytokine production) are preferentially released from the basolateral side of renal proximal tubular epithelial cells under inflammatory conditions (Wang et al. 2021a). Since exosomes are enriched in urine (known as "urine exosomes"), apical exosome release may be involved in the release of cellular wastes. Future extensive investigation of the content and function of apical and basolateral exosomes released from polarized epithelial cells (or different exosome subtypes, from nonpolarized cells) will be necessary to fully understand the significance of heterogeneous exosomes. Moreover, since exosomes from cancer cells are known to promote cancer development (Xu et al. 2018), in the future it will be interesting to investigate whether a specific subtype of cancer exosomes, not all exosomes, mediates cancer development.

Another important goal of research will be the identification of exosome subtypes released from other polarized cells besides epithelial cells. More specifically, neurons that have a unique axon–dendrite polarity (Fig. 6) are generally known to release exosomes, but nothing is known about the existence of axonal and dendritic exosomes or their cargo differences. Future isolation and characterization of axonal and dendritic exosomes by using a special microfluid device that can fluidically isolate axons from dendrites and soma will clarify the existence of exosome subtypes in neurons.

In addition, neuronal exosomes are known to be associated with neurodegenerative diseases, including Parkinson's disease (PD) (Hill 2019; Pinnell et al. 2021). Some studies have reported a protective role of exosomes in neurodegenerative diseases, while others have shown that exosomes mediate the propagation of toxic protein (e.g., α-synuclein) aggregates that cause neurodegenerative diseases (Hill 2019). However, it is still unknown whether the opposing functions of exosomes are directly related to the difference in exosomes subtypes in neurons. Analysis of exosome subtypes in neurons described above will be necessary to resolve this issue. Since loss of Rab39B, an abundant Rab39 isoform expressed in the brain, is known to cause early-onset PD and X-linked non-specific intellectual disability (Giannandrea et al. 2010; Koss et al. 2021; Wilson et al. 2014) and to specifically mediate basolateral exosome release from polarized epithelial cells (Matsui et al. 2023), the results of future investigations of Rab39B-mediated exosome release from a specific domain

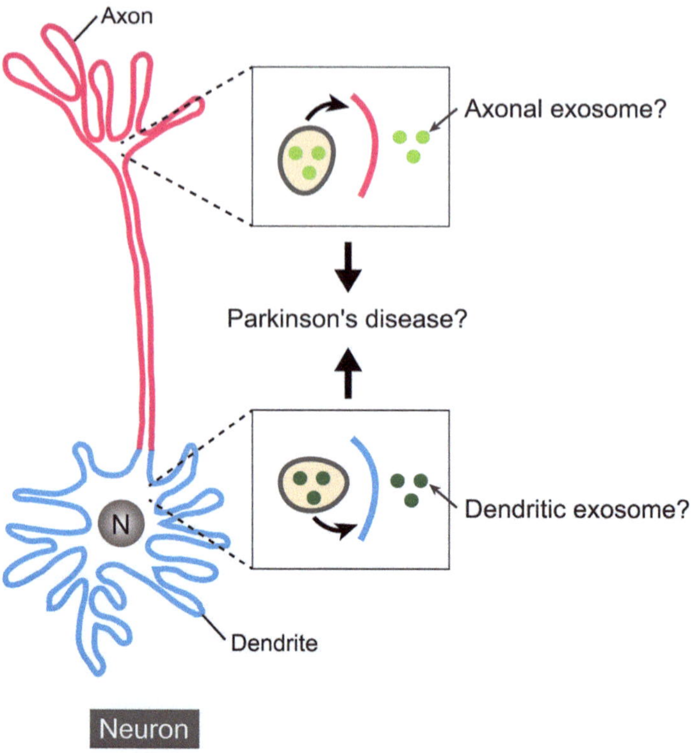

Fig. 6 Neuronal axon–dendrite polarity, and the possible existence of axonal and dendritic exosomes. The same as polarized epithelial cells, neurons possess two membrane domains, an axonal domain and dendritic domain, and thus it is quite likely that neurons secrete axonal exosomes and dendritic (or somatodendritic) exosomes having distinct characters. Rab39B, an important regulator of basolateral exosomes in epithelial cells, is highly expressed in neurons, and its gene mutations are known to be associated with early-onset PD and X-linked non-specific intellectual disability (Giannandrea et al. 2010; Koss et al. 2021; Wilson et al. 2014). Our understanding of the relationship between axonal/dendritic exosomes and PD awaits further investigation

of neurons will provide important clues to understanding the physiological role of exosomes in bran functions as well as the phathogenesis of PD. A long and winding road still lies ahead in our attempt to fully understand the positive and negative roles of exosomes in human diseases.

Acknowledgements This work was supported in part by Grant-in-Aid for Scientific Research (B) 22H02613 from the Ministry of Education, Culture, Sports, Science and Technology (MEXT) of Japan, Japan Science Technology Agency (JST) CREST Grant JPMJCR17H4, and The Uehara Memorial Foundation.

References

Baietti MF et al (2012) Syndecan-syntenin-ALIX regulates the biogenesis of exosomes. Nat Cell Biol 14(7):677–685.

Bonifacino JS, Neefjes J (2017) Moving and positioning the endolysosomal system. Curr Opin Cell Biol 47:1–8.

Colombo M et al (2013) Analysis of ESCRT functions in exosome biogenesis, composition and secretion highlights the heterogeneity of extracellular vesicles. J Cell Sci 126(24):5553–5565.

D'Angelo G et al (2023) Protrusion-derived vesicles: new subtype of EVs? Nat Rev Mol Cell Biol 24(2):81–82.

Davies BA et al (2020) Polarized human cholangiocytes release distinct populations of apical and basolateral small extracellular vesicles. Mol Biol Cell 31(22):2463–2474.

Dixson AC et al (2023) Context-specific regulation of extracellular vesicle biogenesis and cargo selection. Nat Rev Mol Cell Biol 24(7):454–476.

Fader CM et al (2009) TI-VAMP/VAMP7 and VAMP3/cellubrevin: two v-SNARE proteins involved in specific steps of the autophagy/multivesicular body pathways. Biochim Biophys Acta 1793(12):1901–1916.

Fukuda M (2013) Rab27 effectors, pleiotropic regulators in secretory pathways. Traffic 14(9):949–963.

Giannandrea M et al (2010) Mutations in the small GTPase gene *RAB39B* are responsible for X-linked mental retardation associated with autism, epilepsy, and macrocephaly. Am J Hum Genet 86(2):185–195.

Hessvik NP et al (2023) siRNA screening reveals that SNAP29 contributes to exosome release. Cell Mol Life Sci 80(7):177.

Hill AF (2019) Extracellular vesicles and neurodegenerative diseases. J Neurosci 39(47):9269–9273.

Homma Y et al (2019) Comprehensive knockout analysis of the Rab family GTPases in epithelial cells. J Cell Biol 218(6):2035–2050.

Homma Y et al (2021) Rab family of small GTPases: an updated view on their regulation and functions. FEBS J 288(1):36–55.

Hong W (2005) SNAREs and traffic. Biochim Biophys Acta 1744(2):120–144.

Hurley JH (2015) ESCRTs are everywhere. EMBO J 34(19):2398–2407.

Hyenne V et al (2015) RAL-1 controls multivesicular body biogenesis and exosome secretion. J Cell Biol 211(1):27–37.

Jahn R, Scheller RH (2006) SNAREs–engines for membrane fusion. Nat Rev Mol Cell Biol 7(9):631–643.

Jeppesen DK et al (2019) Reassessment of exosome composition. Cell 177(2):428–445.

Kalluri R, LeBleu VS (2020) The biology, function, and biomedical applications of exosomes. Science 367(6478):eaau6977.

Koss DJ et al (2021) Dysfunction of RAB39B-mediated vesicular trafficking in Lewy body diseases. Mov Disord 36(8):1744–1758.

Kowal J et al (2016) Proteomic comparison defines novel markers to characterize heterogeneous populations of extracellular vesicle subtypes. Proc Natl Acad Sci USA 113(8):E968–E977.

Liu C et al (2023) Identification of the SNARE complex that mediates the fusion of multivesicular bodies with the plasma membrane in exosome secretion. J Extracell Vesicles 12(9):e12356.

Mathieu M et al (2019) Specificities of secretion and uptake of exosomes and other extracellular vesicles for cell-to-cell communication. Nat Cell Biol 21(1):9–17.

Mathieu M et al (2021) Specificities of exosome versus small ectosome secretion revealed by live intracellular tracking of CD63 and CD9. Nat Commun12(1):4389.

Matsui T et al (2021) ALIX and ceramide differentially control polarized small extracellular vesicle release from epithelial cells. EMBO Rep 22(5):e51475.

Matsui T et al (2022) Rab39 and its effector UACA regulate basolateral exosome release from polarized epithelial cells. Cell Rep 39(9):110875.

Matsui T et al (2023) VAMP5 and distinct sets of cognate Q-SNAREs mediate exosome release. Cell Struct Funct 48(2):187–198.
Messenger SW et al (2018) A Ca^{2+}-stimulated exosome release pathway in cancer cells is regulated by Munc13-4. J Cell Biol, 217(8):2877–2890.
Mitani F et al (2022) SNAP23-mediated perturbation of cholesterol-enriched membrane microdomain promotes extracellular vesicle production in Src-activated cancer cells. Biol Pharm Bull 45(10):1572–1580.
Mori Y et al (2013) Small GTPase Rab39A interacts with UACA and regulates the retinoic acid-induced neurite morphology of Neuro2A cells. Biochem Biophys Res Commun 435(1):113–119.
Ostrowski M et al (2010) Rab27a and Rab27b control different steps of the exosome secretion pathway. Nat Cell Biol 12(1):19–30.
Peak TC et al (2020) Syntaxin 6-mediated exosome secretion regulates enzalutamide resistance in prostate cancer. Mol Carcinog 59(1):62–72.
Pinnell JR et al (2021) Exosomes in Parkinson disease. J Neurochem 157(3):413–428.
Shelke GV et al (2023) Inhibition of endolysosome fusion increases exosome secretion. J Cell Biol 222(6):e202209084.
Stenmark, H (2009) Rab GTPases as coordinators of vesicle traffic. Nat Rev Mol Cell Biol 10(8):513–525.
Trajkovic K et al (2008) Ceramide triggers budding of exosome vesicles into multivesicular endosomes. Science 319(5867):1244–1247.
Verweij FJ et al (2018) Quantifying exosome secretion from single cells reveals a modulatory role for GPCR signaling. J Cell Biol 217(3)1129–1142.
Vietri M et al (2020) The many functions of ESCRTs. Nat Rev Mol Cell Biol 21(1):25–42.
Wang X et al (2021a) Molecular and functional profiling of apical versus basolateral small extracellular vesicles derived from primary human proximal tubular epithelial cells under inflammatory conditions. J Extracell Vesicles 10(4):e12064.
Wang Y et al (2021b) Syntaxin 2 promotes colorectal cancer growth by increasing the secretion of exosomes. J Cancer 12(7):2050–2058.
Wei Y et al (2017) Pyruvate kinase type M2 promotes tumour cell exosome release via phosphorylating synaptosome-associated protein 23. Nat Commun 8:14041.
Wilson GR et al (2014) Mutations in *RAB39B* cause X-linked intellectual disability and early-onset Parkinson disease with α-synuclein pathology. Am J Hum Genet 95(6):729–735.
Xu R et al (2018) Extracellular vesicles in cancer–implications for future improvements in cancer care. Nat Rev Clin Oncol 15(10):617–638.
Zhang H et al (2018) Identification of distinct nanoparticles and subsets of extracellular vesicles by asymmetric flow field-flow fractionation. Nat Cell Biol 20(3):332–343.

Control of the Particle Trafficking and Dynamics of the Lymphatic System and of the Cellular Microenvironment

Hiroki Tanaka and Hidetaka Akita

1 Introduction: The Lymphatic System as a Target of Drug Delivery

The lymphatic system is a vascular system that runs throughout the body as a second-tier bodily fluid transport system alongside the blood circulatory system. Peripheral lymphatic vessels originate from the subcutaneous tissue and connect to lymph nodes. After antigen-presenting cells (APCs) such as dendritic cells and macrophages in the lymph nodes capture the antigens from pathogens/cancers that flow in through the lymphatic vessels, the APCs display these antigens to the lymphocytes (T cells and B cells) to activate the immunity necessary for defense (Feeney et al. 2020). The lymphatic system is composed of lymph nodes and lymph vessels. This system plays a central role as a site for immune regulation (Louie and Liao 2019), and is a lesion for diseases such as lymph node metastasis and lymphedema (Alitalo 2011). Thus, this system is an intrinsically crucial target for drug discovery.

The lymph fluid flowing through the lymph vessels is generally colorless and transparent, which creates difficulties in visually identifying the lymphatic system. This is one of the reasons that research on lymphatic drug delivery has lagged behind that related to the blood circulatory system. With recent developments in the technology of molecular biology, however, substances such as lymphatic vessel endothelial hyaluronan receptor 1 (LYVE-1) and podoplanin (PDPN) have been identified as marker proteins for lymphatic endothelial cells (LECs) (Breiteneder-Geleff et al.

H. Tanaka (✉) · H. Akita (✉)
Laboratory of DDS Design and Drug Disposition, Graduate School of Pharmaceutical Sciences, Tohoku University, 6-3 Aoba, Aramaki, Aoba-Ku, Sendai City 980-8578, Miyagi, Japan
e-mail: hiroki.tanaka.e1@tohoku.ac.jp

H. Akita
e-mail: hidetaka.akita.a4@tohoku.ac.jp

Y. Baba et al. (eds.), *Extracellular Fine Particles*,
https://doi.org/10.1007/978-981-97-7067-0_4

1999; Banerji et al. 1999). These achievements aided in the identification and isolation of lymphatic vessels from tissues, and research on the functional analysis of LECs in physiological and disease states continues to grow (Oliver et al. 2020).

While blood vessels are closed structures consisting of veins and arteries that circulate blood, capillary lymph vessels in peripheral tissues are finally connected to the subclavian vein via a collection of lymph vessels. The lymph vessels have an open structure with a unidirectional flow of lymph fluid into the subclavian vein. The lymph vessels and the lumen of lymph nodes are formed by LECs, which form structures that differ from those of blood vessel walls (Breslin et al. 2018). Peripheral lymph vessels have no basement membranes formed by collagen and are not supported by lining cells such as the pericytes that are generally found in blood vessels. Additionally, there are gaps between the LECs that act as windows because these adhere to each other through discontinuous connections. Due to these structural characteristics, in addition to draining unnecessary metabolites from tissue fluids and cells in peripheral tissues and organs, the lymphatic system collects substances such as antigens and nanoparticles that have a relatively large molecular weight. Topically administered nanoparticles larger than 3.5 nm cannot cross the blood vessel walls and therefore do not migrate into blood circulation, but they can migrate to the lymphatic system through the gaps in the LECs, and then are distributed to the lymphatic system (Trevaskis et al. 2015). For this reason, substances such as lipid nanoparticles (LNPs) are suitable modalities for targeting the lymphatic system.

Herein, we briefly summarize our recent challenges in researching the lymphatic system: (1) control of nanoparticle trafficking in the lymphatic system; (2) the development of RNA vaccines for topical administration; and, (3) the development of a LEC-targeting nucleic acid delivery system via the use of LNPs (Fig. 1). To help attenuate these challenges, we synthesized the copy of a key component of LNPs, which we refer to as SS-cleavable pH-Activated Lipid-like Material (ssPalm) (Akita et al. 2013; Ukawa et al. 2014; Tanaka et al. 2014). This material contains two hydrophobic scaffolds. The material is functionalized by dual intracellular environment-sensing motifs (Fig. 2). The first scaffold is a tertiary amine that acquires a positive charge in response to an acidic component in endosomes and lysosomes that promotes membrane destabilization and subsequent endosomal escape. The second scaffold is a disulfide bond that can be cleaved in response to a reductive environment (cytoplasm) for spontaneous collapse and cytoplasmic release of RNA molecules (Schafer and Buettner 2001).

2 Control of Nanoparticle Trafficking in the Lymphatic System

To control the function of lymphatic systems, the efficient delivery of drugs to the lymph nodes is a prerequisite. Recently, we comprehensively analyzed the relationship between the physicochemical characteristics of the nanoparticles (liposomes)

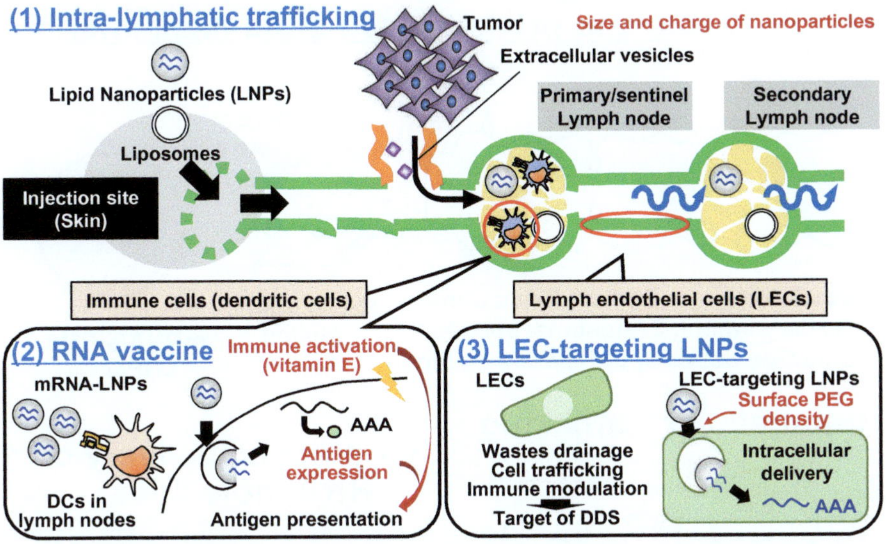

Fig. 1 Drug Delivery Systems (DDS) for lymph. The lymphatic system is a secondary vascular system in the body. The research into a DDS for the lymph has lagged behind that for blood circulation since the flow of colorless lymphatic fluid is difficult to identify. However, current advances in molecular biology has enabled the detailed investigation of lymphatic systems. Three research topics of a DDS for lymph are summarized in this review: (1) the control of nanoparticle trafficking in the lymphatic system (Intra-lymphatic trafficking); (2) the development of RNA vaccine for topical administration (RNA vaccine); and, (3) the development of a LEC-targeting nucleic acid delivery system using LNPs (LEC-targeting LNPs)

and their intra-lymphatic pharmacokinetics (lymph node-to-lymph node trafficking) (Gomi et al. 2021). In normal mice, quantitative analysis of the lymphatic transport is difficult since the lymph vessels run deep into the body. Therefore, to analyze the long-distance transport of the nanoparticles in the lymphatic system, a lymphatic flow-modified (LFM) mouse model was used for the investigation, in which popliteal lymph nodes and the vessels surrounding it were resected (Fig. 3a) (Yamaji et al. 2018). In this mouse model, nanoparticles injected into the footpad are drained into the inguinal lymph node instead of the popliteal lymph node, and then they flow into the axillary lymph node via the lymph vessels that run just beneath the surface of the body.

Fluorescence-labeled liposomal particles with different surface charges and sizes were used as model nanoparticles. To prepare the differently charged liposomes, the cholesterol in the lipid composition was replaced either with cholesteryl hemisuccinate (an anionic derivative) or with DC-cholesterol (a cationic derivative). By using an extruder, small, medium, and large liposomal particles (70 nm, 130 nm, or >300 nm) were prepared. These liposomal particles were injected into the footpad and time-dependent intra-lymphatic trafficking was analyzed via an In Vivo Imaging System (IVIS) for as long as 48 h.

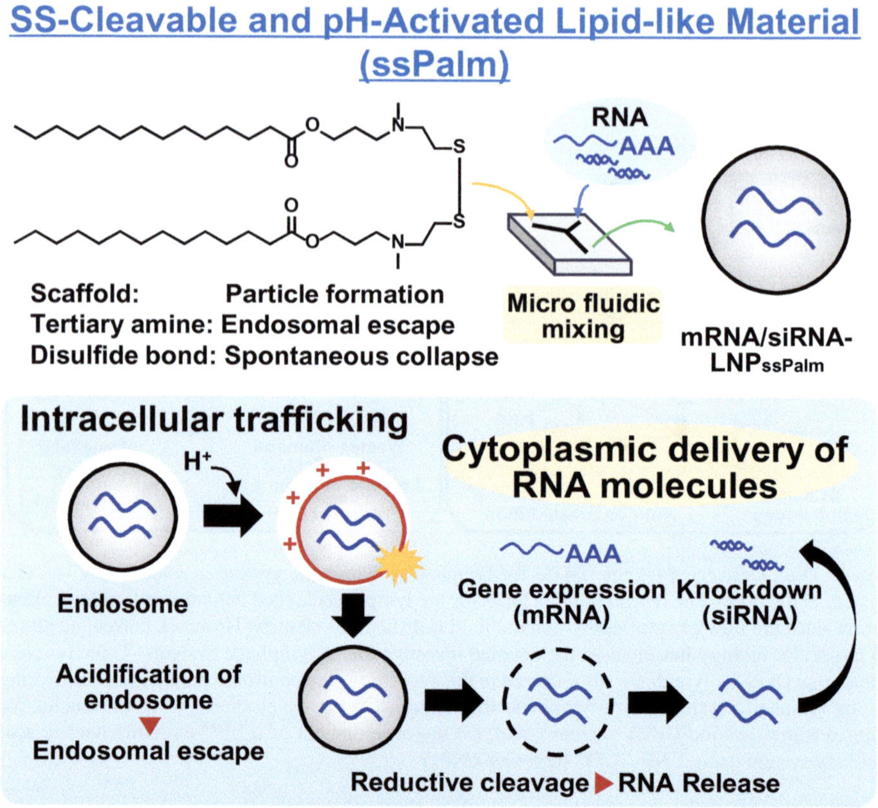

Fig. 2 Concept of SS-Cleavable and pH-Activated Lipid-Like Material (ssPalm). ssPalm contains tertiary amine groups and a disulfide bond as environment-sensing units. The LNPs were formulated using microfluidic mixing technology. Once the LNPssPalm is taken up by cells, the LNPssPalm is exposed to the acidic environment in maturating endosomes. The tertiary amine groups develop a positive charge according to the acidification of the endosomes and then trigger endosomal escape of the LNPssPalm. After the endosomal escape, the disulfide bond is cleaved in the reducing environment in the cytoplasm. This cleavage induces the collapse of the LNPssPalm and promotes intracellular release of the RNA molecules

Based on the relative location of the injection site, the inguinal lymph node and axillary lymph node were defined as the primary lymph node and secondary lymph node, respectively. The accumulation of each fluorescence-labeled particle at the primary and secondary lymph nodes was quantified. This investigation revealed that small and medium anionic liposomal particles efficiently drain into the lymph vessels and accumulate in the lymph nodes. It is noteworthy that medium anionic liposomal particles selectively accumulate in the primary lymph nodes, while the small anionic liposomal particles further pass into the efferent lymphatic flow and into secondary lymph nodes (Fig. 3b). Flow cytometry analysis revealed that the anionic liposomal particles are dominantly taken up by the $CD169^+$ macrophages lining the sinus in the

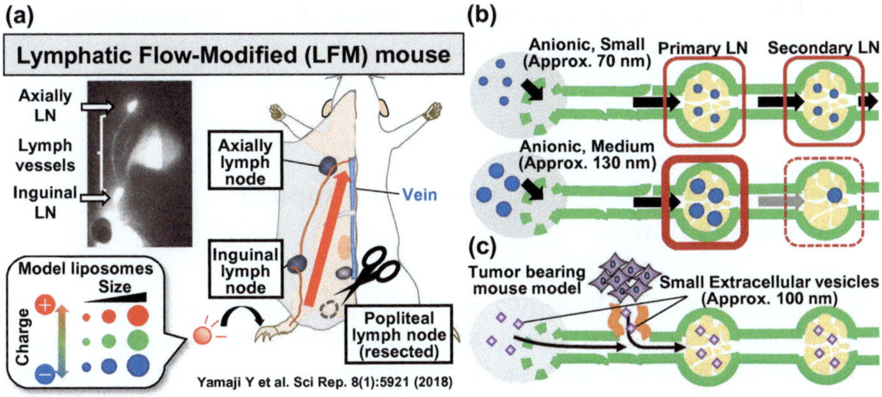

Fig. 3 Analysis of the intra-lymphatic trafficking of nanoparticles. Intra-lymphatic trafficking of artificial nanoparticles (liposomes) and natural nanoparticles (small extracellular vesicles) was investigated. **a** Schematic illustration of a Lymphatic-Flow-Modified (LFM) mouse model. The intra-lymphatic flow was visualized by modifying the flow of lymph fluid by resecting the popliteal lymph node. In LFM mice, the lymph fluid flows just beneath the skin. **b** Intra-lymphatic trafficking of liposomes. Anionic liposomes accumulate in the lymph nodes. Smaller sized particles distribute to multiple lymph nodes while medium sized particles selectively accumulate in the primary lymph node. **c** Intra-lymphatic trafficking of smaller extracellular vesicles. In the cases of both intra-tumoral and subcutaneous injections, the small extracellular vesicles derived from cancer cells are distributed to the lymphatic system

lymph nodes (Junt et al. 2007; Gray and Cyster 2012). These observations suggest that the anionic surface charge is a prerequisite for lymph node targeting, and the size is a key factor for the retention of particles in the lymph nodes.

Small extracellular vesicles (sEVs) are particles of approximately 100 nm in size that are secreted from the cells. These vesicles serve as a natural container for delivering messenger RNA, microRNA, and other mediators. The pharmacokinetics of the sEVs in blood circulation has been generally investigated via intravenous administration (Gudbergsson et al. 2019; Kang et al. 2021). However, particularly in the case of cancer-derived sEVs, they are expected to drain into the lymphatic system. Based on this assumption, the intra-lymphatic trafficking of the sEVs was investigated following topical administration (Sakurai et al. 2023a). Consistent with the previous observations, intravenously administrated B16F0 melanoma cell-derived sEVs mainly accumulated in the liver and spleen. On the other hand, subcutaneously injected sEVs showed preferential accumulation in the primary lymph nodes. Analysis of the intra-lymphatic trafficking using LFM mice has revealed that sEVs injected into the footpad accumulate in inguinal lymph nodes and axillary lymph nodes as in the case of the model liposomal particles. One of the important observations is that sEVs directly injected into a tumor also show an accumulation in the primary lymph nodes (Fig. 3c). These results suggest that natural or artificial nanoparticles that are either in situ-produced or s.c.-administrated primarily flow into the lymphatic vessels. The anionic charge is an important characteristic of the particles that efficiently flow through the lymphatic system. Also, the control of the particle size is one of the strategies that is used to modify lymph node-retention properties.

3 Development of an RNA Vaccine for Topical Administration

RNA vaccines against SARS-CoV-2 are the world's first drug based on an LNP-based mRNA formulation (Kim et al. 2021; Carvalho et al. 2021). By designing the sequence of the mRNA, an RNA vaccine against novel pathogenic microorganisms was rapidly developed once the platform was established for drug delivery system (DDS) technology (Pardi et al. 2018). The management of LNPs is a key technology for RNA vaccines that deliver mRNA into the body as a source of antigens. An ionizable lipid with a tertiary amine moiety is a main component of LNPs. The pH-sensitive characteristics of ionizable lipids confer endosomal escape activity to the LNPs, while the neutral surface of the LNPs in the physiological environment is beneficial for avoiding unspecific electrostatic interactions with biomaterials (Sun and Lu 2023).

Another advantage of RNA vaccines is their ability to activate cellular immunity, which is important for eliminating mutated or infected cells from the body (Sahin et al. 2020; Liu et al. 2021). Since the mRNA is translated to the proteins in the cytoplasm, we assume that the antigens are processed via the proteasome pathway and presented mainly on major histocompatibility class 1 antigens (MHC-I), which is a crucial process in the activation of $CD8^+$ T cells. Therefore, RNA vaccines are considered promising modalities for vaccines against cancers and infectious diseases caused by intracellular parasitic microorganisms (Rojas et al. 2023; Cafri et al. 2020; Ganley et al. 2023; Hayashi et al. 2022). Research has shown that α-tocopherol (Vitamin E) has an adjuvant activity that can further promote the activation of cellular immunity (Kawai et al. 2018; Maeta et al. 2020; Tateshita et al. 2019). This section describes how vitamin E was incorporated into the structure of ssPalm (ssPalmE-P4C2, Fig. 4a) (Oyama et al. 2023).

The mRNA-LNPs composed of the ssPalmE-P4C2 ($LNP_{ssPalmE}$) were prepared using an iLiNP microfluidic device (Kimura et al. 2018). $LNP_{ssPalmE}$ has been used as an RNA vaccine against cancers. The optimal form of mRNA-$LNP_{ssPalmE}$ was identified from the screening of 43 LNPs with different lipid compositions. The optimal lipid composition of mRNA-LNPs is an ssPalmE-P4C2/dioleoyl-sn-glycero phosphoethanolamine (DOPE)/cholesterol/polyethylene glycol conjugated-lipid in a ratio of 60/30/10/3. The optimal mRNA-$LNP_{ssPalmE}$ has shown therapeutic anti-tumor activity against cancer models expressing ovalbumin (E.G.7-OVA) and prophylactic anti-tumor activity against cancer models expressing a human antigen (CT26-hNY-ESO-1). The optimal mRNA-$LNP_{ssPalmE}$ also shows prophylactic activity against

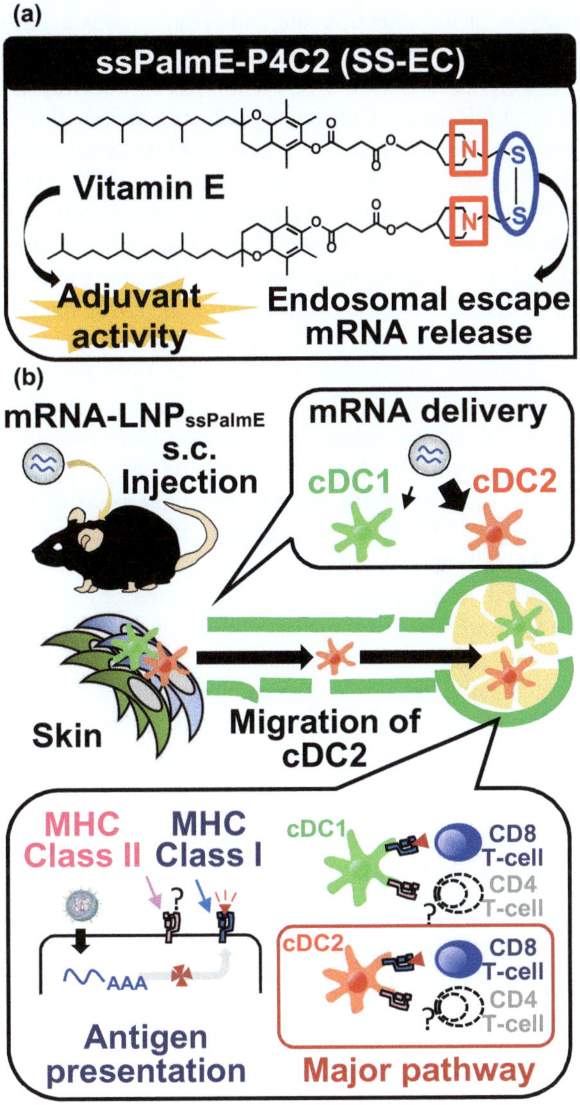

Fig. 4 Development of vitamin E-based ionizable lipids for RNA vaccine. RNA vaccines against cancer and parasites are developed using an ionizable lipid with a vitamin E scaffold. **a** The chemical structure of ssPalmE-P4C2 (SS-EC). The hydrophobic scaffold of the ssPalm was replaced with vitamin E, which has a strong adjuvant effect. **b** The mechanism of action for LNPssPalmE as an RNA vaccine. The LNPssPalmE was taken up by the cDC2 at the injection site. The cells then migrated to the lymph node and activated the $CD8^+$ T cells via antigen presentation

lethal infections of *Toxoplasma gondii* when an antigen derived from this protozoa is coded in the mRNA.

Analysis of the activation of OVA-specific $CD8^+$ T cells (OT-I cells) revealed that mRNA-$LNP_{ssPalmE}$ strongly induces the proliferation and differentiation of these cells. One of the most important observations is that this adjuvant activity depends not on the presence of mRNA molecule, but on the chemical structure of the ionizable lipids. When the vitamin E scaffolds were replaced either with myristic acid scaffolds or oleic acid scaffolds, the adjuvant activity completely disappeared. RNA-seq analysis of the injection site and lymph nodes again revealed the scaffold-specific adjuvant activity of the $LNP_{ssPalmE}$. Pathway analysis of differentially expressed genes indicates that the type-I interferon pathway is key for the adjuvant activity of the $LNP_{ssPalmE}$. Consistent with this observation, treatment with a neutralizing antibody against the receptor of type-I interferon impairs the activation of OT-I cells.

An immunization experiment using a zDC-DTR bone marrow chimeric mice model revealed that conventional dendritic cells (cDCs) are responsible for the antigen presentation. However, the contribution of cDC1s was investigated using XCR-DTR bone marrow chimeric mice and found to be negligible in the case of the mRNA-$LNP_{ssPalmE}$ (Yamazaki et al. 2013; Kitano et al. 2016). This result was unexpected since the cDC1s possess strong cross-presentation activity in which extracellular antigens are processed and presented on the MHC-I (Embgenbroich and Burgdorf 2018). The low contribution of the cDC1s suggests that the proteins processed for the antigen presentation were probably intracellular antigens in the cells other than the cDC1s. The analysis of gene expression suggested that migratory cDC2s at the injection site are potential targets of the mRNA-$LNP_{ssPalmE}$ (Fig. 4b) (Idoyaga et al. 2013). The antigen presentation by the DCs in lymph nodes was successfully visualized via microscopy investigation.

Based on these observations, it seemed obvious that mRNA-LNPs composed of the ssPalmE-P4C2 are potent anti-cancer/anti-parasite vaccines. While the antigen presentation from the APCs to the T cells was observed in the lymph nodes, cDC2s located at the injection site took up the mRNA-LNPs. Thereafter, the cells moved to the lymph nodes via the lymph vessels for the antigen presentation.

4 Using LNPs to Develop LEC-Targeting of the Nucleic Acid Delivery System

Since LECs play a key role in biological processes such as the trafficking of dendritic cells and the migration of tumor cells (Karkkainen et al. 2002; Randolph et al. 2005; Takeda et al. 2023), LECs are attractive targets for nucleic acid delivery in order to control their biological function. LNPs formulated with ssPalm were applied to achieve efficient nucleic acid delivery to the LECs. The efficiency of the intracellular release of nucleic acids is an important property of the LNPs since the activity of a nucleic acid is elicited only when it interacts with appropriate cellular proteins (Tanaka et al. 2020a). Therefore, the biodegradable nature of the nucleic acid vector is one of the important properties.

Analysis of the disulfide cleavage in the ssPalm molecules unexpectedly revealed that the thiol groups produced by the reducing reaction act as nucleophiles that drive hydrolysis reactions in the intraparticle spaces of nanoparticles. Based on this intraparticle hydrolytic reaction, a self-degradable ionizable lipid was designed (Tanaka et al. 2020b). Phenyl ester, a potential substrate for intraparticle hydrolysis, was inserted as a linker between amine groups and hydrophobic scaffolds. The ssPalm molecule with the phenyl ester and oleic acid scaffolds (ssPalmO-Phe) showed self-degradation activity in 6 h when treated with glutathione (Fig. 5a). This self-degradation reaction triggered the release of nucleic acid cargo from the LNPs. The mRNA expression of the $LNP_{ssPalmO\text{-}Phe}$ in the liver following intravenous injection was superior to that of its non-degradable counterpart as well as to that of ionizable lipids that are used in siRNA therapeutics (Akinc et al. 2019). No abnormalities were observed in single-dose toxicity tests on rats.

The ssPalmO-Phe was then used for in vivo siRNA delivery to the LECs (Sakurai et al. 2023b). In targeting LECs, we focused on the mechanism of hepatic uptake by LNPs following *i.v.* administration whereupon apolipoprotein E (ApoE) is absorbed by the surface of the LNPs, and this complex is then recognized by low-density lipoprotein receptors (LDLRs) (Fig. 5b) (Akinc et al. 2010). Since the ApoE concentration was also maintained in the lymph fluid and in the LDLRs existing on the surface of LECs (Milasan et al. 2016), we expected that a similar strategy would be applicable for the targeting of LECs (Fig. 5c). To control the interaction between the surface of LNPs and proteins in the lymph fluid, the amount of PEG-lipids on the LNPs was modified. This revealed that a high density of PEG-lipids with short fatty acid chains (1,2-dimyristoyl-rac-glycero-3-methoxypolyethylene glycol-2000 (DMG-PEG2000)) effectively promotes the siRNA delivery to the LECs. In particular, a modification with 6% of DMG-PEG2000 promoted the delivery of small-interference RNA (siRNA) to the LECs in terms of the gene knockdown activity of VEGFR3. The contribution of ApoE was confirmed using an $ApoE^{-/-}$ mice model. In this transgenic mouse model, the uptake of the LNPs by the LECs was decreased compared with that of normal mice. Pre-incubation with the ApoE improved the uptake of the LNPs by LECs in the $ApoE^{-/-}$ mice. Collectively, the LECs could be targeted in vivo in the ApoE-dependent manner that intrinsically exists in the lymphatic system.

5 Conclusions

In this review, we summarized the effect that physicochemical properties exert on the lymphatic trafficking of liposomes. In particular, ~100 nm anionic particles highly accumulate in the regional lymph nodes and/or in sentinel lymph nodes. Also, in an attempt to develop ssPalm as a platform for the RNA and LEC-targeting of siRNA delivery, a series of derivatives was applied: vitamin E-scaffold materials (ssPalmE) and self-degradable linker-inserted materials (ssPalmO-Phe). The ssPalm material is now commercially available from the NOF CORPORATION. In some cases, however, a large hurdle remains to preparing the LNPs for the life sciences researchers who are tasked with finding a therapeutic target, since they generally possess neither the equipment required for LNP formulation nor the ability to measure the physicochemical properties. However, we recently developed a novel Ready-to-Use-type formulation, which enables the preparation of mRNA-encapsulating LNPs via the simple mixing of an mRNA water solution to create a lyophilized ssPalm

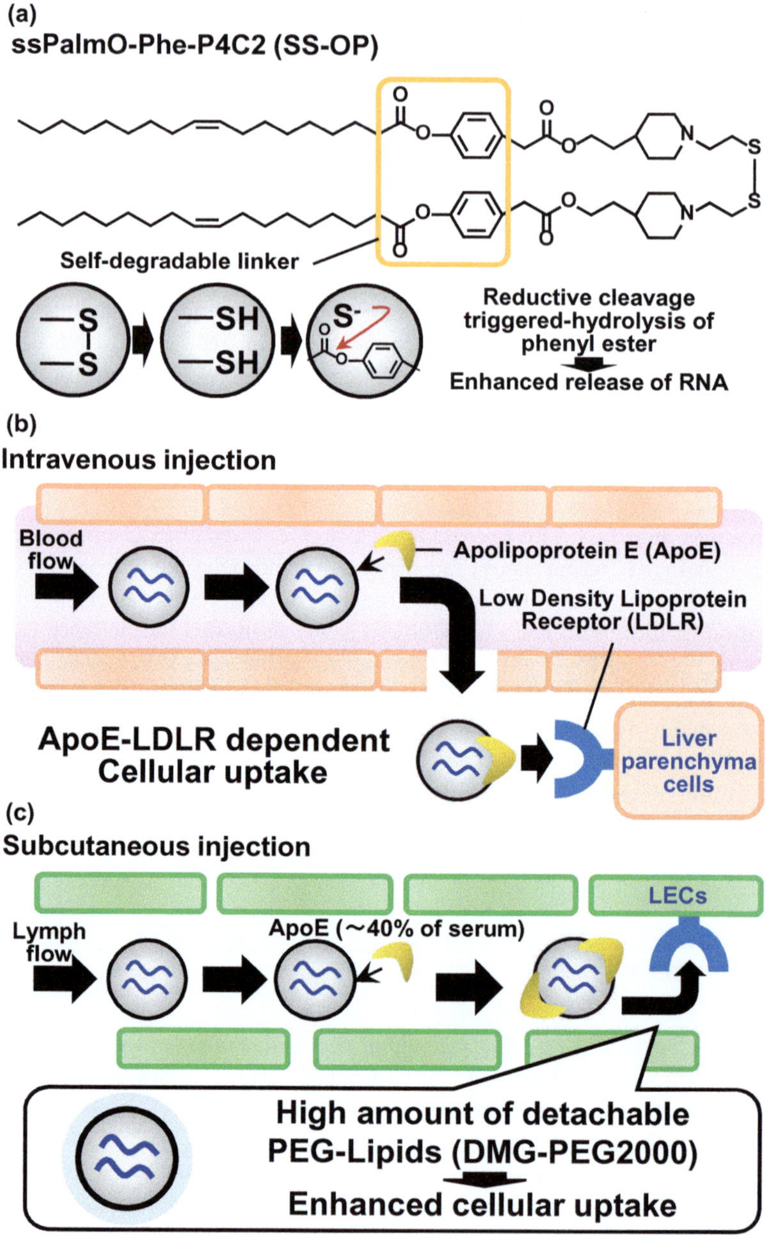
(a)
ssPalmO-Phe-P4C2 (SS-OP)
Self-degradable linker
Reductive cleavage triggered-hydrolysis of phenyl ester
Enhanced release of RNA
(b)
Intravenous injection
Blood flow
Apolipoprotein E (ApoE)
Low Density Lipoprotein Receptor (LDLR)
ApoE-LDLR dependent Cellular uptake
Liver parenchyma cells
(c)
Subcutaneous injection
LECs
Lymph flow
ApoE (~40% of serum)
High amount of detachable PEG-Lipids (DMG-PEG2000)
Enhanced cellular uptake

◀**Fig. 5** Targeting of LECs by self-degradable ionizable lipids. LECs are potential targets of DDS because of their immune-modulating properties. **a** The chemical structure of ssPalmO-Phe-P4C2 (SS-OP). When the LNPssPalmO-Phe is exposed to the reducing environment, a thiol group attacks the phenyl ester and induces hydrolysis of the ionizable lipid. **b** Schematic illustration of the pharmacokinetics of LNPs in the blood. LNPs interact with Apolipoprotein E (ApoE) in the blood. This protein acts as an intrinsic ligand for Low Density Lipoprotein Receptors (LDLRs) and promotes liver delivery of the LNPs. **c** Schematic illustration of the pharmacokinetics of LNPs in the lymphatic fluid. The concentration of ApoE in the lymphatic fluid reaches approximately 40% of that in the blood and LECs express LDLRs. The assumption is that the ApoE-LDLR interaction could be applicable to the targeting of LECs. Higher amounts of detachable PEG-lipids enhance the targeting of LECs via the ApoE-LDLR interaction

cake (Tanaka et al. 2023). We expect this technology to accelerate the innovation of RNA-based therapeutics.

Acknowledgements The authors wish to thank Dr. James McDonald for his helpful advice in writing the English manuscript.

Author Contributions HT and HA designed the review work and wrote the manuscript.

Funding This work was supported in part by JST CREST (JPMJCR17H1).

Competing Interests
H. Tanaka and H. Akita hold the patents on ssPalm chemicals. Also, both authors received joint research funding from the NOF CORPORATION.

References

Akinc A, Querbes W, De S, Qin J, Frank-Kamenetsky M, Jayaprakash KN, et al. Targeted delivery of RNAi therapeutics with endogenous and exogenous ligand-based mechanisms. Mol Ther. 2010;18(7):1357–64. https://doi.org/10.1038/mt.2010.85.

Akinc A, Maier MA, Manoharan M, Fitzgerald K, Jayaraman M, Barros S, et al. The Onpattro story and the clinical translation of nanomedicines containing nucleic acid-based drugs. Nat Nanotechnol. 2019;14(12):1084–7. https://doi.org/10.1038/s41565-019-0591-y.

Akita H, Ishiba R, Hatakeyama H, Tanaka H, Sato Y, Tange K, et al. A neutral envelope-type nanoparticle containing pH-responsive and SS-cleavable lipid-like material as a carrier for plasmid DNA. Adv Healthc Mater. 2013;2(8):1120–5. https://doi.org/10.1002/adhm.201200431.

Alitalo K. The lymphatic vasculature in disease. Nat Med. 2011;17(11):1371–80. https://doi.org/10.1038/nm.2545.

Banerji S, Ni J, Wang SX, Clasper S, Su J, Tammi R, et al. LYVE-1, a new homologue of the CD44 glycoprotein, is a lymph-specific receptor for hyaluronan. J Cell Biol. 1999;144(4):789–801. https://doi.org/10.1083/jcb.144.4.789.

Breiteneder-Geleff S, Soleiman A, Kowalski H, Horvat R, Amann G, Kriehuber E, et al. Angiosarcomas express mixed endothelial phenotypes of blood and lymphatic capillaries: podoplanin as a specific marker for lymphatic endothelium. Am J Pathol. 1999;154(2):385–94. https://doi.org/10.1016/s0002-9440(10)65285-6.

Breslin JW, Yang Y, Scallan JP, Sweat RS, Adderley SP, Murfee WL. Lymphatic Vessel Network Structure and Physiology. Compr Physiol. 2018;9(1):207–99. https://doi.org/10.1002/cphy.c180015.

Cafri G, Gartner JJ, Zaks T, Hopson K, Levin N, Paria BC, et al. mRNA vaccine-induced neoantigen-specific T cell immunity in patients with gastrointestinal cancer. J Clin Invest. 2020;130(11):5976–88. https://doi.org/10.1172/jci134915.

Carvalho T, Krammer F, Iwasaki A. The first 12 months of COVID-19: a timeline of immunological insights. Nat Rev Immunol. 2021;21(4):245–56. https://doi.org/10.1038/s41577-021-00522-1.

Embgenbroich M, Burgdorf S. Current Concepts of Antigen Cross-Presentation. Front Immunol. 2018;9:1643. https://doi.org/10.3389/fimmu.2018.01643

Feeney OM, Gracia G, Brundel DHS, Trevaskis NL, Cao E, Kaminskas LM, et al. Lymph-directed immunotherapy - Harnessing endogenous lymphatic distribution pathways for enhanced therapeutic outcomes in cancer. Adv Drug Deliv Rev. 2020;160:115–35. https://doi.org/10.1016/j.addr.2020.10.002.

Ganley M, Holz LE, Minnell JJ, de Menezes MN, Burn OK, Poa KCY, et al. mRNA vaccine against malaria tailored for liver-resident memory T cells. Nat Immunol. 2023;24(9):1487–98. https://doi.org/10.1038/s41590-023-01562-6.

Gomi M, Sakurai Y, Okada T, Miura N, Tanaka H, Akita H. Development of Sentinel LN Imaging with a Combination of HAase Based on a Comprehensive Analysis of the Intra-lymphatic Kinetics of LPs. Mol Ther. 2021;29(1):225–35. https://doi.org/10.1016/j.ymthe.2020.09.014.

Gray EE, Cyster JG. Lymph node macrophages. J Innate Immun. 2012;4(5-6):424–36. https://doi.org/10.1159/000337007.

Gudbergsson JM, Jønsson K, Simonsen JB, Johnsen KB. Systematic review of targeted extracellular vesicles for drug delivery - Considerations on methodological and biological heterogeneity. J Control Release. 2019;306:108–20. https://doi.org/10.1016/j.jconrel.2019.06.006.

Hayashi CTH, Cao Y, Clark LC, Tripathi AK, Zavala F, Dwivedi G, et al. mRNA-LNP expressing PfCSP and Pfs25 vaccine candidates targeting infection and transmission of Plasmodium falciparum. NPJ Vaccines. 2022;7(1):155. https://doi.org/10.1038/s41541-022-00577-8.

Idoyaga J, Fiorese C, Zbytnuik L, Lubkin A, Miller J, Malissen B, et al. Specialized role of migratory dendritic cells in peripheral tolerance induction. J Clin Invest. 2013;123(2):844–54. https://doi.org/10.1172/jci65260.

Junt T, Moseman EA, Iannacone M, Massberg S, Lang PA, Boes M, et al. Subcapsular sinus macrophages in lymph nodes clear lymph-borne viruses and present them to antiviral B cells. Nature. 2007;450(7166):110–4. https://doi.org/10.1038/nature06287.

Kang M, Jordan V, Blenkiron C, Chamley LW. Biodistribution of extracellular vesicles following administration into animals: A systematic review. J Extracell Vesicles. 2021;10(8):e12085. https://doi.org/10.1002/jev2.12085.

Karkkainen MJ, Mäkinen T, Alitalo K. Lymphatic endothelium: a new frontier of metastasis research. Nat Cell Biol. 2002;4(1):E2–5. https://doi.org/10.1038/ncb0102-e2.

Kawai M, Nakamura T, Miura N, Maeta M, Tanaka H, Ueda K, et al. DNA-loaded nano-adjuvant formed with a vitamin E-scaffold intracellular environmentally-responsive lipid-like material for cancer immunotherapy. Nanomedicine. 2018;14(8):2587–97. https://doi.org/10.1016/j.nano.2018.08.006.

Kim J, Eygeris Y, Gupta M, Sahay G. Self-assembled mRNA vaccines. Adv Drug Deliv Rev. 2021;170:83–112. https://doi.org/10.1016/j.addr.2020.12.014.

Kimura N, Maeki M, Sato Y, Note Y, Ishida A, Tani H, et al. Development of the iLiNP Device: Fine Tuning the Lipid Nanoparticle Size within 10 nm for Drug Delivery. ACS Omega. 2018;3(5):5044–51. https://doi.org/10.1021/acsomega.8b00341.

Kitano M, Yamazaki C, Takumi A, Ikeno T, Hemmi H, Takahashi N, et al. Imaging of the cross-presenting dendritic cell subsets in the skin-draining lymph node. Proc Natl Acad Sci U S A. 2016;113(4):1044–9. https://doi.org/10.1073/pnas.1513607113.

Liu G, Zhu M, Zhao X, Nie G. Nanotechnology-empowered vaccine delivery for enhancing CD8(+) T cells-mediated cellular immunity. Adv Drug Deliv Rev. 2021;176:113889. https://doi.org/10.1016/j.addr.2021.113889.

Louie DAP, Liao S. Lymph Node Subcapsular Sinus Macrophages as the Frontline of Lymphatic Immune Defense. Front Immunol. 2019;10:347. https://doi.org/10.3389/fimmu.2019.00347.

Maeta M, Miura N, Tanaka H, Nakamura T, Kawanishi R, Nishikawa Y, et al. Vitamin E Scaffolds of pH-Responsive Lipid Nanoparticles as DNA Vaccines in Cancer and Protozoan Infection. Mol Pharm. 2020;17(4):1237–47. https://doi.org/10.1021/acs.molpharmaceut.9b01262.

Milasan A, Dallaire F, Mayer G, Martel C. Effects of LDL Receptor Modulation on Lymphatic Function. Sci Rep. 2016;6:27862. https://doi.org/10.1038/srep27862

Oliver G, Kipnis J, Randolph GJ, Harvey NL. The Lymphatic Vasculature in the 21(st) Century: Novel Functional Roles in Homeostasis and Disease. Cell. 2020;182(2):270–96. https://doi.org/10.1016/j.cell.2020.06.039.

Oyama R, Ishigame H, Tanaka H, Tateshita N, Itazawa M, Imai R, et al. An Ionizable Lipid Material with a Vitamin E Scaffold as an mRNA Vaccine Platform for Efficient Cytotoxic T Cell Responses. ACS Nano. 2023;17(19):18758–74. https://doi.org/10.1021/acsnano.3c02251.

Pardi N, Hogan MJ, Porter FW, Weissman D. mRNA vaccines - a new era in vaccinology. Nat Rev Drug Discov. 2018;17(4):261–79. https://doi.org/10.1038/nrd.2017.243.

Randolph GJ, Angeli V, Swartz MA. Dendritic-cell trafficking to lymph nodes through lymphatic vessels. Nat Rev Immunol. 2005;5(8):617–28. https://doi.org/10.1038/nri1670.

Rojas LA, Sethna Z, Soares KC, Olcese C, Pang N, Patterson E, et al. Personalized RNA neoantigen vaccines stimulate T cells in pancreatic cancer. Nature. 2023;618(7963):144–50. https://doi.org/10.1038/s41586-023-06063-y.

Sahin U, Muik A, Derhovanessian E, Vogler I, Kranz LM, Vormehr M, et al. COVID-19 vaccine BNT162b1 elicits human antibody and T(H)1 T cell responses. Nature. 2020;586(7830):594–9. https://doi.org/10.1038/s41586-020-2814-7.

Sakurai Y, Ohtani A, Nakayama Y, Gomi M, Masuda T, Ohtsuki S, et al. Logistics and distribution of small extracellular vesicles from the subcutaneous space to the lymphatic system. J Control Release. 2023;361:77–86. https://doi.org/10.1016/j.jconrel.2023.07.043.

Sakurai Y, Yoshikawa K, Arai K, Kazaoka A, Aoki S, Ito K, et al. siRNA delivery to lymphatic endothelial cells via ApoE-mediated uptake by lipid nanoparticles. J Control Release. 2023;353:125–33. https://doi.org/10.1016/j.jconrel.2022.11.036.

Schafer FQ, Buettner GR. Redox environment of the cell as viewed through the redox state of the glutathione disulfide/glutathione couple. Free Radic Biol Med. 2001;30(11):1191–212. https://doi.org/10.1016/s0891-5849(01)00480-4.

Sun D, Lu ZR. Structure and Function of Cationic and Ionizable Lipids for Nucleic Acid Delivery. Pharm Res. 2023;40(1):27–46. https://doi.org/10.1007/s11095-022-03460-2.

Takeda A, Salmi M, Jalkanen S. Lymph node lymphatic endothelial cells as multifaceted gatekeepers in the immune system. Trends Immunol. 2023;44(1):72–86. https://doi.org/10.1016/j.it.2022.10.010.

Tanaka H, Akita H, Ishiba R, Tange K, Arai M, Kubo K, et al. Neutral biodegradable lipid-envelope-type nanoparticle using vitamin A-Scaffold for nuclear targeting of plasmid DNA. Biomaterials. 2014;35(5):1755–61. https://doi.org/10.1016/j.biomaterials.2013.11.016.

Tanaka H, Sakurai Y, Anindita J, Akita H. Development of lipid-like materials for RNA delivery based on intracellular environment-responsive membrane destabilization and spontaneous collapse. Adv Drug Deliv Rev. 2020;154–155:210–26. https://doi.org/10.1016/j.addr.2020.07.001.

Tanaka H, Takahashi T, Konishi M, Takata N, Gomi M, Shirane D, et al. Self-Degradable Lipid-Like Materials Based on "Hydrolysis accelerated by the intra-Particle Enrichment of Reactant (HyPER)" for Messenger RNA Delivery. Advanced Functional Materials. 2020;30(34):1910575. https://doi.org/10.1002/adfm.201910575

Tanaka H, Hagiwara S, Shirane D, Yamakawa T, Sato Y, Matsumoto C, et al. Ready-to-Use-Type Lyophilized Lipid Nanoparticle Formulation for the Postencapsulation of Messenger RNA. ACS Nano. 2023;17(3):2588–601. https://doi.org/10.1021/acsnano.2c10501.

Tateshita N, Miura N, Tanaka H, Masuda T, Ohtsuki S, Tange K, et al. Development of a lipoplex-type mRNA carrier composed of an ionizable lipid with a vitamin E scaffold and the KALA peptide for use as an ex vivo dendritic cell-based cancer vaccine. J Control Release. 2019;310:36–46. https://doi.org/10.1016/j.jconrel.2019.08.002.

Trevaskis NL, Kaminskas LM, Porter CJ. From sewer to saviour - targeting the lymphatic system to promote drug exposure and activity. Nat Rev Drug Discov. 2015;14(11):781–803. https://doi.org/10.1038/nrd4608.

Ukawa M, Akita H, Hayashi Y, Ishiba R, Tange K, Arai M, et al. Neutralized nanoparticle composed of SS-cleavable and pH-activated lipid-like material as a long-lasting and liver-specific gene delivery system. Adv Healthc Mater. 2014;3(8):1222–9. https://doi.org/10.1002/adhm.201300629.

Yamaji Y, Akita S, Akita H, Miura N, Gomi M, Manabe I, et al. Development of a mouse model for the visual and quantitative assessment of lymphatic trafficking and function by in vivo imaging. Sci Rep. 2018;8(1):5921. https://doi.org/10.1038/s41598-018-23693-9.

Yamazaki C, Sugiyama M, Ohta T, Hemmi H, Hamada E, Sasaki I, et al. Critical roles of a dendritic cell subset expressing a chemokine receptor, XCR1. J Immunol. 2013;190(12):6071–82. https://doi.org/10.4049/jimmunol.1202798.

Elucidation of the Mechanisms that Regulate the Quantity and Quality of Exosomes In Cancer

Chitose Oneyama

Abstract Communication among pre-neoplastic cells, malignant cells, and cancer-associated cells in tumor is critical for cancer development and progression. Transmission of messages between cells facilitates the remodeling of the microenvironment and influences tumor growth and invasion. In addition, intercellular communication not only within a tumor but also to distant tissues promotes metastasis of cancer.

1 Introduction

Communication among pre-neoplastic cells, malignant cells, and cancer-associated cells in tumor is critical for cancer development and progression. Transmission of messages between cells facilitates the remodeling of the microenvironment and influences tumor growth and invasion. In addition, intercellular communication not only within a tumor but also to distant tissues promotes metastasis of cancer. Such signaling is mediated through the exchange of extracellular vesicles (EVs) as well as soluble factors (Kalluri and McAndrews 2023).

EVs include diverse vesicles, but can be categorized into subsets such as exosomes and ectosomes/microvesicles, primarily based on their origins. Several proteins such as tetraspanins (CD9, CD63, and CD81), TSG101, Alix, flotillin, and integrins are enriched in EVs and considered as marker molecules of EVs. Although the expression of some of these marker molecules varies among EV-secreting cells, ectosomes arising from plasma membrane budding are enriched in CD9 and CD81, while exosomes are thought to be enriched in CD9, CD63, CD81, syntenin, and Alix (Kugeratski et al. 2021; Kowal et al. 2016; Mathieu et al. 2021).

Exosomes are secretory membrane vesicles originated from the endocytosis pathway (Johnstone et al. 1987). During the maturation process of endosomes, late endosomes become multivesicular bodies (MVBs) harboring intraluminal vesicles (ILVs) formed by the inward plunging of their own membrane. ILVs contain

C. Oneyama (✉)
Division of Cancer Cell Regulation, Aichi Cancer Center Research Institute, Nagoya, Japan
e-mail: coneyama@aichi-cc.jp

Y. Baba et al. (eds.), *Extracellular Fine Particles*,
https://doi.org/10.1007/978-981-97-7067-0_5

various molecules, including proteins, RNA, DNA, lipids and metabolites (Kalluri and LeBleu 2020). When the MVB fuses with the plasma membrane, the internal ILVs are secreted as exosomes (Sung et al. 2020). Depending on the type of cell and the process of MVB formation, the molecules present inside and on the surface of the exosome vary.

In various cells, exosome production is regulated according to their physiological state, and exosomes with unique lipid, protein, and nucleic acid compositions are released. The exosome surface molecules might be customized for exosome-mediated intercellular communication in a tissue- and cell-specific manner. Thus, physiological, pathological, and therapeutic conditions modulate the composition and quantity of secreted exosomes (Arya et al. 2023). In particular, in cancer cells, secretion and cargo-loading of EVs such as exosomes are significantly altered in association with the dysregulation of various signaling pathways which govern cell survival, proliferation, and motility.

From this point of view, research for diagnostic use of EVs obtained from body fluid has been promoted (Zhao et al. 2019). However, the details of the molecular mechanism by which the "quantity" or "quality" of EVs, i.e., their cargo molecules, changes in diseases are still remain unclear.

In exploring the mechanisms that regulate exosome production, we have specifically focused on how cancer signaling affects the formation and fate determination of MVBs. In this article, we review the molecular mechanisms underlying the activation of Src and MEK signaling, which are important for the expression of cancer traits in cells, and their relationship with exosome production, including our results.

Exosomes may be referred as sEVs (small EVs) owing to their size, whereas not all sEVs are exosomes. Supernatants of cell cultures and samples derived from body fluids yield divergent populations of EVs whose origin is unknown (Jeppesen et al. 2019). Currently, it is not easy to purify specific types of vesicles from them or to attribute their effects on recipient cells to specific vesicle types. Therefore, guidelines proposed by the ISEV (MISEV 2018) recommend that the term "extracellular vesicles" be used collectively for particles released outside the cell (Thery et al. 2018). MISEV is a set of guidelines developed by researchers in the field to present standard methods for EV studies and boost the reproducibility of studies conducted at different institutions.

We have developed and used a system in which luminescent proteins are fused to tetraspanins such as CD63, a marker molecule for extracellular vesicles, in order to study extracellular vesicles efficiently (Hikita et al. 2018) (Fig. 1). The majority of particles detected by this system are considered to be derived from exosomes, but strictly speaking, they are CD63-positive sEVs. Our studies presented in this review are concerned with the regulatory mechanisms of endosome-mediated vesicle formation, and the changes in EVs detected that depend on these mechanisms can be attributed to exosomes. Note that we often use the term "exosomes" hereafter, since the description is mainly about the biogenesis of MVB-derived EVs, not the properties and functions of released EVs.

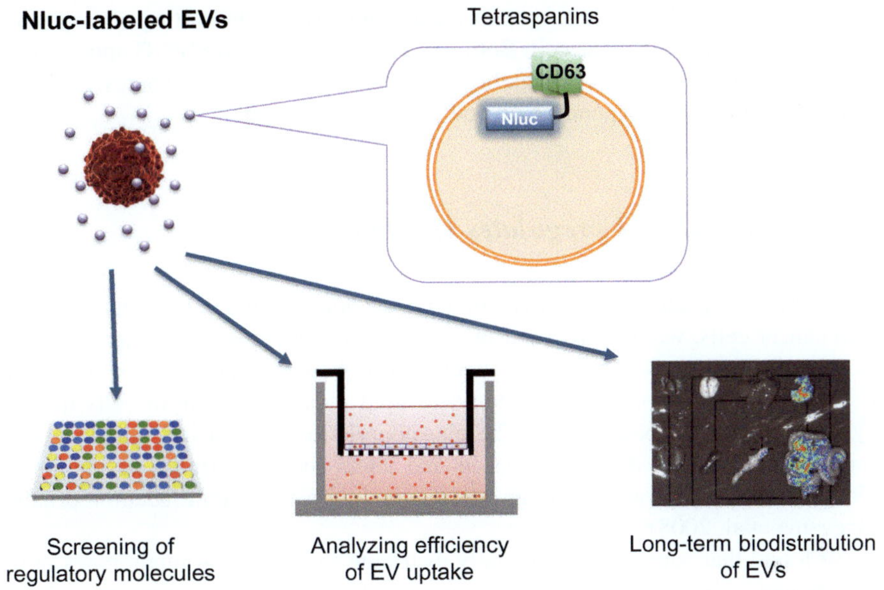

Fig. 1 Application of Nanoluc-labeled exosomes from cells. Cells introduced with tetraspanin-conjugated Nanoluc are useful for screening of regulatory molecules and/or drugs, analyzing uptake efficiency of exosomes, and long-term biodistribution of exosomes. Adapted from the figure in Sci Rep., 9(1):3265, 2019

2 ILV Formation and Src Signaling

2.1 Mechanism of ILV Formation

The formation of ILVs (budding from endosomal membranes), which are essentially exosomes themselves, relies on ESCRT-dependent and -independent mechanisms (Colombo et al. 2013; Stuffers et al. 2009). ESCRTs are complexes involved in membrane remodeling, in which ESCRT-0 and ESCRT-I accumulate with their cargo molecules in lipid microdomains of the endosomal membrane. Accordingly, ESCRT-III is recruited via ESCRT-II. These complexes function together with a number of accessory proteins to bend the membrane and to induce the inward budding of membrane vesicles (Hurley 2008). Furthermore, the ESCRT apparatus and associated proteins such as syntenin, Alix, and CHMP4 (VPS32) are important factors in ESCRT-dependent exosome formation (Baietti et al. 2012).

On the other hand, it has also been shown that MVBs are generated even when the components of the ESCRT complex are deficient, suggesting the mechanism of membrane budding independent of ESCRT (Stuffers et al. 2009). Ceramide molecule produced by enzymatic reaction of neutral type II sphingomyelinase alters the curvature of the membrane. In addition, it has been pointed out that clustering of tetraspanins may induce membrane curvature via their conical structures and budding

(Zimmerman et al. 2016). Tetraspanins are also involved in the sorting of cargo such as integrins into MVBs in the cell. Both ESCRT-dependent and ESCRT-independent mechanisms are thought to be functioned in the formation of exosomes, and their contribution may vary depending on cell type and condition.

2.2 Src Signaling Upregulates ILV Formation

To explore the signaling pathways involved in the upregulation of exosomes secreted from cancer cells, we have focused on the relationship between the mechanism of exosome formation and Src, which is reported to be upregulated in a wide variety of human cancer cells. Src is anchored to lipid membranes within cells through myristoylation of its N-terminus. The phosphorylation state of its regulatory tyrosine residue changes its conformation and alters the localization to membrane microdomains (lipid rafts) through interactions with membrane adaptor proteins (Oneyama et al. 2008).

We found that activated Src partially distribute to endosomal membranes in the transformed cells, which release large amount of exosomes containing Src with its interacting molecules including its substrates (Hikita et al. 2019). Although Src at the plasma membrane, primarily in focal adhesion, is known to be critical in activation of downstream signals required for cancer progression, the role of Src on the endosomal membrane is poorly understood. We also identified Alix, which is thought to facilitate the nipping off the endosomal membrane during ILV formation, as a factor involved in Src-dependent exosome formation.

In human cancer cells, the distribution of Src to the endosomal membrane was unaffected by Alix. This suggests that activated Src is recruited to the endosomal membrane independently of the Alix, where it interacts with Alix. In this process, the proline-rich sequence of Alix is recognized by the SH3 domain of Src, which requires Src to be activated to an SH3-exposed conformation. In addition to the SH3 domain, the kinase activity of Src was required to promote exosome secretion (Fig. 2).

Src has been reported to directly phosphorylate syndecans and syntenins. Syndecans are a family of proteins that interact with various signaling and adhesion molecules via extracellular heparan sulfate chains. Syntenin is an adaptor molecule that binds to the intracellular domain of syndecans and also binds to Alix, and during ILV budding, syndecan-syntenin complexes accumulated at MVB are thought to contribute to the recruitment of the ESCRT machinery (Baietti et al. 2012).

In various human cancers with enhanced Src activation, Src is thought to promote ILV formation and lead to increased exosome secretion through multiple mechanisms, including activation of the syndecan-syntenin complex by its kinase activity and its activation by binding to Alix.

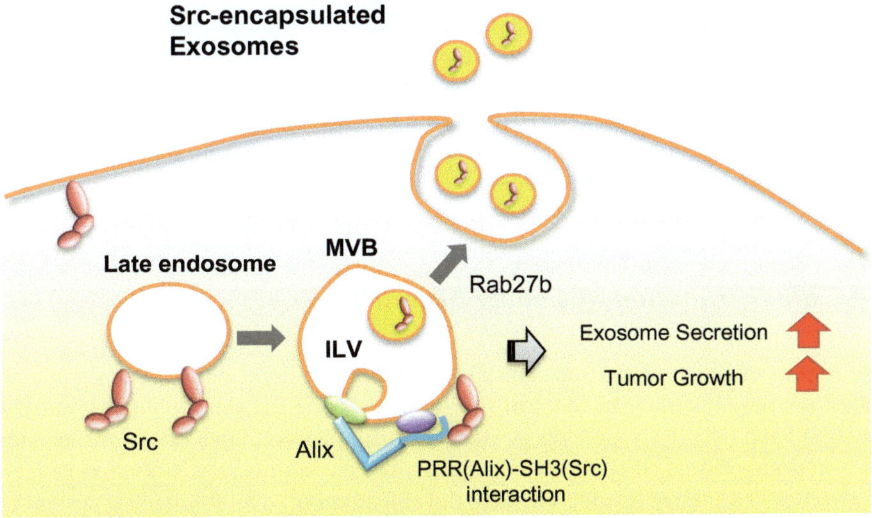

Fig. 2 Src signaling upregulates exosome secretion through promoting ILV formation. In cancer cells, active Src localizes to the endosomal membrane and Src interacts with Alix, promoting ESCRT-dependent ILV formation and exosome secretion

2.3 *Significance of Exosome Upregulation in Src-Activated Cancer Cells*

Although the physiological function of exosome biogenesis enhanced by Src requires further study, it has been reported that exosomes released from cells in which Src activity is inhibited by drugs have a weakened effect on the recipient cells (Imjeti et al. 2017). It is also possible that excess Src is sequestered by MVB/ILV and released into exosomes, thereby maintaining the appropriate level of Src signaling activity in exosome-releasing cells. Indeed, aberrant expression of oncogenes, such as the highly active *v-src*, has often been observed to cause cell death in the early stages of cellular oncogenesis. We have found that knockdown of Alix in Src-transformed cells suppresses the oncogenic phenotypes while decreasing exosome secretion; the Src-Alix interaction not only enhances exosome secretion, but is also important for maintaining the tumorigenic potential of human cancer cells with upregulated Src expression. Notably, the expression of Alix in several tumor tissues correlates with prognosis.

Interestingly, knockdown of Alix in PANC-1 and MIAPaCa-2 pancreatic cancer cell lines suppressed their tumorigenicity. Suppression of cancer phenotypes associated with downregulation of exosome secretion is also caused by knockdown of Rab27b, which is involved in vesicle fusion, and by the neutral sphingomyelinase (N-SMase) inhibitor GW4869, which is widely used to inhibit exosome production. These results imply that defective exosome secretion inhibits cell transformation, regardless of the cause (Hikita et al. 2019). On the other hand, suppression of exosome secretion had little effect on normal cell proliferation. Cancer cell-specific correlation between inhibition of exosome secretion and suppression of cell proliferation was also observed in MEK/ERK signaling-dependent exosome secretion, which will be discussed later. In cancer cells in which intracellular signaling regulation has been disrupted, exosome secretion may be important to maintain cellular homeostasis by discarding excess signaling molecules.

2.4 Biodistrubution of Cancer Secreted EVs

The impact of secreted exosomes or EVs, whether Src-dependent or not, on cancer pathology remains unclear. Most of the studies reported to date to understand EV function have used EVs isolated in vitro and administered large doses of them for short periods of time. It is difficult to discuss the physiological distribution of EVs in vivo based on these results. We have developed an in vivo bioimaging system to monitor EVs circulating in the blood and accumulating in specific organs over time in a subcutaneous xenograft tumor model mouse by utilizing CD63 fused with the high-intensity luciferase Nanoluc (Hikita et al. 2018). We further modified this method and introduced a gene fused to CD63 and BRET reporter Antares2 into tumor cells, and constructed an in vivo imaging system to monitor circulating EVs in the blood and their accumulation in specific organs (Fig. 3). In a mouse model of subcutaneous xenograft tumors, we observed that the administration of a low concentration of the Src inhibitor dasatinib greatly reduce the accumulation of exosomes circulating in the blood from tumor tissue and in organs (Hikita et al. 2020).

Such a system can be used to track the dynamics of cancer-derived EVs and determine the intracellular signals that are important for EV regulation in vivo. Several factors involved in the regulation of EV production have been reported in vitro, but

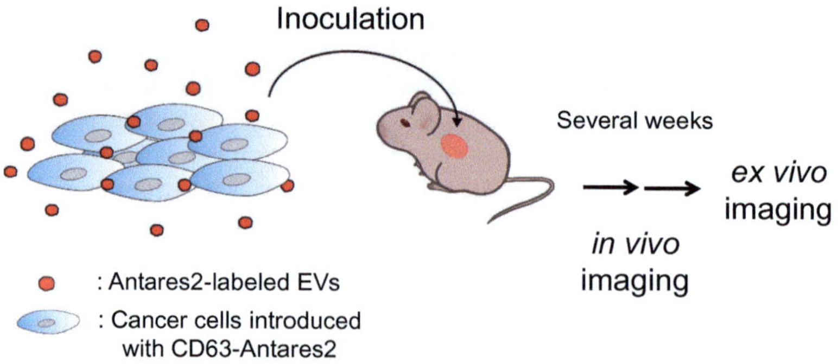

Fig. 3 The tetraspanin-Antares2-expressing cancer cells for in vivo imaging. By transplanting CD63-Antares2-expressing cancer cells into mice, the long-term homing behavior of cancer-derived exosomes released into the bloodstream from the primary tumor to target organs and tissues can be visualized

the extent to which these functions contribute to EV production in vivo and which types of EVs are involved is unknown. Identifying factors involved in specific EV populations would facilitate to clarify the contribution of EVs to cancer progression.

3 Regulation of Contents of EV in Cancer Cells

During the formation of MVB, diverse biomolecules, including proteins and nucleic acids, are incorporated into ILVs. Lipid microdomains enriched with tetraspanins, such as CD9, CD63, and CD81 on exosomes would facilitate protein incorporation into exosomes (Niel et al. 2011; Perez-Hernandez et al. 2013). Certain RNAs are found to be condensed in EVs relative to their producing cells, indicating that a specific RNA loading mechanism regulates the RNAs encapsulated into EVs (Kalluri and McAndrews 2023).

Several factors have been reported as mediators of sorting of protein and RNA into EVs. Recently, it has been suggested that the contact site between the endoplasmic reticulum (ER) and the late endosomal membrane regulates the maturation of endosome, and its association with small GTPases affects the fusion of MVB with the plasma membrane, i.e., exosome secretion (Verweij 2022). Malignant cells usually secrete great amount of EVs than non-malignant cells, which is probably mediated by calcium mobilization from the ER (Taylor et al. 2020). Furthermore, activation of p53 in response to DNA damage is related with increased EV release via TSAP6 (Yu et al. 2006).

Driver genes involved in oncogenesis may affect EV secretion, encapsulation of cargo molecules, and even uptake by recipient cells. AURKB, MYC, and HRAS G12V have been reported to change the size and composition of EVs (Kilinc et al. 2021). Mutant RASs induce EV uptake in cancer cells through micropinocytosis (Choi et al. 2021). Mutant KRAS inhibits Ago2 accumulation in MVBs and EVs and alters miRNA packaging in EVs (McKenzie et al. 2016). Many oncogenic miRNAs have been found to play important roles in tumor development. It has been reported that miRNAs matured from precursor miRNAs in EVs secreted from breast cancer can promote transformation of non-neoplastic epithelial cells (Melo et al. 2014). Although the roles of the DNA carried by EVs in recipient cells are not elucidated, nor is it known whether such transport is functional in vivo, release of EVs containing oncogenic DNA from transformed cell by oncogenic HRAS has been observed (Lee et al. 2014).

These studies suggest that EV secretion driven by oncogenes may increase the risk of cancer development. Future work to identify EV-related factors specific to tumor-forming cells will make it possible to assess the direct involvement of EVs to the risk of cancer progression and will provide an early biomarker for cancer (Kalluri and McAndrews 2023).

4 Fate Determination of MVB and MEK/ERK Signaling

4.1 Fate Determination of MVB and Lysosomes

The formation of MVBs and their fusion with lysosomes is a well-known protein degradation and processing mechanism. For example, certain plasma membrane receptors and MHC class II molecules, when ubiquitinated, are sequestered into ILVs via an ESCRT-dependent mechanism. The MVB containing them will fuse with lysosomes and be degraded. Thus, for ILVs to be released as exosomes, such fusion with lysosomes must be avoided and MVBs must be guided to the plasma membrane. However, how the fate determination of MVB, i.e., the balance between degradation and secretion, is regulated remains unclear (Niel et al. 2018). The inhibition of exosome secretion upon ISGylation of TSG101, a component of ESCRT-I, indicates a mechanism of MVB fate determination that depends on the state of the ESCRT machinery (Villarroya-Beltri et al. 2016). Exosome secretion is increased by simply inhibiting the vacuolar-type H + ATPase (V-ATPase) proton pump then reducing lysosomal activity (Papandreou and Tavernarakis 2017). This suggests that the number and activation state of lysosomes contribute significantly to the fate determination of MVBs.

4.2 Regulation of Lysosomal Activity by MEK/ERK Signaling

To elucidate the signaling pathways involved in exosome biogenesis, we have examined the effects of known cell signaling inhibitors on EV secretion. In this work, we have constructed cancer cell lines that secrete luminescent EVs by fusing CD63 with luciferase (Hikita et al. 2018). Using this system, a good correlation was observed between the luminescence intensity in the cell culture medium and the number of secreted EV particles, making it easy to evaluate the effect of inhibitors on EV secretion.

We found that the MEK inhibitors U0126 and trametinib inhibited EV secretion ftom human colon cancer cells (Hikita et al. 2022). Mutant small G-protein Ras and Raf kinases, which are found in several cancers frequently, promote tumor growth, invasion, and metastasis by abnormally activating the MEK/ERK pathway (Leevers et al. 1994). Therefore, the molecular mechanisms of regulation of the MEK/ERK signaling pathway have been extensively studied since its discovery (Samatar and Poulikakos 2014). However, few studies have yet analyzed the effects of the MEK/ERK pathway in terms of enhanced EV production.

We examined a group of genes whose expression changes upon alteration of MEK/ERK pathway activity and found that inhibition of MEK increases the expression of lysosome-associated genes. Although the activation state of lysosomes in cancer varies among cancer types and cell lines, in certain types of cancer cells, activation of the MEK/ERK pathway reduces the expression of V-ATPase subunit proteins, which inactivate lysosomes. Consequently, MVBs escaped from degradation fuse to the plasma membrane and ILVs are secreted as exosomes. We found that transcription factors of the MiT/TFE family, master regulators of lysosome-associated genes, are involved in this process.

In normal cells, transcription of lysosome-associated genes is maintained by the action of nuclear-localized MiT/TFE family members. Expression of lysosome-associated genes drives lysosome formation and accelerates degradation of MVB. Accordingly, ILVs secreted as exosomes decrease. However, in cancer cells, activation of the MEK/ERK pathway changes the localization of some MiT/TFEs from the nucleus to the cytoplasm. Then, the expression of lysosome-associated genes is suppressed and lysosome activity is reduced. As a result, MVBs accumulate intracellularly and may fuse with the plasma membrane, leading to increased secretion of exosomes (Fig. 4).

The localization of MiT/TFE has been reported to be regulated by direct phosphorylation by ERK or mTOR (Napolitano and Ballabio 2016). Inhibition of mTOR by rapamycin has also been reported to inhibit exosome production, which is thought to be due to MiT/TFE-mediated enhancement of lysosome activity (Gao et al. 2020). However, in our cell system, introduction of MiT/TFE with a mutation in the phosphorylation site did not result in its nuclear transfer or lysosomal activation. However, MEK/ERK activation enhanced the expression of the oncogene MYC, resulting in

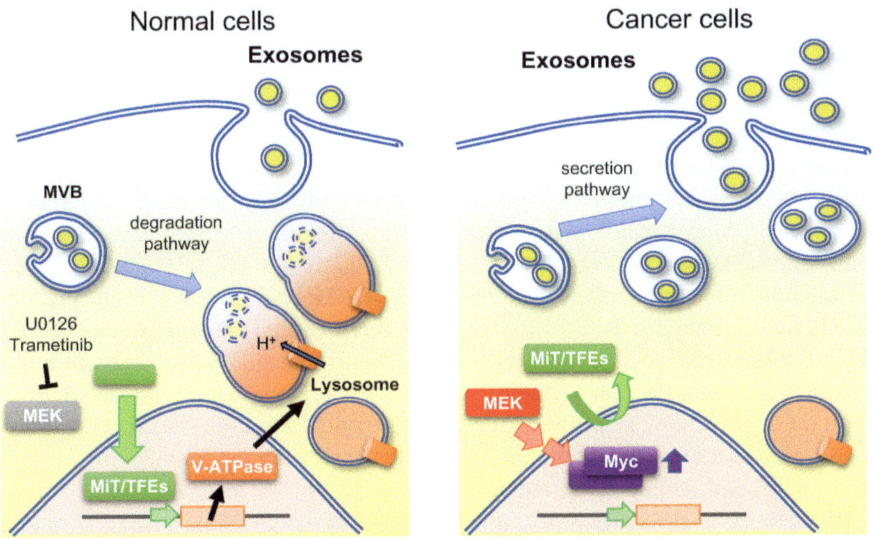

Fig. 4 MEK/ERK pathway regulates the fate of MVBs and control exosome secretion. In normal cells, the MEK/ERK pathway is inactive and maintains lysosomal function. MVBs are degraded and exosome secretion is suppressed. MEK/ERK pathway is activated and disrupts transcription of lysosomal genes and disrupts lysosome function in cancer cells. MVBs fuses with the plasma membrane and promotes exosome secretion. Adapted from the figure in Cancer Sci., 113(4):1264–1276, 2022

nuclear localization of MYC, which in turn suppressed the nuclear localization of MiT/TFE and decreased the expression of lysosome-associated genes. Furthermore, in MEK/ERK-activated renal cancer cells/tissues, upregulation of MYC correlates with downregulation of lysosomal genes (Hikita et al. 2022). These findings support the idea that the MEK/ERK/MYC pathway controls the fate of MVB and promotes exosome secretion from human cancer cells by suppressing lysosomal function.

4.3 Inhibition of Exosome Secretion *via* Lysosome Activation

Because of the potential contribution of EVs to tumor development, EV inhibitors are attracting attention as novel therapeutic agents. However, the mechanisms underlying the signaling pathways that control EV formation remain unclear, limiting the discovery of potent inhibitors of EVs, including exosomes. Several EV inhibitors have been described so far (Zhang et al. 2020; Catalano and O'Driscoll 2020). Most of them act as inhibitors of Rab27A. GW4869, a neutral sphingomyelinase (nSMase) inhibitor, and Manumycin A, an FTase inhibitor, also exert EV inhibitory effects. Moreover, inhibitor screening has demonstrated different mechanisms of EV inhibition, including sulfisoxazole (a sulfonamide antimicrobial agent), H+, K+-ATPase inhibitors (e.g., lansoprazole), and simvastatin (which lowers lipid levels). However, the efficacy and selectivity of such EV inhibitors are not considered sufficient, and the effort to find novel EV inhibitors with different mechanisms of action is still in progress.

We screened fungal extracts isolated from marine organisms using the previously described cell-based high-throughput EV quantification with NanoLuc-labeled CD63. As a result, asteltoxin, a mitochondrial inhibitor, which is secreted by *Aspergillus ochraceopetaliformis*, was found to inhibit EV production (Mitani et al. 2022). At low concentrations, asteltoxin inhibited EV production without causing mitochondrial damage. The reduction of ATP levels by asteltoxin activated AMPK and inhibited mTORC1 activation. Consequently, activation of mTORC1 induced nuclear translocation of the MiT/TFE family and stimulated transcription of lysosome-associated genes, including V-ATPase. Upregulation of lysosomal genes promotes degradation of MVB by inducing lysosome formation. Then, depletion of MVB suppresses exosome secretion. Asteltoxin shows a unique potential to affect the fate of MVB and inhibits exosome production (Fig. 5). The mechanism of exosome suppression by asteltoxin indicates that stress-induced activation of AMPK triggers suppression of exosome secretion and that stress-induced AMPK activation effectively suppresses cancer-derived exosome secretion by depleting glucose instead of asteltoxin.

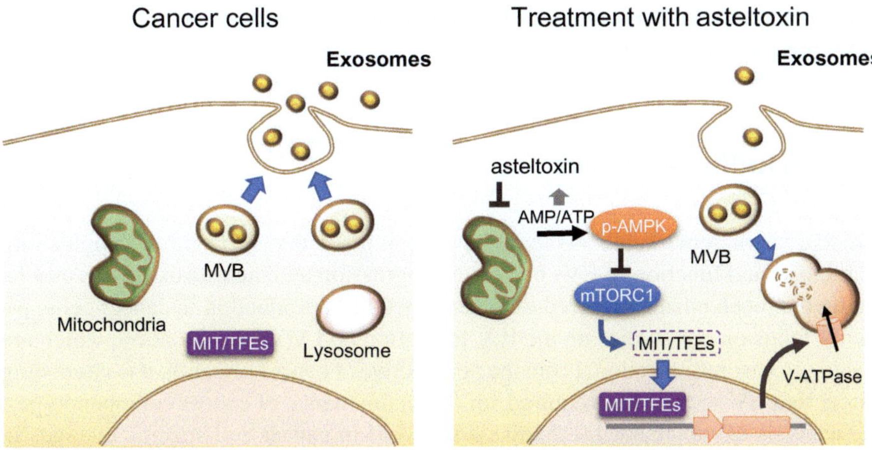

Fig. 5 Asteltoxin controls exosome production by regulating the fate of MVBs, which fuse with the plasma membrane to secrete ILVs as exosomes. Increase of AMP/ATP ratio by treatment with asteltoxin induces AMPK-mediated inactivation of mTORC1. Suppression of mTORC1 activates lysosomal function and MVBs are degraded, resulting in suppression of exosome secretion. Adapted from the figure in Sci Rep., 12(1):6674, 2022

4.4 Exosome Release from MVB

Exosome secretion is a very complex process. The traffic of MVB involving Rab proteins and various cytoskeletal and molecular motors, the fusion process with the plasma membrane involving SNARE proteins, and the regulation of MVB membrane properties by cholesterol and lipid molecules are important, and many findings are being accumulated.

Rab GTPases are a family of small GTPases that regulate the vesicular trafficking process and are also important in MVB fate determination. Rab7, which plays pivotal roles in late endosomal trafficking and fusion with lysosomes, promotes MVB degradation and decreases exosome secretion (Bucci et al. 2000). The mechanism relies on the complex of Coro1a and NEDD8 on the MVB membrane, which interacts to Rab7. By contrast, Rab31 promotes exosome secretion by recruiting the GTPase-activating protein TBC1D2B. Various other Rabs have been reported to be involved in the trafficking of MVB to the plasma membrane, but the details vary among species and cell types (Wei et al. 2021).

MVB on the membrane fusion pathway is transported from the center to the peripheral regions of the cell. MVB moves along the cytoskeleton, including actin filaments and microtubules, and motor proteins such as dynein and kinesin drive MVB movement directly (Stenmark 2009). Alix and clathrin are involved in the process of guiding MVB to the plasma membrane, presumably in cooperation with the plasma membrane-associated actin network (Cabezas et al. 2005). Cortactin, an actin-regulated protein, is a motor protein that promotes exosome secretion. In cancer cells, knockdown of cortactin results in decreased exosome secretion, while overexpression has the opposite effect (Sinha et al. 2016; Calabia-Linares et al. 2011).

5 Concluding Remarks

In recent years, remarkable advances have been made that allow novel insights into the biology and function of EVs in cancer progression and metastasis. In this article, among the mechanisms involved in the enhanced EV production in cancer cells, we mainly focus on our studies on the ILV formation and MVB degradation, which are related to exosome secretion. From the perspective of cancer research, it is interesting to note that EV secretion is required for the maintenance of cancer cell phenotypes. The analysis of molecular mechanisms involved in cancer cell-specific changes in the secretion of EVs is expected to lead to the development of new molecular targets for cancer diagnosis, treatment, and prevention.

On the other hand, the regulatory factors involved in EV secretion are extremely diverse, and the contribution of these factors varies in different cell types (Krylova and Feng 2023). In addition, emerging single-EV analysis methodology have revealed that individual EVs exhibit heterogeneity in their size and marker expression (Mizenko et al. 2021; Banijamali et al. 2022). The majority of researches on the role of EVs in cancer, including our study, remains based on crude EVs collected by methods that acquire heterogeneous EV mixtures. However, it is possible that there are in fact different functional subsets of EVs, each with distinct regulation and function. Recently, EV technology has rapidly evolved to include analysis and isolation of single EVs with size- and charge-based methods, making profiling EV subsets and assessing their functional roles. Future work will elucidate the details of each step involved in the production of the various EVs, and clarify their coordination and dynamics, thereby providing a more detailed picture of the regulation of EV production in cancer and other disease states. This will also help us to better understand the role of EVs in normal tissue homeostasis, a role that remains largely unexplored.

References

Arya SB, Collie SP, & Parent CA (2023) The ins-and-outs of exosome biogenesis, secretion, and internalization. *Trends Cell Biol.*

Baietti MF, *et al.* (2012) Syndecan-syntenin-ALIX regulates the biogenesis of exosomes. *Nat Cell Biol* 14(7):677–685.

Banijamali M, *et al.* (2022) Characterizing single extracellular vesicles by droplet barcode sequencing for protein analysis. *J Extracell Vesicles* 11(11):e12277.

Bucci C, Thomsen P, Nicoziani P, McCarthy J, & van Deurs B (2000) Rab7: a key to lysosome biogenesis. *Mol Biol Cell* 11(2):467–480.

Cabezas A, Bache KG, Brech A, & Stenmark H (2005) Alix regulates cortical actin and the spatial distribution of endosomes. *J Cell Sci* 118(Pt 12):2625–2635.
Calabia-Linares C, *et al.* (2011) Endosomal clathrin drives actin accumulation at the immunological synapse. *J Cell Sci* 124(Pt 5):820–830.
Catalano M & O'Driscoll L (2020) Inhibiting extracellular vesicles formation and release: a review of EV inhibitors. *J Extracell Vesicles* 9(1):1703244.
Choi D, *et al.* (2021) Oncogenic RAS drives the CRAF-dependent extracellular vesicle uptake mechanism coupled with metastasis. *J Extracell Vesicles* 10(8):e12091.
Colombo M, *et al.* (2013) Analysis of ESCRT functions in exosome biogenesis, composition and secretion highlights the heterogeneity of extracellular vesicles. *J Cell Sci* 126(Pt 24):5553–5565.
Gao J, *et al.* (2020) Hepatic stellate cell autophagy inhibits extracellular vesicle release to attenuate liver fibrosis. *J Hepatol* 73(5):1144–1154.
Hikita T, Miyata M, Watanabe R, & Oneyama C (2018) Sensitive and rapid quantification of exosomes by fusing luciferase to exosome marker proteins. *Sci Rep* 8(1):14035.
Hikita T, Kuwahara A, Watanabe R, Miyata M, & Oneyama C (2019) Src in endosomal membranes promotes exosome secretion and tumor progression. *Sci Rep* 9(1):3265.
Hikita T, Miyata M, Watanabe R, & Oneyama C (2020) In vivo imaging of long-term accumulation of cancer-derived exosomes using a BRET-based reporter. *Sci Rep* 10(1):16616.
Hikita T, *et al.* (2022) MEK/ERK-mediated oncogenic signals promote secretion of extracellular vesicles by controlling lysosome function. *Cancer Sci* 113(4):1264–1276.
Hurley JH (2008) ESCRT complexes and the biogenesis of multivesicular bodies. *Curr Opin Cell Biol* 20(1):4–11.
Imjeti NS, *et al.* (2017) Syntenin mediates SRC function in exosomal cell-to-cell communication. *Proc Natl Acad Sci U S A* 114(47):12495–12500.
Jeppesen DK, *et al.* (2019) Reassessment of Exosome Composition. *Cell* 177(2):428–445 e418.
Johnstone RM, Adam M, Hammond JR, Orr L, & Turbide C (1987) Vesicle formation during reticulocyte maturation. Association of plasma membrane activities with released vesicles (exosomes). *J Biol Chem* 262(19):9412–9420.
Kalluri R & LeBleu VS (2020) The biology, function, and biomedical applications of exosomes. *Science* 367(6478).
Kalluri R & McAndrews KM (2023) The role of extracellular vesicles in cancer. *Cell* 186(8):1610–1626.
Kilinc S, *et al.* (2021) Oncogene-regulated release of extracellular vesicles. *Dev Cell* 56(13):1989–2006 e1986.
Kowal J, *et al.* (2016) Proteomic comparison defines novel markers to characterize heterogeneous populations of extracellular vesicle subtypes. *Proc Natl Acad Sci U S A* 113(8):E968–977.
Krylova SV & Feng D (2023) The Machinery of Exosomes: Biogenesis, Release, and Uptake. *Int J Mol Sci* 24(2).
Kugeratski FG, *et al.* (2021) Quantitative proteomics identifies the core proteome of exosomes with syntenin-1 as the highest abundant protein and a putative universal biomarker. *Nat Cell Biol* 23(6):631–641.
Lee TH, *et al.* (2014) Oncogenic ras-driven cancer cell vesiculation leads to emission of double-stranded DNA capable of interacting with target cells. *Biochem Biophys Res Commun* 451(2):295–301.
Leevers SJ, Paterson HF, & Marshall CJ (1994) Requirement for Ras in Raf activation is overcome by targeting Raf to the plasma membrane. *Nature* 369(6479):411–414.
Mathieu M, *et al.* (2021) Specificities of exosome versus small ectosome secretion revealed by live intracellular tracking of CD63 and CD9. *Nat Commun* 12(1):4389.
McKenzie AJ, *et al.* (2016) KRAS-MEK Signaling Controls Ago2 Sorting into Exosomes. *Cell Rep* 15(5):978–987.
Melo SA, *et al.* (2014) Cancer exosomes perform cell-independent microRNA biogenesis and promote tumorigenesis. *Cancer Cell* 26(5):707–721.

Mitani F, *et al.* (2022) Asteltoxin inhibits extracellular vesicle production through AMPK/mTOR-mediated activation of lysosome function. *Sci Rep* 12(1):6674.
Mizenko RR, *et al.* (2021) Tetraspanins are unevenly distributed across single extracellular vesicles and bias sensitivity to multiplexed cancer biomarkers. *J Nanobiotechnology* 19(1):250.
Napolitano G & Ballabio A (2016) TFEB at a glance. *J Cell Sci* 129(13):2475–2481.
van Niel G, *et al.* (2011) The tetraspanin CD63 regulates ESCRT-independent and -dependent endosomal sorting during melanogenesis. *Dev Cell* 21(4):708–721.
van Niel G, D'Angelo G, & Raposo G (2018) Shedding light on the cell biology of extracellular vesicles. *Nat Rev Mol Cell Biol* 19(4):213–228.
Oneyama C, *et al.* (2008) The lipid raft-anchored adaptor protein Cbp controls the oncogenic potential of c-Src. *Mol Cell* 30(4):426–436.
Papandreou ME & Tavernarakis N (2017) Autophagy and the endo/exosomal pathways in health and disease. *Biotechnol J* 12(1).
Perez-Hernandez D, *et al.* (2013) The intracellular interactome of tetraspanin-enriched microdomains reveals their function as sorting machineries toward exosomes. *J Biol Chem* 288(17):11649–11661.
Samatar AA & Poulikakos PI (2014) Targeting RAS-ERK signalling in cancer: promises and challenges. *Nat Rev Drug Discov* 13(12):928–942.
Sinha S, *et al.* (2016) Cortactin promotes exosome secretion by controlling branched actin dynamics. *J Cell Biol* 214(2):197–213.
Stenmark H (2009) Rab GTPases as coordinators of vesicle traffic. *Nat Rev Mol Cell Biol* 10(8):513–525.
Stuffers S, Sem Wegner C, Stenmark H, & Brech A (2009) Multivesicular endosome biogenesis in the absence of ESCRTs. *Traffic* 10(7):925–937.
Sung BH, *et al.* (2020) A live cell reporter of exosome secretion and uptake reveals pathfinding behavior of migrating cells. *Nat Commun* 11(1):2092.
Taylor J, Azimi I, Monteith G, & Bebawy M (2020) Ca(2+) mediates extracellular vesicle biogenesis through alternate pathways in malignancy. *J Extracell Vesicles* 9(1):1734326.
Thery C, *et al.* (2018) Minimal information for studies of extracellular vesicles 2018 (MISEV2018): a position statement of the International Society for Extracellular Vesicles and update of the MISEV2014 guidelines. *J Extracell Vesicles* 7(1):1535750.
Verweij FJ, *et al.* (2022) ER membrane contact sites support endosomal small GTPase conversion for exosome secretion. *J Cell Biol* 221(12).
Villarroya-Beltri C, *et al.* (2016) ISGylation controls exosome secretion by promoting lysosomal degradation of MVB proteins. *Nat Commun* 7:13588.
Wei D, *et al.* (2021) RAB31 marks and controls an ESCRT-independent exosome pathway. *Cell Res* 31(2):157–177.
Yu X, Harris SL, & Levine AJ (2006) The regulation of exosome secretion: a novel function of the p53 protein. *Cancer Res* 66(9):4795–4801.
Zhang H, Lu J, Liu J, Zhang G, & Lu A (2020) Advances in the discovery of exosome inhibitors in cancer. *J Enzyme Inhib Med Chem* 35(1):1322–1330.
Zhao Z, *et al.* (2019) Extracellular vesicles as cancer liquid biopsies: from discovery, validation, to clinical application. *Lab Chip* 19(7):1114–1140.
Zimmerman B, *et al.* (2016) Crystal Structure of a Full-Length Human Tetraspanin Reveals a Cholesterol-Binding Pocket. *Cell* 167(4):1041–1051 e1011.

Membrane Dynamics of Exosomes as Revealed by Single-Molecule Imaging

Kenichi G. N. Suzuki, Koichiro M. Hirosawa, Tatsuki Isogai, Tomokazu Yasuda, and Shinya Hanashima

Abstract Exosomes or small extracellular vesicles (sEVs) play a pivotal role in diverse intercellular communications, notably in the selective metastasis of cancer cells. The membranes of sEVs exhibit abundance in raft-associated molecules such as sphingolipids. A recent elucidation revealed that sEV membranes manifested a composition characterized by the coexistence of raft-like and nonraft-like domains as discerned by the assessment of fluorescence anisotropy and lifetimes. Furthermore, it turned out that sEVs retain some asymmetry in the inner and outer leaflets. In light of the unequivocal heterogeneity inherent in sEVs, it is imperative to meticulously scrutinize the behaviors exhibited by each subtype of sEV particles. This review proposes a strategy to identify the subtypes of sEV, encapsulating extant findings and prospects.

Keywords Small extracellular vesicles · Fluorescence anisotropy · Fluorescence lifetime · Single-particle tracking · Single-molecule imaging

---

K. G. N. Suzuki (✉) · K. M. Hirosawa
Institute for Glyco-Core Research (iGCORE), Gifu University, Gifu 501-1193, Japan
e-mail: suzuki.kenichi.b7@f.gifu-u.ac.jp

K. G. N. Suzuki · T. Isogai
United Graduate School of Agricultural Science, Gifu University, Gifu 501-1193, Japan

K. G. N. Suzuki
Division of Advanced Bioimaging, National Cancer Center Research Institute, Tokyo 104-0045, Japan

T. Yasuda · S. Hanashima (✉)
Graduate School of Science, Osaka University, Osaka 560-0043, Japan
e-mail: hanashima@tottori-u.ac.jp

S. Hanashima
Graduate School of Engineering, Tottori University, Tottori 680-8550, Japan

Y. Baba et al. (eds.), *Extracellular Fine Particles*,
https://doi.org/10.1007/978-981-97-7067-0_6

1 Introduction

Exosomes or small extracellular vesicles (sEVs) with a size range of 40–100 nm have garnered considerable attention as they are involved in near- or far-cell communications in many cell types (Kalluri and LeBleu 2020). sEVs contain proteins, lipids, and nucleic acids such as mRNA, microRNA, and other small noncoding RNA derived from the cells that secreted them (Valadi et al. 2007; Kowal et al. 2016). Due to these characteristics, sEVs are often associated with diverse diseases such as cancer (Kosaka et al. 2016), viral infection (van der Grein et al. 2018), neurodegenerative diseases (Soria et al. 2017), and cardiovascular diseases (Yuan et al. 2018). Recent research has revealed that tumor-derived sEVs intricately bind to and are internalized by recipient cells, leading to an exchange of the cargo between the two (Hoshino et al. 2015; Zomer et al. 2015). This phenomenon induces the formation of a premetastasis niche and an environment conducive to the facile metastasis of cancer cells (Ono et al. 2014). Therefore, the characterization of sEVs becomes increasingly crucial in their therapeutic utility, such as diagnostic biomarkers (Ibsen et al. 2017) and drug-delivery vehicles (Alvarez-erviti et al. 2011).

EVs are known to be generated in several ways. Microvesicles (MVs) with a size range of 100–1,000 nm are generated through the shedding of the cell membrane, while sEVs originate from intraluminal vesicles generated through the invagination of endosomal membranes (Colombo et al. 2014; Hessvik and Llorente 2018). sEV membranes are enriched in cholesterol, sphingolipids, and phosphatidylserine, which are different in the composition of cell plasma membranes (PMs) (Llorente et al. 2013). Furthermore, sEVs contain abundant raft-associated molecules such as glycosylphosphatidylinositol (GPI)-anchored proteins and flotillin. Therefore, the membranes of sEVs are assumed to be rich in raft domains. However, the physicochemical properties of EV membranes, such as raft domains and asymmetry of EV membrane bilayers, have been hardly investigated.

It has recently come to light that membrane composition and encapsulating contents of sEVs are predominantly contingent upon the employed isolation methodologies. For example, the molecular composition of sEVs, obtained by sedimentation of the culture supernatant using an ultracentrifuge, diverged from that achieved by using affinity beads coated with tetraspanin marker proteins, such as CD63 (Jeppesen et al. 2019). Therefore, these results imply the potential existence of distinct EV subtypes. Since the methodologies for sEV isolation, including the duration and velocity of centrifugation, are contingent upon literature (Deun et al. 2017), an analysis of the entire range of sEVs released from cells precludes the elucidation of the distinctive behaviors inherent to each subtype of sEV particles (Deun et al. 2017).

This review summarizes and discusses recent studies regarding the characterization of EVs and their membrane structure to solve the issues mentioned above, employing fluorescence imaging techniques.

2 Characterization of the Lipid Bilayers of sEV Membranes

As described previously, sEV membranes include raftophilic molecules, but the presence of raft domains has never been verified. Furthermore, it is unknown if the asymmetric structures in the inner and outer leaflets are also retained in sEV membranes. This section introduces the fluorescence spectroscopic studies used to solve these issues.

2.1 Detection of Lipid Domains in the sEV Membranes

In a recent work by our group that examined the membrane packing in sEVs (Yasuda et al. 2022), TMA-DPH was incorporated into the membranes of sEVs derived from bovine milk, human prostate cancer PC3 cells, human lung cancer A549 cells, and human embryonic kidney cells 293 (HEK293). The membrane packing in sEVs was examined by measuring the fluorescence anisotropy of TMA-DPH. TMA-DPH is anchored to the outer leaflet of sEV membranes and tends to partition slightly more into the liquid-disordered (L_d) than the liquid-ordered (L_o) phase (St Clair et al. 2020; Beck et al. 1993). The fluorescence anisotropy of TMA-DPH predominantly reports the orders of the lipid chain in the outer leaflet of sEV membranes. The anisotropy values of TMA-DPH in sEVs derived from the cells were comparable to those of asymmetric large unilamellar vesicles (aLUVs), of which the outer leaflet consists of 1-palmitoyl-2-oleoyl phosphatidylcholine (POPC):palmitoyl sphingomyelin (PSM):cholesterol (Chol)(Fig. 1), which is similar to those of the rafts in PMs. Therefore, L_o and L_d phases coexist in the EV membranes derived from the cells. Furthermore, the anisotropy of sEVs derived from PC3 cells isolated by the pellet-down method was similar to that obtained by affinity capture with TIM-4 beads (Yasuda et al. 2022).

Unlike TMA-DPH, the generalized polarization values for C-laurdan indicate the local polarity of the headgroup region of sEV membranes, which suggested that the order and polarity of the outer leaflet of sEV membranes reduced due to the occurrence of a significant amount of L_d lipids, leading to phase separation (Yasuda et al. 2022).

TEMPO is known to partition into the Ld phase in Lo-Ld phase-separated membranes favorably, and its radicals quench TMA-DPH distributed with a slightly higher relative preference for the L_d phase (Suga and Umakoshi 2013). Therefore, TEMPO radicals barely quench the TMA-DPH in the L_o phase, and the remaining TMA-DPH emits fluorescence. The ratio of TMA-DPH fluorescence in the presence and absence of TEMPO in EV membranes and LUVs revealed that 40–50% of the area of sEV membranes can be assumed to consist of the L_o domains (Fig. 2) (Yasuda et al. 2022).

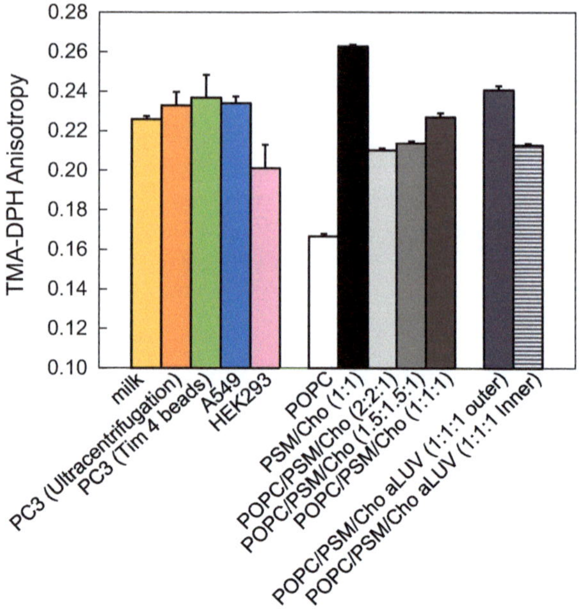

Fig. 1 TMA-DPH fluorescence anisotropy of sEVs and LUVs at 30 °C. aLUVs have a lipid composition of POPC/PSM/Chol in a 1:1:1 ratio, where PSM is the major phospholipid in the outer leaflet of the bilayers. Adapted with permission from Yasuda et al. (2022). ©2022 American Chemical Society

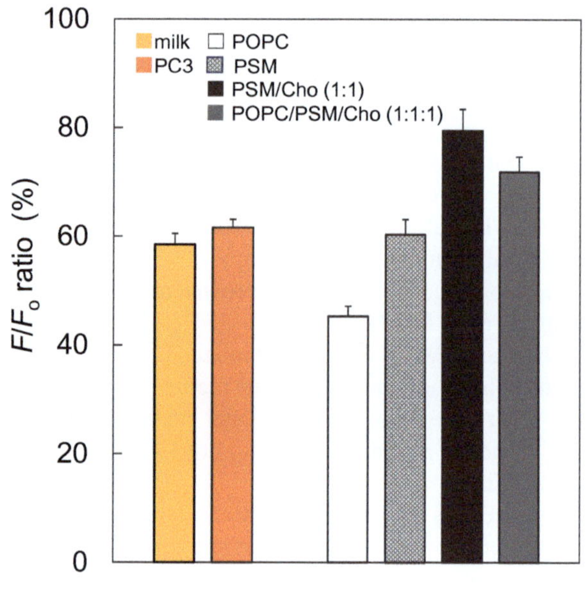

Fig. 2 Residual TMA-DPH fluorescence of sEVs and LUVs by TEMPO quenching at 23 °C. The F/F_0 ratio was determined from the fluorescence intensity of TMA-DPH in the presence (F) and absence (F_0) of TEMPO. Adapted with permission from Yasuda et al. (2022). ©2022 American Chemical Society

2.2 *Detection of Asymmetric Distribution in the Outer and Inner Leaflets of EV Membranes*

The differences in the extended fluorescence lifetime component of C6-NBD-PC in the model and cell PMs accurately indicated the relative lateral packing properties (Stöckl et al. 2008). C6-NBD-PC was incorporated into the EV membranes and asymmetric LUVs at 4 °C, and the fluorescence lifetime of the probe in the outer leaflet was measured immediately (Yasuda et al. 2022). After incubation of the sample for 24 h at 37 °C, a proportion of the C6-NBD-PC flip-flopped and relocated in the inner leaflet of the membranes, and C6-NBD-PC in the outer leaflet of the PMs was subsequently quenched by dithionite ($S_2O_4^{-2}$). Under these conditions, the extended fluorescence lifetime component of C6-NBD-PC could be measured only in the inner leaflet and compared with that in the outer leaflet. The fluorescence lifetime in the outer leaflets was longer than those in the inner leaflets of EV membranes derived from three kinds of cells: PC3, A549, and HEK293 (Fig. 3). These results explicitly exhibit that the outer leaflet of EV membranes was more tightly packed than the inner leaflet. The lipid bilayers of the EV membranes were asymmetric. The remaining activity level of the enzyme responsible for the flip-flop mechanism in the sEV membranes was unclear. However, asymmetric lipid distribution may be partially maintained because of the very slow desymmetrization of phospholipid bilayers, which usually takes more than a few days (Watanabe et al. 2023).

3 Single-Particle Tracking of EVs

To examine molecular interactions between sEVs and molecules in the PMs of the recipient cells or identify the membrane structures that uptake the sEVs, it is necessary to observe the behavior of individual EV particles on cell PMs by single-particle tracking (Suzuki et al. 2005). However, for these observations, sEVs should be first labeled with fluorophores. In this section, the methods used for the labeling of EVs by employing lipid-like probes are introduced.

3.1 *Fluorescent Labeling of sEVs with Lipid-Like Probes*

A diverse array of lipid-like fluorescent probes with a range of excitation (ex) and emission (em) wavelengths such as DiO ($\lambda_{ex/em}$ = 484 nm, 504 nm) (Hood et al. 2011), DiD ($\lambda_{ex/em}$ = 644 nm, 665 nm) (Yang et al. 2017), DiR ($\lambda_{ex/em}$ = 750 nm, 782 nm) (Wiklander et al. 2015), or PKH67 ($\lambda_{ex/em}$ = 490 nm, 502 nm) (Verdera et al. 2017), PKH26 ($\lambda_{ex/em}$ = 551 nm, 567 nm) (Dominkus et al. 2018) have been used for the staining of EVs. These fluorescent probes are carbocyanine-based, which are incorporated into the lipid bilayers of EVs. The multiple types of EVs were labeled

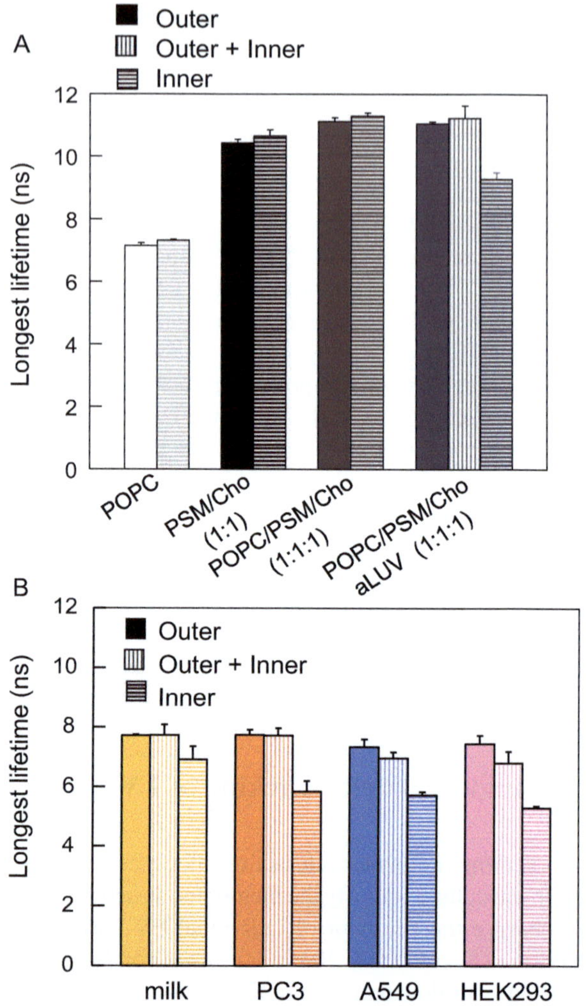

Fig. 3 Long fluorescence lifetime of C6-NBD-PC distributed in each leaflet of the bilayers of symmetric and asymmetric LUVs (**a**) and sEVs (**b**) at 23 °C. Adapted with permission from Yasuda et al. (2022). ©2022 American Chemical Society

with different colors, and the dual-color, single-particle tracking of these EVs was performed simultaneously. Notably, these probes are almost insoluble in water and form micelles and/or aggregates (Dominkus et al. 2018). Therefore, even if EVs are isolated carefully after labeling with lipid-like probes, micelles and/or aggregates of lipid-like probes that remain with EVs may be observed with higher probability. Furthermore, since these probes can transfer between membranes, some would move from the labeled EV membranes to the cell PMs when observing cell–cell interactions (Dominkus et al. 2018). These factors should considered when conducting such experiments. Since the recently developed ExoSparkler series of dyes (Shimomura et al. 2021) do not aggregate and minimally affect the properties of EVs, the dynamics of EVs can be observed more accurately. To overcome the issues associated with fluorescent labeling of EVs with lipid-like molecular probes, more specific fluorescent labeling of EVs via marker proteins such as CD63 and single particle tracking of its behavior has increased in recent years.

3.2 Fluorescent Labeling of Marker Proteins in EVs

Many marker proteins for EVs have been identified. For example, CD63, CD9, and CD81 are three transmembrane (tetraspanin) proteins that form complexes with integrins and growth factor receptors and are known to modulate cellular functions. Other examples include syndecan-1, flotillin-2, and ALIX. One method for fluorescent labeling of EV marker proteins is purifying them and then labeling them with dye-conjugated antibodies. However, it is difficult to accurately quantify marker proteins or determine the ratios of their presence due to the limited detection sensitivity of antibodies, steric hindrance, and detection of apparent colocalization due to aggregation. For example, the labeling efficiency was >90% when the EVs on glass were stained with antibodies (Han et al. 2021). However, this method could not indicate the number of fluorescent molecules per antibody molecule or the fluorescence intensity per particle of EVs, making it impossible to distinguish between one EV particle with a high expression of the marker protein or an aggregate of EVs. In other words, the aggregation of EVs and the colocalization of molecules cannot be distinguished. Steric hindrance between antibodies and aggregation can occur during the staining of marker proteins on EVs using antibodies (Mizenko et al. 2021).

As an alternative to labeling EVs with dye-conjugated antibodies, fluorescent proteins such as GFP or molecules fused with tags such as Halo7 are expressed in cells, and EVs derived from those cells have been isolated. For example, the contents of various marker proteins fused with GFP in sEVs, GFP-fused ALIX, and flotillin-2 were rarely included in sEVs, unlike tetraspanins such as CD63, CD9, and CD81 after labeling with GFP (Corso et al. 2019). Therefore, the sEV marker proteins should be carefully selected when using this technique. In the same study, only 20–30% of CD63-mCherry and CD9-GFP or CD81-GFP were detected in the same sEV. However, only ~40% of the positive control CD63-mCherry and CD63-GFP were on the same sEV. These observations suggest an underestimation of colocalization, which may be due to (a) the low sensitivity of the confocal fluorescence or epi-fluorescence microscopy-based observation and (b) although HEK293T cells were transfected with two marker proteins and transiently overexpressed, cells in which both proteins were coexpressed could not be fractionated, but many cells could express only one of them.

3.3 Single-Particle Imaging of sEVs at the Sensitivity of Detection of a Single-Molecule

In the scenario of two subtypes: sEVs containing all three membrane proteins (sEV-1, Fig. 4) and sEVs containing only one protein (sEV-2, Fig. 4), immunoprecipitation using antibodies against these proteins can detect all the proteins but not subtypes. On the other hand, the presence of sEV subtypes can be verified by single-particle imaging, which determined the coexistence of these proteins in the sEVs. For this, sEVs must be isolated from cells that express all these proteins conjugated with fluorescent proteins such as GFP or with tags such as Halo7-tag. In addition, all the proteins to be examined for colocalization must be fluorescently labeled at a molar ratio 1:1 with a high labeling efficiency. The methodology for fluorescent labeling with a high efficiency has been described previously (Suzuki et al. 2012, 2013). Furthermore, since the presence of proteins in sEV membranes must be detected at the single-molecule level, the fluorescently labeled proteins in individual sEV particles must be observed at the detection sensitivity level of a single-molecule. For this purpose, the most appropriate observation system would be a simultaneous, two-color, single-molecule observation using total internal reflection microscopy (TIRFM).

Here, we introduce our microscopic system, Olympus IX83 or Nikon Ti2 that enables such observations. The system consisted of sCMOS cameras (Hamamatsu Photonics) or high-speed CMOS cameras (Photron) and two image intensifiers (Hamamatsu Photonics), which are photon-amplifying optical devices (Fig. 5). In addition, a high numerical aperture objective lens with a NA of 1.50 or 1.49 was used. Using this system, sEVs were simultaneously illuminated using two-color lasers, and fluorescently labeled membrane molecules on sEVs were concurrently observed by TIRFM with two synchronized cameras at a single-molecule detection

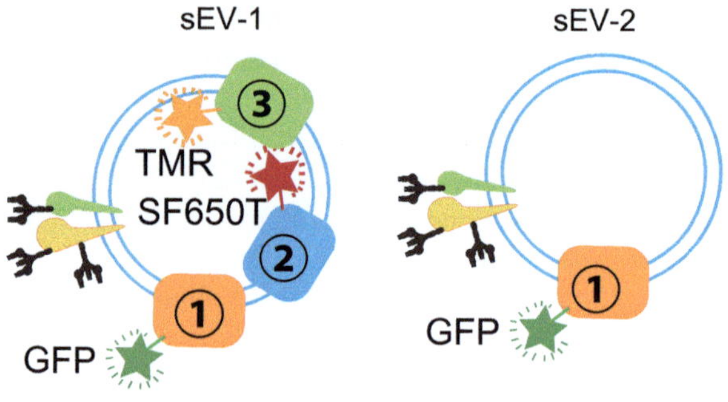

Fig. 4 Schematic drawing of an sEV particle containing three membrane protein (Left) and an sEV particle containing only one protein (Right). These proteins are labeled with GFP, TMR, and SF650T. These two sEV particles can be distinguishable by single-particle imaging but not by immunoprecipitation with antibodies against these proteins

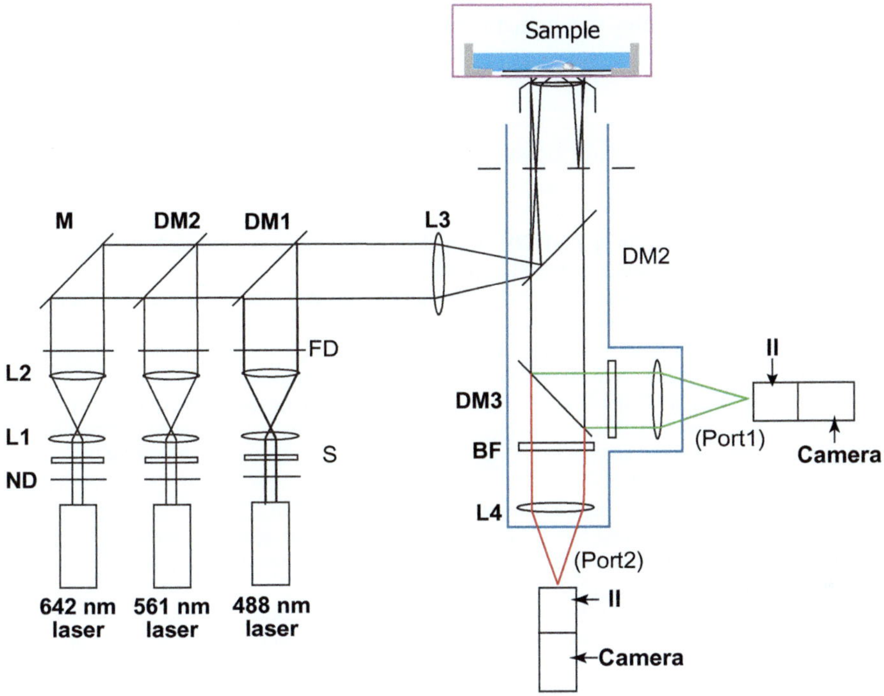

Fig. 5 Schematic diagram of the total internal reflection fluorescence microscopic system to perform simultaneous, dual-color, single-particle tracking and single-molecule imaging. (Left) Optical ray diagram L1 and L2 beam expander, L3 focusing lens, L4 projection lens, S electric shutter, ND neutral density filter, FD field diagram, DM dichroic mirror, BF band-pass filter, II image intensifier. The microscope used was an Olympus IX83 or Nikon ECLIPSE Ti2. The cameras used were ORCA fusion (Hamamatsu Photonics) or FAST CAM AX-I.I. (Photron)

sensitivity. This system enabled the high-speed, dual-color, single-molecule observation of molecules in cell PMs (Komura et al. 2016; Kinoshita et al. 2017; Morise et al. 2019) and to observe the interactions between the sEVs and molecules in PMs.

4 Conclusions

Recent studies in this review revealed that sEV membranes have a mixer of L_o-like and L_d-like phases with structurally asymmetric outer and inner leaflets. Furthermore, recently developed fluorescence imaging techniques can elucidate the presence of sEV subtypes and the dynamic behaviors of sEVs on the PMs of the recipient cells. Future studies using these techniques are expected to (a) identify more detailed sEV subtypes and inclusions, (b) visualize uptake pathways of sEVs by the recipient cells, and (c) unravel the detailed mechanisms of sEV-induced modification of the recipient cells.

References

Alvarez-erviti L, Seow Y, Yin H et al (2011) Delivery of siRNA to the mouse brain by systemic injection of targeted exosomes. Nat Biotechnol 29:341–345.

Beck A, Heissler D, Duportail G (1993) New diphenylhexatriene derivatives as fluorescent membrane probes: Partitioning properties. J Fluoresc 3:145–147.

Colombo M, Raposo G, Thèry C (2014) Biogenesis, secretion, and intercellular interactions of exosomes and other extracellular vesicles. Annu Rev Cell Dev Biol 30:255–289.

Corso G, Heusermann W, Trojer D et al (2019) Systematic characterization of extracellular vesicle sorting domains and quantification at the single molecule – Single vesicle level by fluorescence correlation spectroscopy and single particle imaging. J Extracell Vesicles 8:1663043.

Deun JV, Metsdagh P, Agostinis P et al (2017) EV-TRACK: Transparent reporting and centralizing knowledge in extracellular vesicle research. Nat Methods 14:228–232.

Dominkus PP, Stenovec M, Sitar S et al (2018) PKH26 labeling of extracellular vesicles: Characterization and cellular internalization of contaminating PKH26 nanoparticles. Biochim Biophys Acta Biomembr 1860:1350–1361.

Van der Grein SG, Defourny KAY, Slot EFJ et al (2018) Intricate relationships between naked viruses and extracellular vesicles in the crosstalk between pathogen and host. Semin Immunopathol 40:491–504.

Han C, Kang H, Yi J et al (2021) Single-vesicle imaging and co-localization analysis for tetraspanin profiling of individual extracellular vesicles. J Extracell Vesicles 10:e12047.

Hessvik NP, Llorente A (2018) Current knowledge on exosome biogenesis and release. Cell Mol Life Sci 75:193–208.

Hood JL, San RS, Wickline SA (2011) Exosomes released by melanoma cells prepare sentinel lymph nodes for tumor metastasis. Cancer Res 71:3792–3801.

Hoshino A, Costa-Silva B, Shen TL et al (2015) Tumour exosome integrins determine organotropic metastasis. Nature 527:329–335.

Ibsen SD, Wright J, Lewis JM et al (2017) Rapid isolation and detection of exosomes and associated biomarkers from plasma. ACS Nano 11:6641–6651.

Jeppesen DK, Fenix AM, Franklin JL et al (2019) Reassessment of exosome composition. Cell 177:428–445.e18.

Kalluri R, LeBleu VS (2020) The biology, function, and biomedical applications of exosomes. Science 367:eaau6977.

Kinoshita M, Suzuki KGN, Matsumori N et al (2017) Raft-based sphingomyelin interactions revealed by new fluorescent sphingomyelin analogs. J Cell Biol 216:1183–1204.

Komura N, Suzuki KGN, Ando H et al (2016) Raft-based interactions of gangliosides with a GPI-anchored receptor. Nat Chem Biol 12:402–410. https://doi.org/10.1038/nchembio.2059

Kosaka N, Yoshioka Y, Fujita Y et al (2016) Versatile roles of extracellular vesicles in cancer. J Clin Invest 126:1163–1172.

Kowal J, Arras G, Colombo M et al (2016) Proteomic comparison defines novel markers to characterize heterogeneous populations of extracellular vesicle subtypes. Proc Natl Acad Sci USA 113:E968–E977.

Llorente A, Skotland T, Sylvänne T et al. (2013) Molecular lipidomics of exosomes released by PC-3 prostate cancer cells. Biochim Biophys Acta 1831:1302–1309.

Mizenko RR, Brostoff T, Rojalin T et al (2021) Tetraspanin immunocapture phenotypes extracellular vesicles according to biofluid source but may limit identification of multiplexed cancer biomarkers. BioRxiv, https://doi.org/10.1101/2021.03.02.433595

Morise J, Suzuki KGN, Kitagawa A et al (2019) AMPA receptors in the synapse turnover by monomer diffusion. Nat Commun 10:5245.

Ono M, Kosaka N, Tominaga N et al (2014) Exosomes from bone marrow mesenchymal stem cells contain a microRNA that promotes dormancy in metastatic breast cancer cells. Sci Signal 7:ra63.

Shimomura T, Seino R, Umezaki K, et al (2021) New lipophilic fluorescent dyes for labeling extracellular vesicles: Characterization and monitoring of cellular uptake. Bioconjug Chem 32:680–684.

Soria FN, Pampliega O, Bourdenx M et al (2017) Exosomes, an unmasked culprit in neurodegenerative diseases. Front Neurosci 11:26.

St. Clair J W, Kakuda S, London E (2020) Induction of ordered lipid raft domain formation by loss of lipid asymmetry. Biophys J 119:483–492.

Stöckl M, Plazzo AP, Korte T et al (2008) Detection of lipid domains in model and cell membranes by fluorescence lifetime imaging microscopy of fluorescent lipid analogues. J Biol Chem 283: 30828–30837.

Suga K, Umakoshi H (2013) Detection of nanosized ordered domains in DOPC/DPPC and DOPC/Ch binary lipid mixture systems of large unilamellar vesicles using a TEMPO quenching method. Langmuir 29:4830–4838.

Suzuki K, Ritchie K, Kajikawa E et al (2005) Rapid hop diffusion of a G-protein-coupled receptor in the plasma membrane as revealed by single-molecule techniques. Biophys J 88:3659–3680.

Suzuki KGN, Kasai RS, Hirosawa KM et al (2012) Transient GPI-anchored protein homodimers are units for raft organization and function. Nat Chem Biol 8:774–783.

Suzuki KGN, Kasai RS, Fujiwara K et al (2013) Single-molecule imaging of receptor-receptor interactions. Methods Cell Biol 117:373–390.

Valadi H, Ekstrom K, Bossios A et al (2007) Exosome-mediated transfer of mRNAs and microRNAs is a novel mechanism of genetic exchange between cells. Nat Cell Biol 9:654–659.

Verdera HC, Gitz-Francois JJ, Schiffelers RM et al (2017) Cellular uptake of extracellular vesicles is mediated by clathrin-independent endocytosis and macropinocytosis. J Control Release 28: 100–108.

Watanabe H, Hanashima S, Yano Y, et al (2023) Passive translocation of phospholipids in asymmetric model membranes: Solid-state ^{1}H NMR characterization of flip–flop kinetics using deuterated sphingomyelin and phosphatidylcholine. Langmuir 39.

Wiklander OPB, Nordin JZ, O'Laughlin A et al (2015) Extracellular vesicle in vivo biodistribution is determined by cell source, route of administration and targeting. J Extracell Vesicles 4:26316.

Yang Y, Han Q, Hou Z et al (2017) Exosomes mediate hepatitis B virus (HBV) transmission and NK-cell dysfunction. Cell Mol Immunol 14:465–475.

Yasuda T, Watanabe H, Hirosawa KM et al (2022) Fluorescence spectroscopic analysis of lateral and transbilayer fluidity of exosome membranes. Langmuir 38:14695–14703.

Yuan Y, Du W, Liu J et al (2018) Stem cell-derived exosome in cardiovascular diseases: Macro roles of micro particles. Front Pharmacol 9:547.

Zomer A, Maynard C, Verweij FJ et al (2015) In vivo imaging reveals extracellular vesicle-mediated phenocopying of metastatic behavior. Cell 161:1046–1057.

Glycan Remodeling by Small Extracellular Vesicles

Yasuhiko Kizuka

Abstract Glycosylation of proteins is a fundamental and frequent process that regulates protein functions and is associated with various diseases. Glycan biosynthesis is catalyzed by sequential actions of many glycosyltransferases in the endoplasmic reticulum and Golgi apparatus, but how their intracellular activity is regulated remains largely unresolved. Recently, an increasing number of reports have demonstrated that glycans and related molecules are included in small extracellular vesicles (sEVs), and their physiological and pathological roles in sEVs have been actively studied. Furthermore, a few glycosyltransferases, such as GnT-V (MGAT5), have also been demonstrated to be included in sEVs and transferred from cell to cell via incorporation of these glycosyltransferase-loaded sEVs by recipient cells. Moreover, glycan structures in the recipient cells have been shown to be remodeled by the incorporation of glycosyltransferase-positive sEVs. These findings demonstrate a novel non-genetic sEV-mediated mechanism of glycan remodeling of cells, highlighting a new concept in which glycosyltransferases are transferred among cells via sEVs for regulating glycan expression.

Keywords Glycosylation · Glycosyltransferase · MGAT5 · *N*-Glycan

1 Protein Glycosylation in Mammals

1.1 General Introduction

Glycosylation is one of the most frequent modifications of proteins, and more than 50% of proteins are estimated to be glycosylated in mammals (Apweiler et al. 1999). These glycans have enormous structural diversity and regulate various biological processes, such as development, immunity and learning (Varki 2017; Moremen et al. 2012; Stanley et al. 2015; Zhao et al. 2008). Furthermore, subtle changes

Y. Kizuka (✉)
Institute for Glyco-Core Research, Gifu University, Gifu 501-1193, Japan
e-mail: kizuka.yasuhiko.k8@f.gifu-u.ac.jp

Y. Baba et al. (eds.), *Extracellular Fine Particles*,
https://doi.org/10.1007/978-981-97-7067-0_7

in glycan structure have been reported to be functionally associated with the onset and aggravation of various diseases, including cancer, Alzheimer's disease (AD) and diabetes (Ohtsubo and Marth 2006; Kizuka and Taniguchi 2016; Pinho and Reis 2015; Lauc et al. 1860). These findings underscore the importance of glycosylation for understanding disease mechanisms and developing novel diagnostics and therapeutics.

Glycosylation basically occurs via the secretory pathway (except *O*-GlcNAcylation). The biosynthesis of glycans on proteins starts in the endoplasmic reticulum (ER) lumen and is subsequently matured and completed in the Golgi apparatus (Moremen et al. 2012). Because this process is accomplished by sequential actions of a series of glycosyltransferases (Moremen et al. 2012), the glycan profile in a specific cell largely depends on the expression and activity of glycosyltransferases in the cell. However, it remains largely unclear how the activity and quantity of glycosyltransferases in cells are regulated. Therefore, it is pivotal to understand the regulation mechanisms for glycosyltransferase expression and activity for controlling glycan profiles in cells.

Glycans attached to proteins are classified into several types (Varki 2017). *N*-Glycans are one of the most common and abundant glycans, with more than 7,000 human proteins estimated to be *N*-glycosylated (Sun et al. 2019). The biosynthesis of *N*-glycans is initiated in the ER, and the common oligosaccharide ($Glc_3Man_9GlcNAc_2$) is transferred to the Asn residue in the consensus sequence (N-X-S/T-) (Stanley et al. 2015). After trimming terminal Glc and Man residues, *N*-glycan structures are further modified by various Golgi-resident glycosyltransferases to have diverse structures in protein- and *N*-glycosylation site-specific manners (Nagae et al. 2020).

1.2 N-Glycan Branching and Its Regulation

One of the major structural features of mammalian *N*-glycans is the variable number of GlcNAc branches (Nagae et al. 2020). These branches are biosynthesized by the dedicated glycosyltransferases that are designated as GnT-I–Vb (MGAT1–MGAT5B) (Fig. 1). After the prior actions of GnT-I and GnT-II common to all complex *N*-glycans, the competitive and sequential actions of GnT-III, IVa, IVb, V and Vb (IX) synthesize protein- and glycosylation site-specific branch patterns (Stanley et al. 2015). Functionally, the presence or absence of a particular branch regulates various functions of glycoproteins, including cell adhesion molecules, cell surface receptors and proteases (Moremen et al. 2012; Stanley et al. 2015; Zhao et al. 2008; Kizuka and Taniguchi 2016). Furthermore, knockout (KO) mouse and human clinical studies revealed that alterations in the amount of a branch are profoundly associated with diseases. For example, the mRNA level of *MGAT3* (GnT-III) was found to be upregulated in the brain of AD patients (Akasaka-Manya et al. 2010), and GnT-III KO mice crossing with AD model mice improved AD pathology with a reduction in the deposition of the disease-causative Aβ peptide in the brain (Kizuka et al. 2015, 2016). In addition, intake of a high-fat diet by mice caused a decrease in the mRNA expression of *Mgat4a* (GnT-IVa) in the pancreas (Ohtsubo et al. 2011), and GnT-IVa KO mice displayed diabetic phenotypes, such as elevated blood glucose and impaired insulin secretion (Ohtsubo et al. 2005), indicating that the branch produced by GnT-IVa is involved in type 2 diabetes.

GnT-V (MGAT5) is well known to be involved in cancer malignancy. The mRNA expression of *MGAT5* is driven by the oncogenic Ras-raf-ets pathway (Kang et al. 1996; Buckhaults et al. 1997), and aberrantly high expression of *MGAT5* was reported in many types of cancer cells. Furthermore, GnT-V KO mice exhibited suppression of breast cancer growth in a PyMT-induced model (Granovsky et al. 2000). GnT-V KO tumor cells also showed suppressed in vivo growth upon implantation into mice (Silva et al. 2020). These findings suggest that GnT-V is a promising drug target for cancer therapy, which is expected to be accelerated by the available crystal structures of GnT-V (Nagae et al. 2018; Darby et al. 2020). GnT-V was also reported to regulate T cell functions and immunity. GnT-V KO mice showed severe phenotypes in experimental autoimmune encephalomyelitis (Grigorian and Demetriou 2011) and inflammatory bowel disease (Dias et al. 2018) models, as well as spontaneous development of the lupus nephritis phenotype in the kidney (Alves et al. 2023). These disease-associated phenotypes are considered to be caused by the lowered threshold of the T cell receptor (TCR) and hyperactivated TCR signaling in GnT-V KO mice (Demetriou et al. 2001).

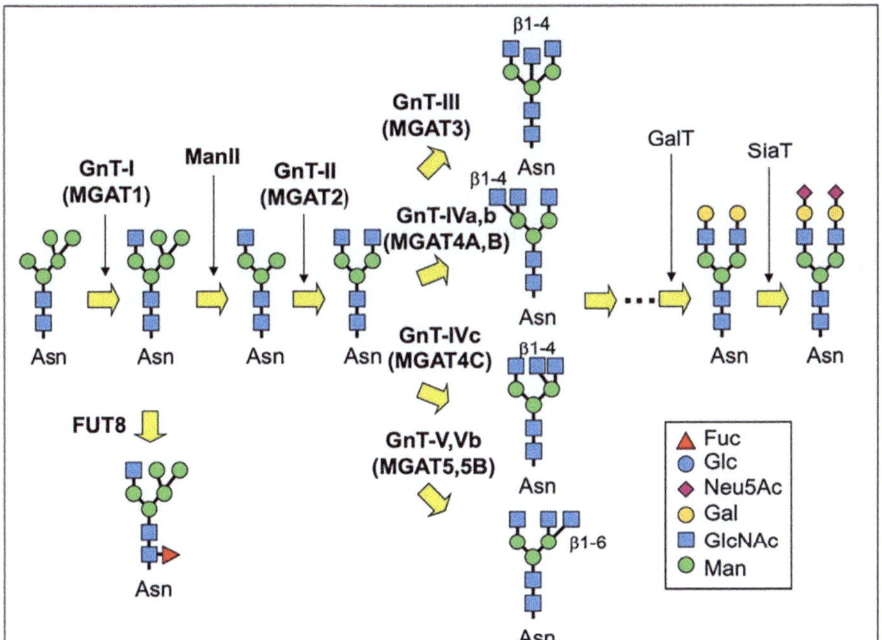

Fig. 1 *N*-Glycan maturation in the Golgi apparatus catalyzed by glycosyltransferases

1.3 *MGAT5, a Unique Enzyme and Its Regulation*

Although the physiological and pathological functions of GnT-V have been well characterized, as described above, regulation mechanisms of GnT-V protein levels and activity in cells remain largely unclear. Recently, SPPL3 was identified as the sheddase of GnT-V and for many other Golgi-resident type II glycosyltransferases (Voss et al. 2014; Kuhn et al. 2015). SPPL3 KO cells accumulated GnT-V and activity in cells, with essentially no secreted GnT-V found in the media. Furthermore, SPPL3 KO cells showed drastic alterations in *N*-glycan structures (Voss et al. 2014). These observations indicate that SPPL3-mediated shedding is a major factor regulating cellular GnT-V activity.

We found that SPPL3-mediated shedding of GnT-V into the extracellular space is significantly suppressed upon treating cells with the *N*-glycan biosynthesis inhibitor kifunensine, which converts most *N*-glycans to immature oligomannose forms (Hirata et al. 2022). Conversely, kifunensine treatment increased the cellular amount and enzymatic activity of GnT-V. Furthermore, similar suppression of GnT-V shedding was found when sialylation or galactosylation was blocked by knocking out the transporter for CMP-Sia or UDP-Gal. Glycomics analysis showed an increase in the GnT-V-produced branch in Gal-depleted cells. These findings suggest that cells sense a reduction in terminal modifications (maturation) of *N*-glycans and upregulate GnT-V activity to produce more branching by stopping GnT-V secretion (Fig. 2) (Hirata et al. 2022). SPPL3 may recognize the glycan structures of GnT-V because GnT-V is *N*-glycosylated. Further analysis is needed to fully understand the mechanism that blocks GnT-V shedding upon reduction in *N*-glycan maturation.

Glycosyltransferases in body fluids were discovered a long time ago (Kim et al. 1972a, 1972b), and many enzymes were reported to maintain their activity in these fluids, including blood. Most Golgi-resident glycosyltransferases are type II membrane proteins with a short cytosolic tail and a large luminal catalytic domain. Thus, it is likely that active glycosyltransferases in body fluids are cleaved beforehand in cells near the transmembrane to become soluble forms or included in membranous vesicles in body fluids as either an intact or cleaved form. We focused on the latter possibility and explored whether glycosyltransferases are included in sEVs, such as exosomes.

2 SEV and Glycosylation

Cells secrete various EVs that are categorized by their size (Kalluri and LeBleu 2020). Exosomes, one type of sEVs, are derived from endosomal membranes and range in diameter from 50 to 200 nm. In addition, novel types of sEVs called exomeres and supermeres were also found recently, which are nanoparticles smaller than exosomes (<50 nm in diameter) that lack a membranous vesicular structure (Zhang et al. 2018, 2021).

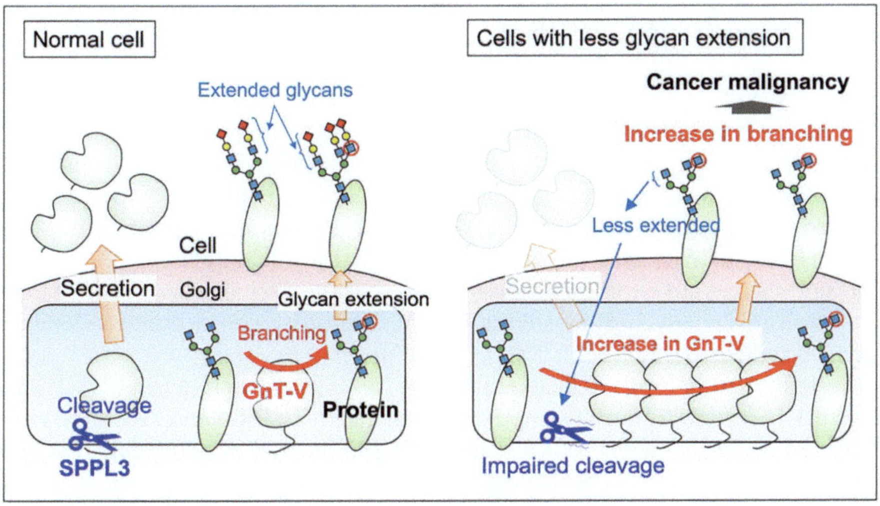

Fig. 2 Regulation of GnT-V secretion. GnT-V is constitutively cleaved by SPPL3 and secreted. SPPL3-mediated cleavage is impaired when the *N*-glycan extension is blocked, thereby increasing the amount and activity of GnT-V in cells, which enhances the biosynthesis of the *N*-glycan branch

EVs contain various cargoes, including DNA, miRNA, proteins, lipids and glycans (Niel et al. 2018; Mathieu et al. 2019; Tkach and Thery 2016). These sEVs are incorporated by recipient cells and function in cell–cell communication under physiological and pathological conditions (Kalluri and LeBleu 2020; Wortzel et al. 2019). In particular, sEVs derived from tumors play significant roles in tumor malignancy, such as promoting proliferation (Hong et al. 2009; Qu et al. 2009), angiogenesis (Zhou et al. 2014; Treps et al. 2017) and metastasis (Hoshino et al. 2015; Zhang et al. 2017; Rodrigues et al. 2019). Glycans and related molecules in sEVs may play key roles in cancer malignancy because specific glycans produced in cells are well known to be deeply related to cancer malignancy.

Glycans in sEVs differ from those in cells and display large structural diversity, which is dependent on the cellular origin (Zhang et al. 2018; Shimoda et al. 2019), and sEV glycans play many biological roles. For example, sEVs with high levels of bisecting GlcNAc, the GnT-III product, in *N*-glycans inhibit the migration of recipient cells by impairing integrin functions (Tan et al. 2020). Glycans in donor cells are also involved in forming sEVs. Blocking *N*-glycosylation in cells was recently found to inhibit the formation of non-exosomal vesicles mainly containing Met (Harada et al. 2020). In addition, exosome biogenesis requires oligomerization of cellular syndecan mediated by heparan sulfate (Baietti et al. 2012). These findings underscore the importance of glycans for the biogenesis and functions of sEVs.

2.1 *Glycosyltransferase in sEV*

2.1.1 MGAT5

As described above, we hypothesized that glycosyltransferases are also included in sEVs. This hypothesis was tested by measuring the activity of glycosyltransferases involved in *N*-glycan branching in sEVs derived from various cancer cells. As a result, GnT-V activity was found to be highly and commonly enriched in sEVs (Fig. 3a), whereas other enzymes showed less activity in sEVs than in cells, indicating that cancer-related glycosyltransferase GnT-V is selectively loaded into sEVs with enzymatic activity (Hirata et al. 2023).

Western blotting revealed that GnT-V from sEVs migrates faster in SDS-PAGE than GnT-V derived from cell samples. The band shift was still observed even after cleavage of *N*-glycans with PNGase F, suggesting that the GnT-V polypeptide in sEVs is a cleaved form (Fig. 3b). Knocking out SPPL3, which is the major sheddase for GnT-V, drastically reduced the amount of GnT-V in sEVs. This observation suggests that GnT-V in sEV is a soluble form and that cleavage by SPPL3 is a prerequisite

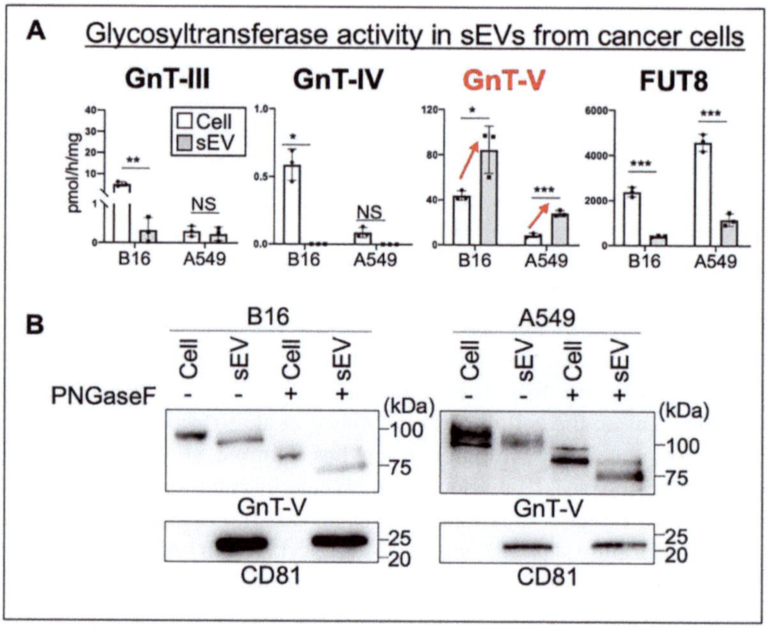

Fig. 3 Activity and state of GnT-V in sEVs. **a** Activity of *N*-glycan branching enzymes in sEVs derived from mouse melanoma B16 and human lung carcinoma A549 cells. GnT-V activity is selectively enriched in the sEV fraction. **b** Western blotting of cellular and sEV proteins from B16 and A549 cells for GnT-V and the sEV marker CD81. PNGaseF is an enzyme that removes *N*-glycans to show clearly the mobility of GnT-V in SDS-PAGE. This figure was modified from a previous paper (Hirata et al. 2023)

for loading GnT-V into sEVs. In addition, the analysis of the average particle size of GnT-V-positive sEVs was larger than typical exosomes. Sucrose-density gradient fractionation experiments further showed that GnT-V-positive sEVs were distributed to fractions different from typical exosomes. These findings indicate that GnT-V is loaded into non-exosomal vesicles in cells after cleavage by SPPL3 (Hirata et al. 2023), although the pathway and the mechanism for this loading process remain to be elucidated.

We further examined whether GnT-V-containing sEVs are incorporated by recipient cells and whether the transferred GnT-V can locally remodel glycans in recipient cells. These questions were tested by incubating GnT-V-positive sEVs with GnT-V KO Hela cells, and enzyme activity and glycans in the recipient cells were then analyzed after washing (Fig. 4a). The results showed that GnT-V activity was detected in cell lysates after incubation with sEVs (Fig. 4b). Furthermore, lectin staining revealed that recipient cells become positive for GnT-V-produced glycans (Fig. 4c) (Hirata et al. 2023). These findings indicate that GnT-V can be transferred from donor to recipient cells via sEVs with intact activity and that the glycans in recipient cells are remodeled, suggesting a novel non-genetic mechanism of glycan-remodeling mediated by sEVs.

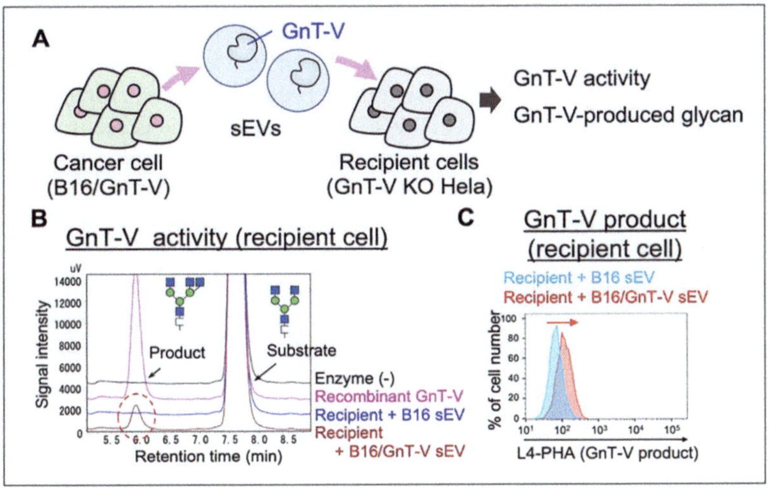

Fig. 4 Cell-to-cell transfer of GnT-V. **a** Experimental scheme. sEVs were collected from B16 cells expressing GnT-V and incubated with GnT-V KO recipient cells. The activity and the levels of GnT-V-produced glycans in the recipient cells were analyzed. **b** The in vitro GnT-V activity assay of recipient cells incubated with sEVs derived from B16 or B16/GnT-V cells. **c** The levels of GnT-V-produced glycans were analyzed by flow cytometry with the L4-PHA lectin. This figure was modified from a previous paper (Hirata et al. 2023)

2.1.2 ST6GAL1

Glycosyltransferase ST6GAL1 is a major sialyltransferase for α2,6-sialic acid in *N*-glycans (Meng et al. 2013) and is involved in sEV functions. In particular, this enzyme was reported to be contained in exomeres (Zhang et al. 2019). Proteomics analysis revealed that the cargo molecules of exosomes and exomeres are largely different, and ST6GAL1 was detected in both of these sEVs. Furthermore, ST6GAL1 in these particles was enzymatically active, and incubating recipient cells with these ST6GAL1-containing particles increased the amount of cellular α2,6-linked sialic acid. This result suggested that sialylated glycoproteins are transferred from particles or that transferred ST6GAL1 locally biosynthesizes α2,6-sialic acids. This report also shows an example of a non-genetic mechanism of glycan remodeling via small particles.

3 Future Perspective

In this chapter, recent findings about cell-to-cell transfer of glycosyltransferases via sEVs were presented. In particular, the incorporation of active GnT-V into sEVs remodeled glycans leads to recipient cells becoming positive for the GnT-V-produced *N*-glycan branch (Fig. 5). This non-genetic glycan remodeling may contribute to enhancing metastasis by GnT-V-positive tumors because GnT-V is highly related to cancer metastasis. In more detail, GnT-V-sEVs from cancer cells are transferred to non-cancer cells to alter the structures of glycans to more cancerous types. However, several questions remain unanswered. These questions include: (i) how is GnT-V loaded into sEVs; (ii) what glycosyltransferases are included in sEVs and how selective is this process; (iii) where do the transferred glycosyltransferases localize in the recipient cells; and (iv) do these transferred glycosyltransferases actively synthesize glycans. Exploring these questions will promote our understanding of how sEVs regulate glycosylation, providing new insights into how glycosylation is regulated in cells.

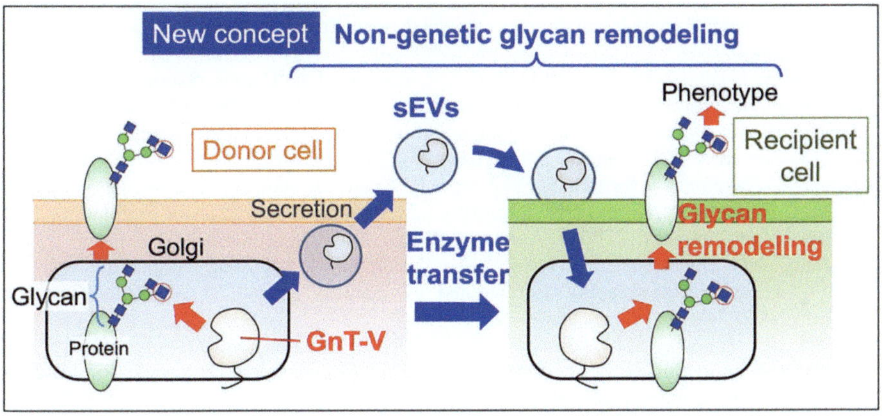

Fig. 5 A new concept showing the non-genetic pathway of glycan remodeling. GnT is included in sEVs derived from donor cells and transferred to recipient cells, where glycans are remodeled

Acknowledgements This work was partially supported by a CREST Grant (18070267) and a FOREST grant (JPMJFR215Z] from Japan Science and Technology Agency (JST) and a Grant from Takeda Science Foundation. We thank Edanz (https://jp.edanz.com/ac) for editing a draft of this manuscript.

References

K. Akasaka-Manya, H. Manya, Y. Sakurai, B.S. Wojczyk, Y. Kozutsumi, Y. Saito, N. Taniguchi, S. Murayama, S.L. Spitalnik, T. Endo, Protective effect of N-glycan bisecting GlcNAc residues on beta-amyloid production in Alzheimer's disease, Glycobiology, 20 (2010) 99–106.

I. Alves, B. Santos-Pereira, N. de la Cruz, A. Campar, V. Pinto, P.M. Rodrigues, M. Araujo, S. Santos, J. Ramos-Soriano, C. Vasconcelos, R. Silva, N. Afonso, F. Mira, C.C. Barrias, N.L. Alves, J. Rojo, L. Santos, A. Marinho, S.S. Pinho, Host-derived mannose glycans trigger a pathogenic gammadelta T cell/IL-17a axis in autoimmunity, Sci Transl Med, 15 (2023) eabo1930.

R. Apweiler, H. Hermjakob, N. Sharon, On the frequency of protein glycosylation, as deduced from analysis of the SWISS-PROT database, Biochim Biophys Acta, 1473 (1999) 4–8.

M.F. Baietti, Z. Zhang, E. Mortier, A. Melchior, G. Degeest, A. Geeraerts, Y. Ivarsson, F. Depoortere, C. Coomans, E. Vermeiren, P. Zimmermann, G. David, Syndecan-syntenin-ALIX regulates the biogenesis of exosomes, Nat Cell Biol, 14 (2012) 677–685.

P. Buckhaults, L. Chen, N. Fregien, M. Pierce, Transcriptional regulation of N-acetylglucosaminyltransferase V by the src oncogene, J Biol Chem, 272 (1997) 19575–19581.

J.F. Darby, A.K. Gilio, B. Piniello, C. Roth, E. Blagova, R.E. Hubbard, C. Rovira, G.J. Davies, L. Wu, Substrate Engagement and Catalytic Mechanisms of N-Acetylglucosaminyltransferase V, ACS Catalysis, 10 (2020) 8590–8596.

M. Demetriou, M. Granovsky, S. Quaggin, J.W. Dennis, Negative regulation of T-cell activation and autoimmunity by Mgat5 N-glycosylation, Nature, 409 (2001) 733–739.

A.M. Dias, A. Correia, M.S. Pereira, C.R. Almeida, I. Alves, V. Pinto, T.A. Catarino, N. Mendes, M. Leander, M.T. Oliva-Teles, L. Maia, C. Delerue-Matos, N. Taniguchi, M. Lima, I. Pedroto, R. Marcos-Pinto, P. Lago, C.A. Reis, M. Vilanova, S.S. Pinho, Metabolic control of T cell immune response through glycans in inflammatory bowel disease, Proc Natl Acad Sci U S A, 115 (2018) E4651–E4660.

M. Granovsky, J. Fata, J. Pawling, W.J. Muller, R. Khokha, J.W. Dennis, Suppression of tumor growth and metastasis in Mgat5-deficient mice, Nat Med, 6 (2000) 306–312.

A. Grigorian, M. Demetriou, Mgat5 deficiency in T cells and experimental autoimmune encephalomyelitis, ISRN Neurol, 2011 (2011) 374314.

Y. Harada, K. Nakajima, T. Suzuki, T. Fukushige, K. Kondo, J. Seino, Y. Ohkawa, T. Suzuki, H. Inoue, T. Kanekura, N. Dohmae, N. Taniguchi, I. Maruyama, Glycometabolic Regulation of the Biogenesis of Small Extracellular Vesicles, Cell Rep, 33 (2020) 108261.

T. Hirata, M. Takata, Y. Tokoro, M. Nakano, Y. Kizuka, Shedding of N-acetylglucosaminyltransferase-V is regulated by maturity of cellular N-glycan, Commun Biol, 5 (2022) 743.

T. Hirata, Y. Harada, K.M. Hirosawa, Y. Tokoro, K.G.N. Suzuki, Y. Kizuka, N-acetylglucosaminyltransferase-V (GnT-V)-enriched small extracellular vesicles mediate N-glycan remodeling in recipient cells, iScience, 26 (2023) 105747.

B.S. Hong, J.H. Cho, H. Kim, E.J. Choi, S. Rho, J. Kim, J.H. Kim, D.S. Choi, Y.K. Kim, D. Hwang, Y.S. Gho, Colorectal cancer cell-derived microvesicles are enriched in cell cycle-related mRNAs that promote proliferation of endothelial cells, BMC Genomics, 10 (2009) 556.

A. Hoshino, B. Costa-Silva, T.L. Shen, G. Rodrigues, A. Hashimoto, M. Tesic Mark, H. Molina, S. Kohsaka, A. Di Giannatale, S. Ceder, S. Singh, C. Williams, N. Soplop, K. Uryu, L. Pharmer, T. King, L. Bojmar, A.E. Davies, Y. Ararso, T. Zhang, H. Zhang, J. Hernandez, J.M. Weiss, V.D. Dumont-Cole, K. Kramer, L.H. Wexler, A. Narendran, G.K. Schwartz, J.H. Healey, P. Sandstrom, K.J. Labori, E.H. Kure, P.M. Grandgenett, M.A. Hollingsworth, M. de Sousa, S. Kaur, M. Jain, K. Mallya, S.K. Batra, W.R. Jarnagin, M.S. Brady, O. Fodstad, V. Muller, K. Pantel, A.J. Minn, M.J. Bissell, B.A. Garcia, Y. Kang, V.K. Rajasekhar, C.M. Ghajar, I. Matei, H. Peinado, J. Bromberg, D. Lyden, Tumour exosome integrins determine organotropic metastasis, Nature, 527 (2015) 329–335.

R. Kalluri, V.S. LeBleu, The biology, function, and biomedical applications of exosomes, Science, 367 (2020).

R. Kang, H. Saito, Y. Ihara, E. Miyoshi, N. Koyama, Y. Sheng, N. Taniguchi, Transcriptional regulation of the N-acetylglucosaminyltransferase V gene in human bile duct carcinoma cells (HuCC-T1) is mediated by Ets-1, J Biol Chem, 271 (1996) 26706–26712.

Y.S. Kim, J. Perdomo, J.S. Whitehead, Glycosyltransferases in human blood.I. Galactosyltransferase in human serum and erythrocyte membranes, J Clin Invest, 51 (1972) 2024–2032.

Y.S. Kim, J. Perdomo, J.S. Whitehead, K.J. Curtis, Glycosyltransferases in human blood. II. Study of serum galactosyltransferase and N-acetylgalactosaminyltransferase in patients with liver diseases, J Clin Invest, 51 (1972) 2033–2039.

Y. Kizuka, N. Taniguchi, Enzymes for N-Glycan Branching and Their Genetic and Nongenetic Regulation in Cancer, Biomolecules, 6 (2016).

Y. Kizuka, S. Kitazume, R. Fujinawa, T. Saito, N. Iwata, T.C. Saido, M. Nakano, Y. Yamaguchi, Y. Hashimoto, M. Staufenbiel, H. Hatsuta, S. Murayama, H. Manya, T. Endo, N. Taniguchi, An aberrant sugar modification of BACE1 blocks its lysosomal targeting in Alzheimer's disease, EMBO Mol Med, 7 (2015) 175–189.

Y. Kizuka, M. Nakano, S. Kitazume, T. Saito, T.C. Saido, N. Taniguchi, Bisecting GlcNAc modification stabilizes BACE1 protein under oxidative stress conditions, Biochem J, 473 (2016) 21–30.

P.H. Kuhn, M. Voss, M. Haug-Kroper, B. Schroder, U. Schepers, S. Brase, C. Haass, S.F. Lichtenthaler, R. Fluhrer, Secretome analysis identifies novel signal Peptide peptidase-like 3 (Sppl3) substrates and reveals a role of Sppl3 in multiple Golgi glycosylation pathways, Mol Cell Proteomics, 14 (2015) 1584–1598.

G. Lauc, M. Pezer, I. Rudan, H. Campbell, Mechanisms of disease: The human N-glycome, Biochim Biophys Acta, 1860 (2016) 1574–1582.

M. Mathieu, L. Martin-Jaular, G. Lavieu, C. Thery, Specificities of secretion and uptake of exosomes and other extracellular vesicles for cell-to-cell communication, Nat Cell Biol, 21 (2019) 9–17.

L. Meng, F. Forouhar, D. Thieker, Z. Gao, A. Ramiah, H. Moniz, Y. Xiang, J. Seetharaman, S. Milaninia, M. Su, R. Bridger, L. Veillon, P. Azadi, G. Kornhaber, L. Wells, G.T. Montelione, R.J. Woods, L. Tong, K.W. Moremen, Enzymatic basis for N-glycan sialylation: structure of rat alpha2,6-sialyltransferase (ST6GAL1) reveals conserved and unique features for glycan sialylation, J Biol Chem, 288 (2013) 34680–34698.

K.W. Moremen, M. Tiemeyer, A.V. Nairn, Vertebrate protein glycosylation: diversity, synthesis and function, Nat Rev Mol Cell Biol, 13 (2012) 448–462.

M. Nagae, Y. Kizuka, E. Mihara, Y. Kitago, S. Hanashima, Y. Ito, J. Takagi, N. Taniguchi, Y. Yamaguchi, Structure and mechanism of cancer-associated N-acetylglucosaminyltransferase-V, Nat Commun, 9 (2018) 3380.

M. Nagae, Y. Yamaguchi, N. Taniguchi, Y. Kizuka, 3D Structure and Function of Glycosyltransferases Involved in N-glycan Maturation, Int J Mol Sci, 21 (2020).

G. van Niel, G. D'Angelo, G. Raposo, Shedding light on the cell biology of extracellular vesicles, Nat Rev Mol Cell Biol, 19 (2018) 213–228.

K. Ohtsubo, J.D. Marth, Glycosylation in cellular mechanisms of health and disease, Cell, 126 (2006) 855–867.

K. Ohtsubo, S. Takamatsu, M.T. Minowa, A. Yoshida, M. Takeuchi, J.D. Marth, Dietary and genetic control of glucose transporter 2 glycosylation promotes insulin secretion in suppressing diabetes, Cell, 123 (2005) 1307–1321.

K. Ohtsubo, M.Z. Chen, J.M. Olefsky, J.D. Marth, Pathway to diabetes through attenuation of pancreatic beta cell glycosylation and glucose transport, Nat Med, 17 (2011) 1067–1075.

S.S. Pinho, C.A. Reis, Glycosylation in cancer: mechanisms and clinical implications, Nat Rev Cancer, 15 (2015) 540–555.

J.L. Qu, X.J. Qu, M.F. Zhao, Y.E. Teng, Y. Zhang, K.Z. Hou, Y.H. Jiang, X.H. Yang, Y.P. Liu, Gastric cancer exosomes promote tumour cell proliferation through PI3K/Akt and MAPK/ERK activation, Dig Liver Dis, 41 (2009) 875–880.

G. Rodrigues, A. Hoshino, C.M. Kenific, I.R. Matei, L. Steiner, D. Freitas, H.S. Kim, P.R. Oxley, I. Scandariato, I. Casanova-Salas, J. Dai, C.R. Badwe, B. Gril, M. Tesic Mark, B.D. Dill, H. Molina, H. Zhang, A. Benito-Martin, L. Bojmar, Y. Ararso, K. Offer, Q. LaPlant, W. Buehring, H. Wang, X. Jiang, T.M. Lu, Y. Liu, J.K. Sabari, S.J. Shin, N. Narula, P.S. Ginter, V.K. Rajasekhar, J.H. Healey, E. Meylan, B. Costa-Silva, S.E. Wang, S. Rafii, N.K. Altorki, C.M. Rudin, D.R. Jones, P.S. Steeg, H. Peinado, C.M. Ghajar, J. Bromberg, M. de Sousa, D. Pisapia, D. Lyden, Tumour exosomal CEMIP protein promotes cancer cell colonization in brain metastasis, Nat Cell Biol, 21 (2019) 1403–1412.

A. Shimoda, S.I. Sawada, Y. Sasaki, K. Akiyoshi, Exosome surface glycans reflect osteogenic differentiation of mesenchymal stem cells: Profiling by an evanescent field fluorescence-assisted lectin array system, Sci Rep, 9 (2019) 11497.

M.C. Silva, A. Fernandes, M. Oliveira, C. Resende, A. Correia, J.C. de-Freitas-Junior, A. Lavelle, J. Andrade-da-Costa, M. Leander, H. Xavier-Ferreira, J. Bessa, C. Pereira, R.M. Henrique, F. Carneiro, M. Dinis-Ribeiro, R. Marcos-Pinto, M. Lima, B. Lepenies, H. Sokol, J.C. Machado, M. Vilanova, S.S. Pinho, Glycans as Immune Checkpoints: Removal of Branched N-glycans Enhances Immune Recognition Preventing Cancer Progression, Cancer Immunol Res, 8 (2020) 1407–1425.

P. Stanley, N. Taniguchi, M. Aebi, N-Glycans, in: 3rd, A. Varki, R.D. Cummings, J.D. Esko, P. Stanley, G.W. Hart, M. Aebi, A.G. Darvill, T. Kinoshita, N.H. Packer, J.H. Prestegard, R.L. Schnaar, P.H. Seeberger (Eds.) Essentials of Glycobiology, Cold Spring Harbor (NY), 2015, pp. 99–111.

S. Sun, Y. Hu, M. Ao, P. Shah, J. Chen, W. Yang, X. Jia, Y. Tian, S. Thomas, H. Zhang, N-GlycositeAtlas: a database resource for mass spectrometry-based human N-linked glycoprotein and glycosylation site mapping, Clin Proteomics, 16 (2019) 35.

Z. Tan, L. Cao, Y. Wu, B. Wang, Z. Song, J. Yang, L. Cheng, X. Yang, X. Zhou, Z. Dai, X. Li, F. Guan, Bisecting GlcNAc modification diminishes the pro-metastatic functions of small extracellular vesicles from breast cancer cells, J Extracell Vesicles, 10 (2020) e12005.
M. Tkach, C. Thery, Communication by Extracellular Vesicles: Where We Are and Where We Need to Go, Cell, 164 (2016) 1226–1232.
L. Treps, R. Perret, S. Edmond, D. Ricard, J. Gavard, Glioblastoma stem-like cells secrete the pro-angiogenic VEGF-A factor in extracellular vesicles, J Extracell Vesicles, 6 (2017) 1359479.
A. Varki, Biological roles of glycans, Glycobiology, 27 (2017) 3–49.
M. Voss, U. Kunzel, F. Higel, P.H. Kuhn, A. Colombo, A. Fukumori, M. Haug-Kroper, B. Klier, G. Grammer, A. Seidl, B. Schroder, R. Obst, H. Steiner, S.F. Lichtenthaler, C. Haass, R. Fluhrer, Shedding of glycan-modifying enzymes by signal peptide peptidase-like 3 (SPPL3) regulates cellular N-glycosylation, EMBO J, 33 (2014) 2890–2905.
I. Wortzel, S. Dror, C.M. Kenific, D. Lyden, Exosome-Mediated Metastasis: Communication from a Distance, Dev Cell, 49 (2019) 347–360.
H. Zhang, T. Deng, R. Liu, M. Bai, L. Zhou, X. Wang, S. Li, X. Wang, H. Yang, J. Li, T. Ning, D. Huang, H. Li, L. Zhang, G. Ying, Y. Ba, Exosome-delivered EGFR regulates liver microenvironment to promote gastric cancer liver metastasis, Nat Commun, 8 (2017) 15016.
H. Zhang, D. Freitas, H.S. Kim, K. Fabijanic, Z. Li, H. Chen, M.T. Mark, H. Molina, A.B. Martin, L. Bojmar, J. Fang, S. Rampersaud, A. Hoshino, I. Matei, C.M. Kenific, M. Nakajima, A.P. Mutvei, P. Sansone, W. Buehring, H. Wang, J.P. Jimenez, L. Cohen-Gould, N. Paknejad, M. Brendel, K. Manova-Todorova, A. Magalhaes, J.A. Ferreira, H. Osorio, A.M. Silva, A. Massey, J.R. Cubillos-Ruiz, G. Galletti, P. Giannakakou, A.M. Cuervo, J. Blenis, R. Schwartz, M.S. Brady, H. Peinado, J. Bromberg, H. Matsui, C.A. Reis, D. Lyden, Identification of distinct nanoparticles and subsets of extracellular vesicles by asymmetric flow field-flow fractionation, Nat Cell Biol, 20 (2018) 332–343.
Q. Zhang, J.N. Higginbotham, D.K. Jeppesen, Y.P. Yang, W. Li, E.T. McKinley, R. Graves-Deal, J. Ping, C.M. Britain, K.A. Dorsett, C.L. Hartman, D.A. Ford, R.M. Allen, K.C. Vickers, Q. Liu, J.L. Franklin, S.L. Bellis, R.J. Coffey, Transfer of Functional Cargo in Exomeres, Cell Rep, 27 (2019) 940–954 e946.
Q. Zhang, D.K. Jeppesen, J.N. Higginbotham, R. Graves-Deal, V.Q. Trinh, M.A. Ramirez, Y. Sohn, A.C. Neininger, N. Taneja, E.T. McKinley, H. Niitsu, Z. Cao, R. Evans, S.E. Glass, K.C. Ray, W.H. Fissell, S. Hill, K.L. Rose, W.J. Huh, M.K. Washington, G.D. Ayers, D.T. Burnette, S. Sharma, L.H. Rome, J.L. Franklin, Y.A. Lee, Q. Liu, R.J. Coffey, Supermeres are functional extracellular nanoparticles replete with disease biomarkers and therapeutic targets, Nat Cell Biol, 23 (2021) 1240–1254.
Y. Zhao, Y. Sato, T. Isaji, T. Fukuda, A. Matsumoto, E. Miyoshi, J. Gu, N. Taniguchi, Branched N-glycans regulate the biological functions of integrins and cadherins, FEBS J, 275 (2008) 1939–1948.
W. Zhou, M.Y. Fong, Y. Min, G. Somlo, L. Liu, M.R. Palomares, Y. Yu, A. Chow, S.T. O'Connor, A.R. Chin, Y. Yen, Y. Wang, E.G. Marcusson, P. Chu, J. Wu, X. Wu, A.X. Li, Z. Li, H. Gao, X. Ren, M.P. Boldin, P.C. Lin, S.E. Wang, Cancer-secreted miR-105 destroys vascular endothelial barriers to promote metastasis, Cancer Cell, 25 (2014) 501–515.

Analysis of Immune Responses Induced by Inhaled Fine Particulates

Etsushi Kuroda

Abstract Inhaled fine particulates, such as particulate matter 2.5 (PM2.5) and sand dust, are closely associated with the onset of respiratory diseases, especially with allergic disorders. These fine particulates travel through the respiratory tract via inhalation and are deposited deep into the lungs. These deposited particulates are engulfed by alveolar macrophages, which serve as sentinel cells of the lungs, and subsequently removed from the lungs, indicating that alveolar macrophages play a crucial role in clearing particulates in the lungs. On the other hand, some particulates stimulate alveolar macrophages to induce inflammatory responses. Thus, the detailed role of alveolar macrophages in response to inhaled fine particulates is still unclear.

Keywords Particulate matter · Allergy · IgE · Adjuvant · Cell death · IL-1α · DAMPs · Alveolar macrophages · iBALT

Abbreviations

PM2.5	Particulate matter 2.5
IgE	Immunoglobulin E
Siglec F	Sialic acid-binding immunoglobulin-like lectin F
CD	Cluster of differentiation
CX3CR1	CX3C chemokine receptor 1
GM-CSF	Granulocyte-monocyte colony-stimulating factor
PPAR-γ	Peroxisome proliferator-activated receptor-γ
TGF-β	Transforming growth factor-β
IL	Interleukin
NLRP	Nucleotide-binding oligomerization domain-like receptor family, pyrin domain-containing
DAMP	Damage-associated molecular pattern

E. Kuroda (✉)
Department of Immunology, Hyogo Medical University School of Medicine, 1-1 Mukogawa-Cho, Nishinomiya 663-8501, Hyogo, Japan
e-mail: kuroetu@hyo-med.ac.jp

Y. Baba et al. (eds.), *Extracellular Fine Particles*,
https://doi.org/10.1007/978-981-97-7067-0_8

NET	Neutrophil Extra Trap
PG	Prostaglandin
WT	Wild type
iBALT	Inducible bronchus-associated lymphoid tissue
Th2	Type 2 T helper

1 Introduction

In daily life, a variety of particulates are present in our environment. Some particulates, such as particulate pollutants, cigarette smoke, and aerosol (organic and inorganic) associated with respiratory diseases, can be harmful to our bodies. These fine particulates travel through the respiratory tracts and are deposited in the lungs. Once deposited, these particulates are engulfed by professional lung phagocytes known as alveolar macrophages. Engulfed particulates are transported to respiratory tracts and then removed from the lungs through the action of cilia. So, the function of alveolar macrophages is indispensable for clearing particulates in the lungs (Ling and Eeden 2009; Oberdörster et al. 2005). However, certain particulates are toxic and can cause death in alveolar macrophages after engulfment (Kuroda et al. 2016). In such cases, inflammatory responses are triggered. Interestingly, many of these particulates are known to induce not only temporary inflammatory responses but also acquired immune responses, such as allergic inflammation characterized by immunoglobulin E (IgE) responses and activation of eosinophils (Kuroda et al. 2013). These particulates are called 'adjuvants,' that are substances to induce acquired immunity. However, mechanisms of action of these particulates in immune responses are still unknown. In this review, we would like to explore the immune responses against fine particulates from the viewpoint of the function of alveolar macrophages in response to fine particulates.

2 Macrophages in the Lungs

It is thought that inhaled fine particulates are engulfed by macrophages. There are two types of macrophages in the lungs: alveolar macrophages and interstitial macrophages. Several studies have described the differences in development and maintenance between these types of macrophages. First, they express different cell surface markers. Alveolar macrophages from mice express sialic acid-binding immunoglobulin-like lectin F (Siglec F) and cluster of differentiation 11c (CD11c), which are known to be expressed in eosinophils and dendritic cells, respectively. On the other hand, interstitial macrophages have CD11b and CX3C chemokine receptor 1 (CX3CR1) as common macrophage markers. Thus, alveolar macrophages are thought to be a unique macrophage phenotype (Aegerter et al. 2022; Kulle et al. 2022; Hou et al. 2021).

Second, required growth factors differ between alveolar and interstitial macrophages. In general, granulocyte-monocyte colony-stimulating factor (GM-CSF) is a well-known cytokine for the differentiation and proliferation of alveolar macrophages. Furthermore, recent studies revealed that the activation of peroxisome proliferator-activated receptor-γ (PPAR-γ) and the stimulation of transforming growth factor-β (TGF-β) are necessary to maintain the characteristics of alveolar macrophages (Shibata et al. 2001; Schneider et al. 2014; Yu et al. 2017). Based on these findings, methods for promoting the proliferation of alveolar macrophages or differentiating alveolar macrophage-like cells from bone marrow cells have been reported (Kuroda et al. 2016; Luo et al. 2021; Kendall et al. 2023). These methods are useful tools for the in vitro analysis of immune responses and immunotoxicity related to particulates.

Lastly, the origin and development differ between alveolar and interstitial macrophages. Experiments using parabiosis, which involves surgical joining of two mice to share the blood circulation, have shown that interstitial macrophages are developed and regenerated from circulating blood monocytes in the same way as macrophages in the skin and intestine. However, alveolar macrophages are maintained by self-renewing and not from blood monocytes in steady state (Tan and Krasnow 2016). Furthermore, fate mapping studies of macrophages demonstrated that alveolar macrophages are mostly derived from fetal liver monocytes (Ginhoux and Guilliams 2016).

These two types of macrophages are considered to have different functions and roles in the lung immune systems. Since alveolar macrophages are thought to be the first cells to come into contact with inhaled fine particulates, we are focusing on alveolar macrophages and their role and responses to particulates.

3 Mechanisms of Adjuvant Activity of Particulates

The underlying mechanisms by which particulates stimulate immune cells and induce inflammation and acquired immune responses remain unclear. The studies on the effect of particulate on immune responses, particularly aluminum salts (alum), a substance used in vaccine adjuvant, have been well researched in immunology. It has long been known that the 'depot effect' is important for the adjuvant effect of alum. The depot effect is a mechanism by which the antigen adsorbed on fine particulate stays at the inoculation site and is gradually released, thereby activating the immune cells through prolonged stimulation by the antigen (Glenny et al. 1926; Harrison 1935). Recently, however, it has been revealed that activation of innate immunity is required for the adjuvant effect, suggesting that an immunological mechanism of action other than the depot effect exists. In experiments using mice, it has been reported that the antigen-specific immune responses remained at normal levels even when the inoculated ear wing was cut off several hours after inoculation with antigen-adsorbed alum to the ear (Hutchison et al. 2011). This result suggests that alum shows adjuvant activity to induce acquired immunity without relying on the depot effect.

The mechanism by which particulates activate innate immunity remained unclear, but in 2008, several papers reported that inorganic particulates stimulate innate immune cells such as macrophages and dendritic cells to induce the activation of the inflammasome (Hornung et al. 2008; Eisenbarth et al. 2008; Dostert et al. 2008). These studies showed that phagocytosis of alum, silica, or asbestos by macrophages and dendritic cells activates the intracellular pattern recognition receptor complex NLRP3 inflammasome, which induces interleukin (IL)-1β and IL-18 in a caspase-1-dependent manner. In addition, caspase-1-dependent cell death, pyrotosis, is also induced by alum, silica or asbestos. It appeared that the activation of the inflammasome might be important in adjuvant activity. However, inflammatory responses induced by alum have been reported in caspase-1- or NLRP3-deficient mice, suggesting the existence of adjuvant effects induced by inflammasome-independent mechanisms (Franchi and Núñez 2008; McKee et al. 2009).

Cell death and damage-associated molecular pattern (DAMP) release are known to be important events for the induction of adjuvant activities. Many fine particulates like alum and silica induce phagocytic cell death. In particular, both uric acid and DNA have been reported to be important DAMPs for adjuvant activities (Kuroda et al. 2013). Uric acid is a catabolic metabolite of nucleic acids and can crystallize to form urate crystals at specific conditions. Urate crystals have been reported to function as potent adjuvants and are thought to be deeply involved in the adjuvant activity of alum (Behrens et al. 2008; Kool et al. 2011, 2008; Kono et al. 2010). However, the mechanism by which urate crystals induce acquired immune responses has not yet been fully clarified. In addition to uric acid, DNA released from dead cells has been suggested to be involved in adjuvant activities. It has been reported that the administration of alum induces the release of DNA from dead cells at the inoculation site, and the released DNA acts as an adjuvant to induce antigen-specific antibodies (Desmet and Ishii 2012; Marichal et al. 2011; McKee et al. 2013). Furthermore, it has also been reported that Neutrophil Extracellular Traps (NETs), DNA released from activated neutrophils that migrate to the local site after inoculation with alum, are involved in antigen-specific antibody responses (Stephen et al. 2017). In general, innate immune cells express various DNA and RNA sensors, which may contribute to the adjuvant activity of particulate like alum and silica, but the detailed mechanism has not been clarified.

In addition to dead cell-derived factors, molecules that recognize dead cells are reported to be involved in the adjuvant effect, such as CD300a, a molecule that binds to phosphatidylserine expressed on dead (apoptotic) cells. CD300a is mainly expressed on inflammatory dendritic cells, which are accumulated at the site of alum inoculation, and it plays an important role in dead cell recognition by inflammatory dendritic cells. CD300a-deficient mice showed decreased production of antigen-specific antibodies, suggesting that activation of inflammatory dendritic cells by the recognition of dead cells might be involved in the induction of acquired immunity

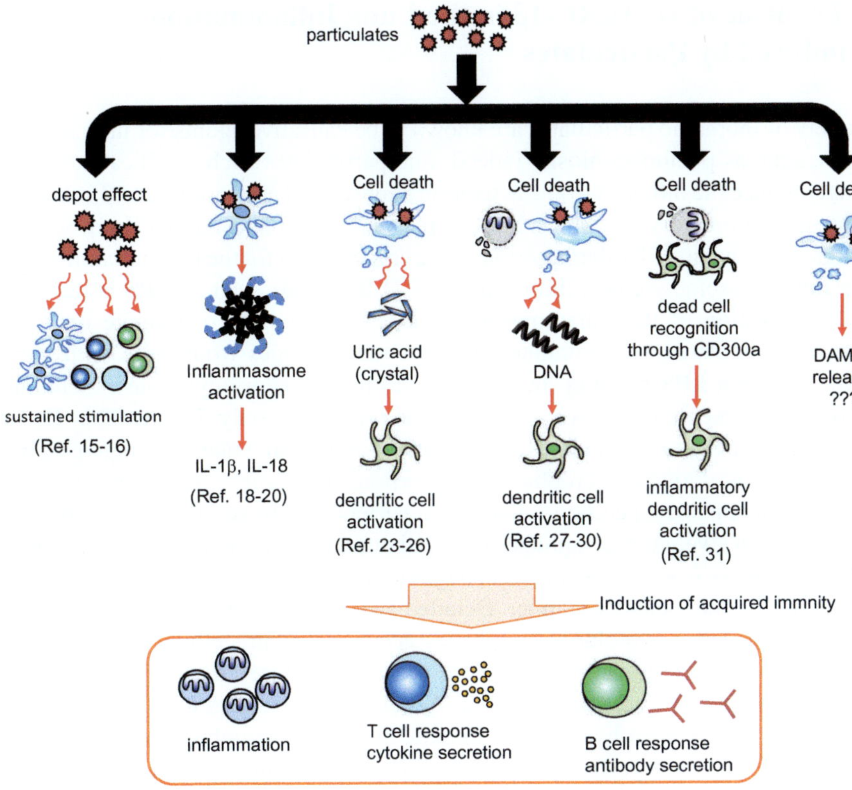

Fig. 1 Model diagrams illustrating the mechanisms of action of the adjuvant effect induced by particulates (alum). Numbers in parentheses indicate relevant references

(Miki et al. 2015). Figure 1 summarizes the mechanisms of action of the adjuvant effect induced by particulates.

The role of prostaglandin E_2 (PGE_2) in adjuvant activity has been reported. PGE_2 is a lipid mediator and released from both live and dead cells. However, PGE_2 act as a modifier of immune responses rather than a functioning as a direct adjuvant (Kuroda et al. 2011). Currently, the mechanisms underlying the adjuvant activities of particulates, including alum, are not fully understood, however, some DAMPs are thought to be involved in the adjuvant activities by particulates.

4 The Role of IL-1α/IL-1β in the Lung Inflammation induced by Particulates

A variety of inorganic particulates are known to be causative agents for lung inflammation, such as pneumoconiosis. Indeed, fine particulates such as silica and alum strongly induce the inflammasome activation as described above, and IL-1β is known as one of inflammatory cytokines. For this reason, it was believed that IL-1β release through the activation of inflammasome was a crucial event for the lung inflammation.

In inflammasome studies, the priming of TLR ligands such as LPS is required for the activation of the inflammasome and IL-1β release. To evaluate only the effect of particulates, we stimulated alveolar macrophages from mice with alum or silica without LPS priming. In this case, no inflammatory cytokines, including IL-1β, were observed, but we observed the release of only IL-1α. This IL-1α release in response to particulates was observed only in alveolar macrophages, not in bone marrow-derived macrophages and peritoneal macrophages, suggesting that this phenomenon was specific to alveolar macrophages. In addition, IL-1α release was detected in alveolar macrophages from inflammasome-deficient mice, indicating that IL-1α release from alveolar macrophages in response to particulates occurs in an inflammasome-independent manner. Detailed analysis of the mechanisms of IL-1α release revealed that alveolar macrophages accumulate intracellular IL-1α, which is subsequently released from cells due to cell death induced by particulate stimulation, as shown in Fig. 2 (Kuroda et al. 2016). Similar studies also demonstrated that silica and beryllium hydroxide stimulate alveolar macrophages to induce cell death and subsequently released intracellular IL-1α (Rabolli et al. 2014; Wade et al. 2018). Furthermore, different from particulate studies, Dagvadorj et al. reported that IL-1α release in the lungs after the intratracheal administration of LPS is mediated by cell death of alveolar macrophages, suggesting that the death of alveolar macrophage followed by IL-1α release is a crucial immune response for host defense against infection and the removal of foreign materials (Dagvadorj et al. 2015).

The phenomenon of dead cell derived IL-1α release is a crucial event in immune responses. Both IL-1α and IL-1β are initially synthesized as precursor forms within cells. Upon cellular activation, pro-IL-1α and pro-IL-1β are cleaved by calpain and caspase-1, respectively, leading to their conversion into mature forms. However, different from IL-1β, pro-IL-1α is the bioactive and released pro-IL-1a, acting as a dead cell factor, can induce inflammation through IL-1 receptor-dependent mechanisms. Thus, IL-1α is categorized as a DAMP, which is an immune-stimulatory factor derived from dead cells (Paolo and Shayakhmetov 2016; Kim et al. 2013; Rider et al. 2011).

Recently, Ikoma et al. reported on the different roles of IL-1α and IL-1β in particulate-induced lung inflammation. After the intratracheal instillation of silica, both IL-1α and IL-1β release and infiltration of neutrophils are observed in the lungs. However, IL-1α release and the number of neutrophils in the lungs were comparable between wild type (WT) mice and inflammasome-deficient mice, despite the absence of IL-1β release. On the other hand, the infiltration of neutrophils in the lungs was

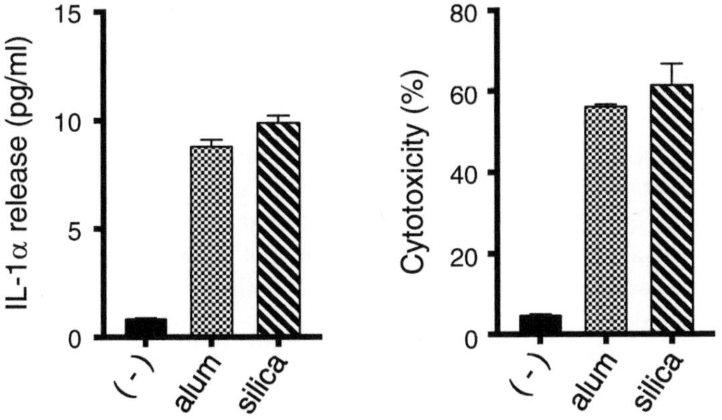

Fig. 2 Inflammatory particulate, alum and silica, induce alveolar macrophage death and the release of intracellular IL-1α. Primary alveolar macrophages were harvested and stimulated with the indicated particulates for 18 h in vitro. Released IL-1α and enzymatic activities of lactose dehydrogenase (as an indicator of cell death) were measured

significantly suppressed by the administration of anti-IL-1α neutralizing antibody to WT mice. These results suggest that silica-induced lung inflammation is triggered by the action of IL-1α release (Ikoma et al. 2022). We also observed similar results using the influenza virus infection model. Both IL-1α and IL-1β release were observed in the lungs after infection with the influenza virus, similar to the results seen with silica administration. Interestingly, IL-1α-deficient mice showed reduced resistance to infection, whereas no changes in susceptibility against influenza virus were observed between WT and inflammasome-deficient mice (Momota et al. 2020). These studies suggest that IL-1α plays a pivotal role in host defenses and inflammatory responses in the lungs, and IL-1α is thought to be released from alveolar macrophages that have undergone cell death due to the stimulation of particulates or viral infection.

5 Inhaled Particulates and Allergic Inflammation

The number of patients with allergic asthma and rhinitis has recently increased, and multiple reports have shown that particulate pollutants such as PM2.5 and sand dust are closely associated with the onset of allergic inflammation (Kanatani et al. 2010; Schwartz et al. 1993; Pulendran and Artis 2012). In this case, it is considered that particulates are involved in the induction of type 2 acquired immunity, characterized by activation of eosinophils and elevation of allergen-specific serum IgE levels. As described above, it is thought that inhaled particulates that induce allergic inflammation also function as adjuvants in the lungs. Not only particulate pollutants, but inorganic fine particulates such as silica and alum are also known to have adjuvant activities to induce acquired immunity. As inflammatory particulates like alum and silica induce cell death and IL-1α release in alveolar macrophages in the lungs, we suggest that the release of IL-1α is a key event for particulate-induced allergic acquired immune responses.

It has long been reported that IL-1 has the ability to induce IgE, which is a mediator of allergic inflammation (Nambu and Nakae 2010). We also observed that antigen-specific IgE is induced after airway sensitization with IL-1α and antigen in mice (Kuroda et al. 2016). In addition, we found that antigen-specific IgE responses induced after airway sensitization with alum and antigen were significantly reduced in IL-1 receptor-deficient mice. These studies indicate that released IL-1α as a DAMP function as an adjuvant to induce type 2 acquired immunity. Furthermore, a recent study has reported that in a mouse model of peanut allergy, inhalation of peanut flour induces IL-1 in the lungs, and the released IL-1 is involved in the induction of serum IgE (Dolence et al. 2018). These studies indicated that particulate- or allergen-induced IL-1α participates in both acute lung inflammation and the induction of allergic acquired immune responses.

6 Particulates and Ectopic Lymphoid Tissue

One of the unique lung immune responses induced after the inhalation of fine particulates is the formation of inducible bronchus-associated lymphoid tissue (iBALT). Multiple papers have reported that iBALT is a tertiary (ectopic) lymphoid tissue, an acquired lymphoid structure formed in the lungs after chronic inflammation (GeurtsvanKessel et al. 2009; Moyron-Quiroz et al. 2004; Rangel-Moreno et al. 2011). Similar to normal lymphoid tissue, iBALT contains T cell areas, B cell areas with germinal centers that generate B cells producing antigen-specific antibodies, and clusters of antibody-producing plasma cells, indicating that iBALT structure functions as lymphoid tissue. So far, iBALT was reported to be generated after influenza virus infection and chronic administration of LPS into the lungs. Additionally, we found that a large number of iBALT structures were induced in the lungs after intratracheal instillation of alum and allergens, as shown in Fig. 3 (Kuroda et al. 2016). In addition, we observed that the number of iBALT structures that were induced by alum was significantly reduced in IL-1 receptor-deficient mice, indicating that IL-1α is a crucial factor for the formation of iBALT as well as the elevation of serum IgE levels. Interestingly, pathogenic memory type 2 T helper (Th2) cells, which are important for chronic allergic inflammation, have also been shown to be involved in iBALT formation, suggesting that iBALT formation is also closely associated with the development and exacerbation of allergic diseases (Chvatchko et al. 1996; Shinoda et al. 2016). There are several interesting reports on iBALT formation. In studies on iBALT formation in humans, it has been reported that iBALT is more developed in infants than in healthy adults, suggesting that iBALT might be involved in asthma in children (Emery and Dinsdale 1973; Tschernig et al. 1995). In addition,

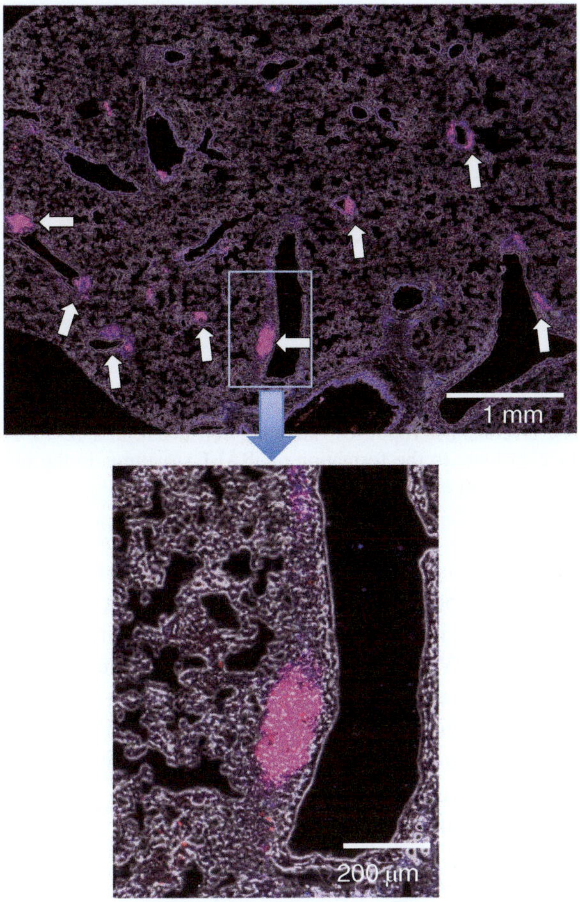

Fig. 3 Formation of iBALT structures in the lungs following airway sensitization to particulates. Mice were sensitized with alum and antigen through intratracheal instillation. One month later, the lungs were collected, and frozen sections were stained with B cell markers, anti-B220 (APC) and anti-CD21 (PE)

chronic exposure of tobacco smoke to mice developed iBALT formation in an IL-1-depenedent manner (Morissette et al. 2014). As tobacco smoke is also categorized as particulates, similar immunological mechanisms might be triggered in response to inhaled particulate matter in the lungs. Interestingly, it has been reported that the administration of steroids, one of the treatments for allergic diseases, hampers iBALT maturation, especially the formation of germinal centers. This phenomenon is intriguing in terms of the anti-allergic effects of steroid drugs (Silina et al. 2018).

7 Conclusion

In this review, we have discussed both the adjuvant activity of particulates and the lung-specific immune response they induce. Specifically, we focused on the lung-specific immune mechanisms such as alveolar macrophage function and iBALT

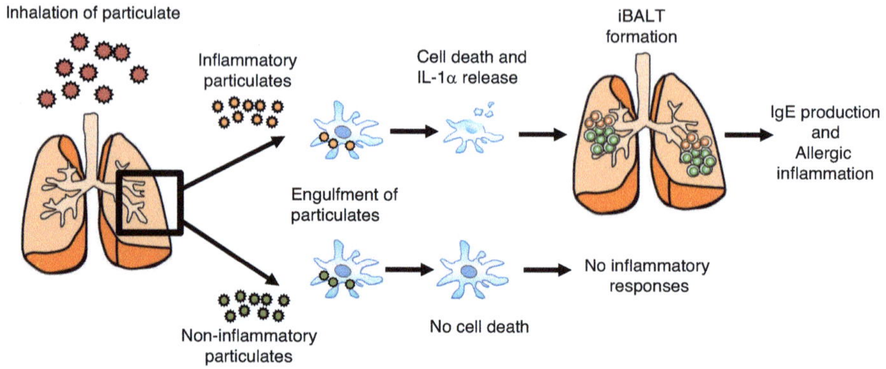

Fig. 4 Model of fine particulate-induced lung inflammation. Inflammatory particulates stimulate alveolar macrophages, inducing cell death, and IL-1α is released as DAMP. Released IL-1α contribute to the formation of iBALT structures in the lungs, which might participate in the production of allergen-specific IgE

formation, as well as the adjuvant effects of IL-1α, as summarized in Fig. 4. So far, most reports have investigated the effect of particulate that cause inflammation on immune responses. However, not all particulates induce similar inflammatory responses. We observed that some particulates do not induce cell death in alveolar macrophages and do not cause inflammation and adjuvant activity. Furthermore, we also found that some particulates induce cell death but do not release IL-1α. What are the difference between particles that cause inflammation and those that do not? It is crucial to analyze the mechanism by which non-inflammatory particulates do not cause inflammation. Therefore, it is crucial to focus on the physicochemical properties of particulates in future research.

Understanding the immune response to particulates with different properties will enhance our knowledge of the various diseases caused by the inhalation of fine particulates and contribute to the development of new therapies.

Acknowledgements This work was supported through a grant of the Japan Science and Technology Agency (JST) CREST (grant number JPMJCR19H3), JST PRESTO (grant number JPMJPR17H4), and Long-range Research Initiative (LRI) by Japan Chemical Industry Association (JCIA).

References

Aegerter H, Lambrecht BN, Jakubzick CV. Biology of lung macrophages in health and disease. Immunity. 2022;55(9):1564–80.

Behrens MD, Wagner WM, Krco CJ, Erskine CL, Kalli KR, Krempski J, et al. The endogenous danger signal, crystalline uric acid, signals for enhanced antibody immunity. Blood. 2008;111:1472–9.

Chvatchko Y, Kosco-Vilbois MH, Herren S, Lefort J, Bonnefoy JY. Germinal center formation and local immunoglobulin E (IgE) production in the lung after an airway antigenic challenge. J Exp Med. 1996;184(6):2353–60.

Dagvadorj J, Shimada K, Chen S, Jones HD, Tumurkhuu G, Zhang W, et al. Lipopolysaccharide Induces Alveolar Macrophage Necrosis via CD14 and the P2X7 Receptor Leading to Interleukin-1alpha Release. Immunity. 2015;42(4):640–53.

Desmet CJ, Ishii KJ. Nucleic acid sensing at the interface between innate and adaptive immunity in vaccination. Nat Rev Immunol. 2012;12:479–91.

Dolence JJ, Kobayashi T, Iijima K, Krempski J, Drake LY, Dent AL, et al. Airway exposure initiates peanut allergy by involving the IL-1 pathway and T follicular helper cells in mice. J Allergy Clin Immunol. 2018;142(4):1144–58 e8.

Dostert C, Petrilli V, Van Bruggen R, Steele C, Mossman BT, Tschopp J. Innate Immune Activation Through Nalp3 Inflammasome Sensing of Asbestos and Silica. Science. 2008;320:674–7.

Eisenbarth SC, Colegio OR, O'Connor W, Sutterwala FS, Flavell RA. Crucial role for the Nalp3 inflammasome in the immunostimulatory properties of aluminium adjuvants. Nature. 2008;453:1122–6.

Emery JL, Dinsdale F. The postnatal development of lymphoreticular aggregates and lymph nodes in infants' lungs. J Clin Pathol. 1973;26(7):539–45.

Franchi L, Núñez G. The Nlrp3 inflammasome is critical for aluminium hydroxide-mediated IL-1β secretion but dispensable for adjuvant activity. Eur J Immunol. 2008;38:2085–9.

GeurtsvanKessel CH, Willart MA, Bergen IM, van Rijt LS, Muskens F, Elewaut D, et al. Dendritic cells are crucial for maintenance of tertiary lymphoid structures in the lung of influenza virus-infected mice. J Exp Med. 2009;206(11):2339–49.

Ginhoux F, Guilliams M. Tissue-Resident Macrophage Ontogeny and Homeostasis. Immunity. 2016;44(3):439–49.

Glenny AT, Pope CG, Waddington H, Wallace U. Immunological notes XVLL.-XXIV. J Pathol Bacteriol. 1926;29(1):31–40.

Harrison WT. Some Observations on the Use of Alum Precipitated Diphtheria Toxoid. Am J Public Health Nations Health. 1935;25:298–300.

Hornung V, Bauernfeind F, Halle A, Samstad EO, Kono H, Rock KL, et al. Silica crystals and aluminum salts activate the NALP3 inflammasome through phagosomal destabilization. Nat Immunol. 2008;9:847–56.

Hou F, Xiao K, Tang L, Xie L. Diversity of Macrophages in Lung Homeostasis and Diseases. Front Immunol. 2021;12:753940.

Hutchison S, Benson RA, Gibson VB, Pollock AH, Garside P, Brewer JM. Antigen depot is not required for alum adjuvanticity. FASEB J. 2011;26:1272–9.

Ikoma K, Takahama M, Kimishima A, Pan Y, Taura M, Nakayama A, et al. Oridonin suppresses particulate-induced NLRP3-independent IL-1alpha release to prevent crystallopathy in the lung. Int Immunol. 2022;34(10):493–504.

Kanatani KT, Ito I, Al-Delaimy WK, Adachi Y, Mathews WC, Ramsdell JW, et al. Desert dust exposure is associated with increased risk of asthma hospitalization in children. Am J Respir Crit Care Med. 2010;182(12):1475–81.

Kendall RL, Ray JL, Hamilton RF, Jr., Holian A. Self-replicating murine ex vivo cultured alveolar macrophages as a model for toxicological studies of particle-induced inflammation. Toxicol Appl Pharmacol. 2023;461:116400.

Kim B, Lee Y, Kim E, Kwak A, Ryoo S, Bae SH, et al. The Interleukin-1alpha Precursor is Biologically Active and is Likely a Key Alarmin in the IL-1 Family of Cytokines. Front Immunol. 2013;4:391.

Kono H, Chen CJ, Ontiveros F, Rock KL. Uric acid promotes an acute inflammatory response to sterile cell death in mice. The Journal of clinical investigation. 2010;120(6):1939–49.

Kool M, Soullie T, van Nimwegen M, Willart MAM, Muskens F, Jung S, et al. Alum adjuvant boosts adaptive immunity by inducing uric acid and activating inflammatory dendritic cells. J Exp Med. 2008;205:869–82.

Kool M, Willart MA, van Nimwegen M, Bergen I, Pouliot P, Virchow JC, et al. An unexpected role for uric acid as an inducer of T helper 2 cell immunity to inhaled antigens and inflammatory mediator of allergic asthma. Immunity. 2011;34(4):527–40.

Kulle A, Thanabalasuriar A, Cohen TS, Szydlowska M. Resident macrophages of the lung and liver: The guardians of our tissues. Front Immunol. 2022;13:1029085.

Kuroda E, Ishii KJ, Uematsu S, Ohata K, Coban C, Akira S, et al. Silica crystals and aluminum salts regulate the production of prostaglandin in macrophages via NALP3 inflammasome-independent mechanisms. Immunity. 2011;34(4):514–26.

Kuroda E, Coban C, Ishii KJ. Particulate adjuvant and innate immunity: past achievements, present findings, and future prospects. International reviews of immunology. 2013;32(2):209–20.

Kuroda E, Ozasa K, Temizoz B, Ohata K, Koo CX, Kanuma T, et al. Inhaled Fine Particles Induce Alveolar Macrophage Death and Interleukin-1alpha Release to Promote Inducible Bronchus-Associated Lymphoid Tissue Formation. Immunity. 2016;45(6):1299–310.

Ling SH, van Eeden SF. Particulate matter air pollution exposure: role in the development and exacerbation of chronic obstructive pulmonary disease. International journal of chronic obstructive pulmonary disease. 2009;4:233–43.

Luo M, Lai W, He Z, Wu L. Development of an Optimized Culture System for Generating Mouse Alveolar Macrophage-like Cells. J Immunol. 2021;207(6):1683–93.

Marichal T, Ohata K, Bedoret D, Mesnil C, Sabatel C, Kobiyama K, et al. DNA released from dying host cells mediates aluminum adjuvant activity. Nature medicine. 2011;17(8):996–1002.

McKee AS, Munks MW, MacLeod MKL, Fleenor CJ, Van Rooijen N, Kappler JW, et al. Alum Induces Innate Immune Responses through Macrophage and Mast Cell Sensors, But These Sensors Are Not Required for Alum to Act As an Adjuvant for Specific Immunity. J Immunol. 2009;183:4403–14.

McKee AS, Burchill MA, Munks MW, Jin L, Kappler JW, Friedman RS, et al. Host DNA released in response to aluminum adjuvant enhances MHC class II-mediated antigen presentation and prolongs CD4 T-cell interactions with dendritic cells. Proceedings of the National Academy of Sciences of the United States of America. 2013;110(12):E1122–31.

Miki H, Nakahashi-Oda C, Sumida T, Shibuya A. Involvement of CD300a Phosphatidylserine Immunoreceptor in Aluminum Salt Adjuvant-Induced Th2 Responses. J Immunol. 2015;194(11):5069–76.

Momota M, Lelliott P, Kubo A, Kusakabe T, Kobiyama K, Kuroda E, et al. ZBP1 governs the inflammasome-independent IL-1alpha and neutrophil inflammation that play a dual role in anti-influenza virus immunity. Int Immunol. 2020;32(3):203–12.

Morissette MC, Jobse BN, Thayaparan D, Nikota JK, Shen P, Labiris NR, et al. Persistence of pulmonary tertiary lymphoid tissues and anti-nuclear antibodies following cessation of cigarette smoke exposure. Respir Res. 2014;15(1):49.

Moyron-Quiroz JE, Rangel-Moreno J, Kusser K, Hartson L, Sprague F, Goodrich S, et al. Role of inducible bronchus associated lymphoid tissue (iBALT) in respiratory immunity. Nat Med. 2004;10(9):927–34.

Nambu A, Nakae S. IL-1 and Allergy. Allergol Int. 2010;59(2):125–35.

Oberdörster G, Oberdörster E, Oberdörster J. Nanotoxicology: An Emerging Discipline Evolving from Studies of Ultrafine Particles. Environmental Health Perspectives. 2005;113(7):823–39.

Di Paolo NC, Shayakhmetov DM. Interleukin 1alpha and the inflammatory process. Nat Immunol. 2016;17(8):906–13.

Pulendran B, Artis D. New paradigms in type 2 immunity. Science. 2012;337(6093):431–5.

Rabolli V, Badissi AA, Devosse R, Uwambayinema F, Yakoub Y, Palmai-Pallag M, et al. The alarmin IL-1alpha is a master cytokine in acute lung inflammation induced by silica micro- and nanoparticles. Part Fibre Toxicol. 2014;11:69.

Rangel-Moreno J, Carragher DM, de la Luz Garcia-Hernandez M, Hwang JY, Kusser K, Hartson L, et al. The development of inducible bronchus-associated lymphoid tissue depends on IL-17. Nat Immunol. 2011;12(7):639–46.

Rider P, Carmi Y, Guttman O, Braiman A, Cohen I, Voronov E, et al. IL-1alpha and IL-1beta recruit different myeloid cells and promote different stages of sterile inflammation. J Immunol. 2011;187(9):4835–43.

Schneider C, Nobs SP, Kurrer M, Rehrauer H, Thiele C, Kopf M. Induction of the nuclear receptor PPAR-gamma by the cytokine GM-CSF is critical for the differentiation of fetal monocytes into alveolar macrophages. Nat Immunol. 2014;15(11):1026–37.

Schwartz J, Slater D, Larson TV, Pierson WE, Koenig JQ. Particulate air pollution and hospital emergency room visits for asthma in Seattle. Am Rev Respir Dis. 1993;147(4):826–31.

Shibata Y, Berclaz PY, Chroneos ZC, Yoshida M, Whitsett JA, Trapnell BC. GM-CSF regulates alveolar macrophage differentiation and innate immunity in the lung through PU.1. Immunity. 2001;15(4):557–67.

Shinoda K, Hirahara K, Iinuma T, Ichikawa T, Suzuki AS, Sugaya K, et al. Thy1+IL-7+ lymphatic endothelial cells in iBALT provide a survival niche for memory T-helper cells in allergic airway inflammation. Proceedings of the National Academy of Sciences of the United States of America. 2016;113(20):E2842–51.

Silina K, Soltermann A, Attar FM, Casanova R, Uckeley ZM, Thut H, et al. Germinal Centers Determine the Prognostic Relevance of Tertiary Lymphoid Structures and Are Impaired by Corticosteroids in Lung Squamous Cell Carcinoma. Cancer Res. 2018;78(5):1308–20.

Stephen J, Scales HE, Benson RA, Erben D, Garside P, Brewer JM. Neutrophil swarming and extracellular trap formation play a significant role in Alum adjuvant activity. npj Vaccines. 2017;2(1).

Tan SY, Krasnow MA. Developmental origin of lung macrophage diversity. Development. 2016;143(8):1318–27.

Tschernig T, Kleemann WJ, Pabst R. Bronchus-associated lymphoid tissue (BALT) in the lungs of children who had died from sudden infant death syndrome and other causes. Thorax. 1995;50(6):658–60.

Wade MF, Collins MK, Richards D, Mack DG, Martin AK, Dinarello CA, et al. TLR9 and IL-1R1 Promote Mobilization of Pulmonary Dendritic Cells during Beryllium Sensitization. J Immunol. 2018;201(8):2232–43.

Yu X, Buttgereit A, Lelios I, Utz SG, Cansever D, Becher B, et al. The Cytokine TGF-beta Promotes the Development and Homeostasis of Alveolar Macrophages. Immunity. 2017;47(5):903–12 e4.

Harnessing DNA and Energy Cargo: Unveiling the Active Biogenesis and Applications of Bacterial Extracellular Vesicles

Sotaro Takano and Akihiro Okamoto

1 Introduction

Human microbiota, akin to human cells releasing exosomes, produce spherical biological nanoparticles, bacterial extracellular vesicles (BEVs). These BEVs are composed of lipid bilayers and encapsulate a variety of biological molecules from their source cells such as signaling molecules, genetic materials, and proteins. BEVs have been known to contribute to diverse biological processes in the human body by mediating both microbe-microbe and host-microbe interactions (Schwechheimer and Kuehn 2015; Cuesta et al. 2021). Yet, while the importance of their cargo is well-recognized, the question remains: do bacteria actively biosynthesize the BEVs to control their cargo on purpose?

Recent studies have been demonstrated that BEVs are not just the by-products of cell-lysis or imbalance in local cell membrane properties but produced via various types of regulations (Fig. 1) (Schwechheimer and Kuehn 2015). Genetic alterations can either enhance or inhibit the formation of BEVs (Kitagawa et al. 2010; Kulp et al. 2015), and evidence points to the selective, rather than random, packaging of certain molecules like proteins and lipids within BEVs (Schwechheimer and Kuehn 2015; Bonnington and Kuehn 2016; Orench-Rivera and Kuehn 2021; Naradasu et al. 2021).

S. Takano · A. Okamoto (✉)
Research Center for Macromolecules and Biomaterials, National Institute for Materials Science, Tsukuba, Japan
e-mail: okamoto.akihiro@nims.go.jp

A. Okamoto
Graduate School of Chemical Sciences and Engineering, Hokkaido University, Sapporo, Japan

Graduate School of Science and Engineering, College of Science and Engineering, University of Tsukuba, Tsukuba, Japan

Institute of Innovation for Future Earth (IRFE) and Laboratory for Integrated Science and Materials (LiSM), Tokyo Institute of Technology, Tokyo, Japan

Y. Baba et al. (eds.), *Extracellular Fine Particles*,
https://doi.org/10.1007/978-981-97-7067-0_9

However, selective packaging into BEVs is still under debate due to the ambiguity of its underlying mechanisms, which could be interpreted as either spontaneous (physical) responses to environmental contexts or as part of a tightly regulated biological process.

In this chapter, we pivot our focus to the functional attributes of BEVs, setting aside the intricacies of their assembly mechanisms, to consider the possibility of an active BEV production strategy. We will focus on "bacterial DNA cargo", which potentially has versatile roles including ecological functions in the context of microbiota, and "electron cargo", crucial for anaerobic metabolism and cell survival. Additionally, we will also showcase the emerging applications of these BEVs' function for diagnosis, novel drug-delivery carriers or vaccines. These recent advances in BEV application have shed light on their great potential for medical applications.

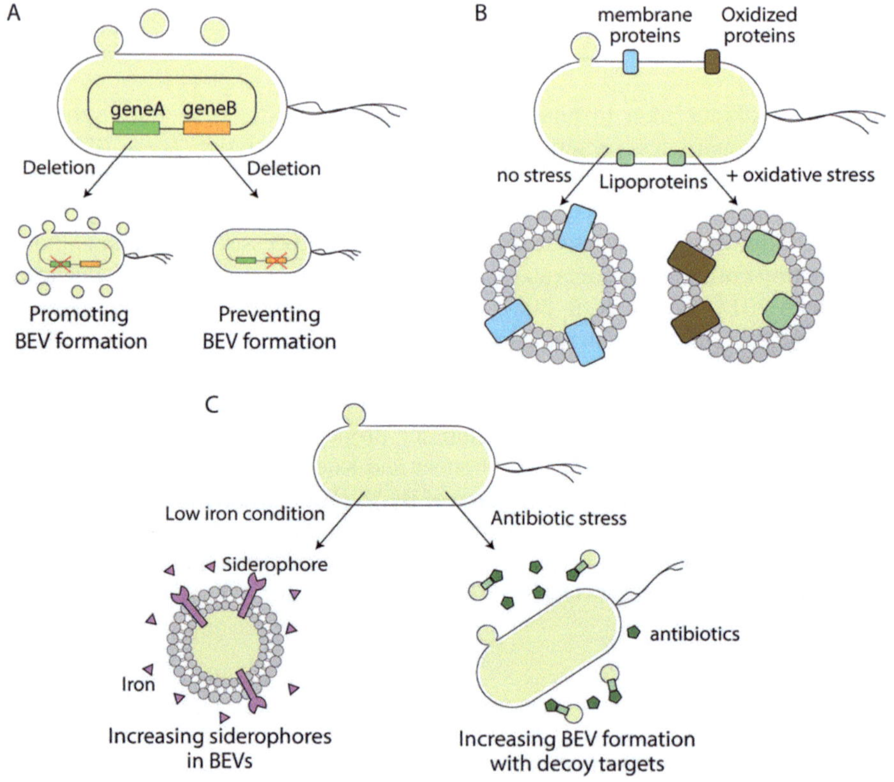

Fig. 1 Observations for BEV biogenesis modified by intracellular and extracellular perturbation. **A** There are cases where genetic disruption promotes or suppresses the BEV-biogenesis. **B** The imposition of stress usually leads to the package of different cellular contents. **C** The environmental changes cause the enhancement of loading specific molecules into BEVs or BEV-biogenesis itself. Some of those work as a benefit for the bacterial population

2 DNA in BEVs as a Key Mediator of Gene Exchange Among Microbiota

Historically, intercellular transfer of genetic materials has been considered as the product of mainly three molecular mechanisms: natural transformation, conjugation, and transduction. Each of those mechanisms required specific molecular machinery, genetic elements, and phenotypic traits, resulting in several "barriers" to gene transfer. Natural transformation is the active process of introducing extracellular free DNA fragments, which are ubiquitously produced by bacterial cells (e.g., cell lysis) and present in the environment. However, the natural transformation usually requires expressions of a set of proteins (e.g., type VI pili) for recipient cells (Mell and Redfield 2014), and the number of bacterial strains that can be transformable via free DNA uptake is limited to date (Johnsborg et al. 2007). Gene transfer via conjugation also needs a set of molecular types of machinery and mobile genetic elements (e.g., plasmids), limiting their prevalence in microbiota and the type of transferrable genes. Transduction via bacteriophages does not require a specific set of machinery for bacterial cells but its efficiency strongly depends on phage-host specificity, which could be very high (i.e., nearly one-to-one basis) (Koskella and Meaden 2013). The formation of "nanotubes", enabling the cells to interexchange cytoplasmic content, was discovered as a new mechanism of gene transfer between bacterial cells, yet it has been observed in a limited number of species (Dubey and Ben-Yehuda 2011).

Compared to these mechanisms, BEVs have different characteristics, and potentially work more efficiently for gene transfer. BEV biogenesis is currently appreciated as a universal mechanism observed among bacteria with several generation routes, including pervasive biological events under stressful conditions such as cell death and prophage induction (Toyofuku et al. 2019). Recently, Tran and Boedicker explored the relationship between the gene transfer efficiency via BEV and donor-recipient relatedness and found that HGT via BEV could occur between phylogenetically distant bacterial species (Tran and Boedicker 2017). Those studies suggest the pervasiveness of BEV biogenesis and extensive targets among microbiota.

Evidence suggests that various organisms within the human microbiome, both harmful and beneficial, produce BEVs (Grenier and Mayrand 1987; Engevik et al. 2021; Liao et al. 2014; Naradasu et al. 2021; Jones et al. 2020; Carvalho et al. 2019). DNA in BEVs plays a crucial role in microbiota such as transferring genetic materials among microbial populations by horizontal gene transfer (HGT). From decades ago, the capability of BEVs for transferring the packaged DNA to target cells was reported (Rumbo et al. 2011; Fulsundar et al. 2014; Klieve et al. 2005; Chiura et al. 2011; Kahn et al. 1983; Dorward et al. 1989; Kadurugamuwa and Beveridge 1995; Kolling and Matthews 1999). Some of those studies observed that BEVs can transfer beneficial genes and provide novel phenotypes to target cells such as antibiotic resistance (Rumbo et al. 2011), degrading macromolecule substrates (Klieve et al. 2005), and virulence (Yaron et al. 2000). These genetic exchanges can lead to significant phenotypic changes and improved survival for the recipient cells, which may, in turn, affect the biological environment of the BEV-producing

A

Examples of selective packaging of DNA in BEVs

Organism	Types or functions of genes	Method
Pseudomonas aerugionosa (26)	Plasmid	PCR
Streptococcus agalactiae (27)	Virulent factor (hemolysin) synthesis	PCR
Prochlorococcus (30)	Miscellaneous	Shotgun metagenomics
Paracoccus denitrificans (29)	Prophage genes	Shotgun metagenomics
Porphyromonas gingivalis	Organization of genomic archtecture, Vitamin synthesis	Droplet sequencing
Alcaligenes faecalis	Virulent factor (LPS) synthesis	Droplet sequencing

Examples of transformation by BEVs

Donor	Phenotypic changes of recipients
Acinetobacter baumannii (26)	Antibiotic resistance
Ruminococcus spp. (27)	Degradation of cellulose
Escherichia coli O157:H7 (30)	Virulence genes

B

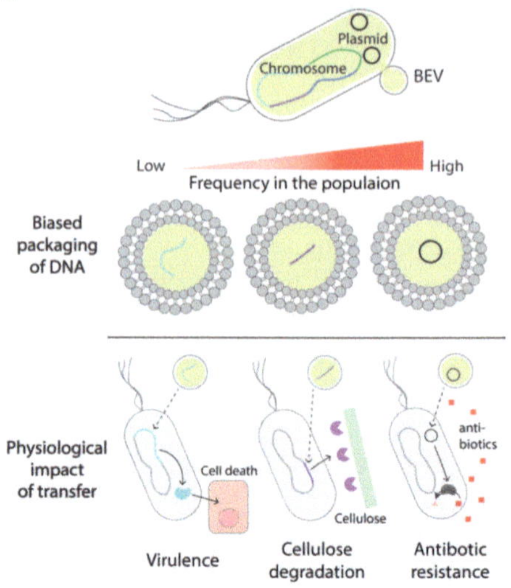

Fig. 2 Selectively packaged DNA in BEVs and its impact on the phenotypic changes via horizontal gene transfer. **A** Examples of selectively packaged DNA in BEVs and resulting phenotypic changes in recipient cells that were reported in previous studies. **B** Schematics of biased packaging of DNA and physiological impacts by the gene transfers. Several studies reported the prevalence of circularized DNA (e.g., plasmid) in BEVs and bias in the loaded genomic regions, which constraints the transferability of genes. The recipient cells of BEVs obtained beneficial functions for their survival such as virulence, macromolecule degradation, and antibiotic resistance

microbes within the microbiome. This suggests that HGT via BEVs is not a random process but is likely regulated to benefit the producing organisms (Fig. 2A).

The preferential packaging of certain genomic regions could restrict HGT. Although the mechanisms for this selective gene presence are unknown, studies have noted a non-uniform distribution of DNA within BEVs (Surve et al. 2016; Bitto et al. 2017; Yasuda et al. 2022; Biller et al. 2014), which could impede the transfer of specific genes. Also, The DNA content within BEVs may present barriers or biases to gene transfer. Research has largely focused on plasmid DNA due to its smaller size and autonomy from chromosomal DNA, which facilitates its incorporation into BEVs (Tran and Boedicker 2017; Rumbo et al. 2011; Fulsundar et al. 2014; Kahn et al. 1983; Dorward et al. 1989; Renelli et al. 2004). Conversely, chromosomal DNA is less frequently reported in BEVs, and its packaging is not well understood, potentially limiting the range of genes transferable via BEVs. More comprehensive and detailed investigations of genetic materials in BEVs would be necessary to estimate the impact of DNA cargo in BEVs on HGT. Also, our knowledge about donor-recipient range of BEV-mediated HGT is still limited.

3 Single Particle DNA Content Support the HGT Function in BEVs

As evidence for horizontal gene transfer (HGT) in pure cultures increases, its prevalence and the particular genes often transferred within the microbiome are still unconfirmed. This uncertainty arises from the limitations of standard metagenomic techniques, which measure the overall DNA content in bacterial extracellular vesicles (BEVs) but fail to identify specific gene sets within individual vesicles. Because the effectiveness of HGT hinges on the quantity of BEVs containing the desired genes, a metagenomic strategy focused solely on total gene quantities cannot sufficiently determine the likelihood of gene transfer in the microbiome.

One possible approach to this end is to analyze each particle by single-cell-genomics techniques such as droplet sequencing methods (Hosokawa et al. 2017; Gole et al. 2013; Macosko et al. 2015). Those methods were recently used for the characterization of a single-cell bacterial genome by whole genome amplification (WGA) with low contamination and biased amplification risk. In the recent pilot studies, such droplet DNA sequencing technique was applied to characterize DNA sequences that were internally stored in a single BEV (Takano et al. 2024). One major advantage of this method is analyzing low-amount DNA in BEVs by whole genome amplification (WGA) that were usually difficult to detect in other approaches. For instance, the current study focusing on BEVs from *Porphyromonas gingivalis* (*P. gingivalis*), a prominent pathogen of periodontitis, revealed that over 60% of BEVs contained chromosomal DNA of parental bacteria and coded ~50 genes on average. This percentage was estimated by staining the BEV-containing droplets with DNA dye after WGA (Takano et al. 2024). Notably, the direct staining of BEVs can detect quite a lower percentage of the positively stained particles by DNA dye. In the case of *P. gingivalis,* only ≈1% of BEVs were estimated as DNA-contained using nanoparticle tracking analysis (Takano et al. 2024). The study using epifluorescence microscopy also detected less than 1% of BEVs with DNA fluorescence in other bacterial cases (Biller et al. 2017). This is reasonable because the length of contained DNA usually affects the results of the fluorescence staining method (Tomaru and Nagasaki 2007). Droplet sequencing detected BEVs containing less than 10 kbp of the genomic regions and the DNA content of such BEVs would be below the detection limit of the direct fluorescence staining. Those results suggest that packaging of chromosomal DNA into BEVs is a more prevalent event than previously reported (Biller et al. 2017).

The droplet sequencing analysis also captured considerable heterogeneity in loaded DNA length and region among the BEV population isolated from the same bacterial culture (Takano et al. 2024). Meanwhile, statistical screening revealed the presence of 10~30 kbp genomic regions that were frequently detected among the BEV population, which indicates that there is enrichment of specific loci in the parental bacterial genome in BEVs. The presence of such a "genetic barcode" was observed in both a bacterial pure culture system and oral microbiota (Takano et al. 2024), suggesting that the bias in the loaded chromosomal region is ubiquitous among

BEVs in nature. Interestingly, the packaged genes showed quite similar functions to each other. In the case of *P. gingivalis*, genes related to DNA integration are significantly enriched (Fig. 2). Since those gene sets directly affect genome organization, the prevalence of those genes in BEVs implicitly explains their association with HGT. The enrichment of gene clusters associated with specific metabolism or virulence was also found in several bacterial cases. Importantly, those genes are located on the chromosome of the original bacterium and not plasmid or mobile genetic element, and thus BEVs can work as mediators of chromosomal gene transfer among bacterial populations, yet the mechanism of the selective packaging of chromosomal DNA is still under debate. Taken together, those observations indicate that packaging of the chromosomal DNA would frequently occur but the propensity of loading into BEVs is different depending on genomic loci, resulting in the functional bias in the DNA cargo of BEVs.

4 DNA Cargo in BEVs for Human Cells: Immunogenic and Potential Carcinogenic Materials

BEVs from the commensal microbiota (e.g., gut microbiota) spread to the whole human body by transmigrating to the bloodstream (Jones et al. 2020; Jang et al. 2015) and can interact with various host human cells via their molecular cargoes (Kim et al. 2017; Shen et al. 2012), and thus have a potential impact on systemic disease (Chronopoulos and Kalluri 2020). This high permeability of BEVs over the epithelial barrier compared to the parental bacterial cells would result from their relatively small size ($\approx$100 nm). Tulkens et al. reported that BEVs produced by gut bacteria residing in the lumen space can cross the epithelial barrier when the barrier is dysfunctionalized by intestinal bowel disease (Tulkens et al. 2020a), and then subsequently interact with wide-range of host immune cells and migrate to other organs (e.g., brain and liver) via systemic circulation. It is usually known that specific types of bacterial DNA (e.g., CpG DNA) possess immunostimulatory properties (Hacker et al. 2002; Paludan et al. 2019), and it is possible that BEVs can modulate the host immune response by transferring such DNA molecules to host cells. Several routes for the entry into host cells have been suggested such as endocytosis and membrane fusion (O'Donoghue and Krachler 2016), suggesting that the translocation of DNA cargo into the host cells is not a rare event. The fact that BEV-associated DNA can translocate into the nucleus of host mammalian cells (Bitto et al. 2017), implicating that those DNA molecules work as carcinogenic materials through the integration into the host chromosome (Chronopoulos and Kalluri 2020; Riley et al. 2013).

Recently, the presence of BEVs in the bloodstream gained great attention for both the causes of diseases and promising targets for diagnosis (Chronopoulos and Kalluri 2020; Tulkens et al. 2020a), and their internal DNA would be also potential biomarkers. Circulating bacterial DNA in the bloodstream has attracted a novel reservoir of biomarkers, and a recent study explored the correlation between profiles of microbial reads and cancer types (Poore et al. 2020), although its origin is still controversial. It is highly possible that some of those microbial reads are derived from circulating BEVs translocated from the gut lumen. Compared to free circulating bacterial DNA, bacterial DNA packaged into BEVs potentially contains more multiparametric information because it implicitly explains the presence of other molecular cargo (e.g., virulence factors) from the same parental bacterial cells, some of which have a greater health impact. Although exploring BEV-derived DNA in the systemic circulation would harbor unique microbiome signatures depending on the host health status, comprehensive analysis by omics approach is technically difficult due to the quite low concentration of BEVs in human blood (10^6 particles/mL in the plasma) (Tulkens et al. 2020b) (Fig. 3).

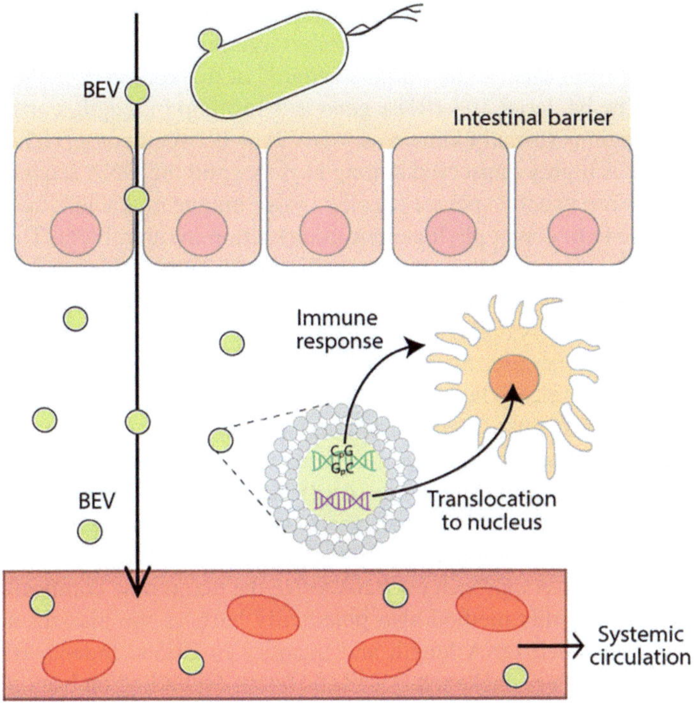

Fig. 3 High permeability of BEVs into intestinal barrier enable direct host-bacteria interactions in whole human body. By the dysfunctionalization of the epithelial barrier, BEVs can migrate into lamina propria and directly interact with immune cells. BEVs can also transmigrate into the systemic circulation and propagate in whole human body. The loaded DNA into BEVs potentially cause the immune response and can translocate into the host nucleus

5 DNA in BEVs for Understanding Their Origin in Microbiota

In the meantime, profiling the origins of BEVs in human body has proven to be as a novel approach for inferring one's health-status. Among the molecular cargo of BEVs, DNA is generally appreciated as a promising target for comprehensive and high-throughput profiling (e.g., next-generation sequencing). Especially, in the case of microbiota, profiling DNA has been a standard approach to characterize its taxonomic composition. With a similar idea to this, the internally encapsulated DNA is one of the most promising targets for understanding the taxonomic origins of BEVs. Recently, many studies have analyzed BEV-associated DNA isolated from human body fluids such as blood, urea, and sputum to characterize the possible origins and further understand their association with systemic diseases (Kim et al. 2020; Lee et al. 2020a, b; Philley et al. 2019; Choi et al. 2017; Samra et al. 2019; Jones et al. 2021). One technical disadvantage of those analyses is targeting only 16S rRNA sequences, which have long been used as the "Gold standard" of marker genes in the microbial ecology field. The bacterial 16S rRNA gene is $\approx$1500 bp in length, consisting of 9 hypervariable regions (usually known as V1–V9) in between conserved sequences. 16S rRNA gene is highly conserved among bacteria, and the DNA sequences of the variable regions are usually species-specific, those unique characteristics enable us to exploit this gene to depict phylogenetic composition in microbiota. The presence of the conserved regions and the comparatively short length of this marker gene also allow low-cost and rapid analysis in combination with next-generation sequencing. However, it is rather likely that vesicles encompassing such a narrow range of genetic regions would be a rare case. Recent studies revealed that specific genomic regions of parent bacteria were enriched in their BEVs (Surve et al. 2016; Bitto et al. 2017; Yasuda et al. 2022). If this is general among bacteria, there would be a bias in the abundance of every coding sequence including 16S rRNA sequences by bacterial species, and taxonomic profiling using 16S rRNA sequences would be potentially limited to the identification of bacterial species whose BEVs frequently contain 16S rRNA. Comprehensive genomic analysis such as shotgun metagenome sequencing would be the alternative approach to characterize the taxonomic composition of the BEV population, but this method also potentially harbors the bias resulting from the difference in loaded DNA length and quantity in BEVs. Indeed, Biller et al. reported that the length of packaged DNA is different depending on bacterial species (Biller et al. 2017). Since those potential heterogeneities in the genetic content in BEVs implicitly explain the risk of the conventional bulk metagenomics approach for profiling the origins of BEVs in microbiota, understanding the fundamental properties of BEV-derived DNA is a key to develop more reliable approach for taxonomic annotation.

While there has been no established method to identify possible BEV producers in the context of microbiota due to the lack of a method to analyze DNA from low-biomass to date, taxonomic annotation of sequence reads from each droplet provided the possible producers of BEVs in human microbiota (Takano et al. 2024), and thus the droplet sequencing method would be one possible approach to overcome such a technical difficulty. In the case of BEVs from the periodontal oral biofilm, a taxonomic profile of BEVs with DNA was quite different from that in the microbiome even though both samples were derived from the same biofilm (Takano et al. 2024). Thus, it is quite possible that the minor bacterial group in the oral biofilm more actively produces BEVs than the major bacterial group. Importantly, there were several oral pathogens that were only detected in the BEV fraction but not in the bacterial cell fraction (Takano et al. 2024). Therefore, taxonomic profiles in the DNA cargo of BEVs could monitor distinct signatures of the host health status from those captured by microbiome analysis. Though packaging of genomic regions on the host chromosome was observed in most of the BEV particles, loading of the 16S rRNA gene was rarely detected in the previous study focusing on the oral microbiota (Takano et al. 2024). These results suggest the technical limitation of taxonomic identification by 16S rRNA-based methods such as amplicon sequencing.

6 DNA Barcodes in BEVs as Possible Novel Biomarkers

The application of biological nanoparticles for health diagnosis recently has been one trending topic (Zhou et al. 2020). Especially, clinical applications for biological nanoparticles such as exosomes have recently attracted great attention (Herrmann et al. 2021). However, more reliable and specific biomarkers are still missing for practical implementation in clinical diagnosis. BEVs are produced by a wide range of bacteria (Toyofuku et al. 2019), and are also ubiquitous nanoparticles in body fluids taking critical health impact (Schwechheimer and Kuehn 2015). A recent study also reported the presence of "DNA barcodes" in BEVs (Takano et al. 2024), which could work as marker genes in the detection of specific BEVs. Given the microflora has a genetic diversity that exceeds that of human cells by two orders of magnitude (Gilbert et al. 2018), BEV-associated biological molecules could bring higher specificity as biomarkers compared to eukaryotic extracellular vesicles.

Recently, differences in BEV-DNA signatures based on health status were explored using human oral saliva (Takano et al. 2024). Around 800 BEVs collected from healthy people and periodontal patients were analyzed by droplet sequencing, and the distinct taxonomic composition of DNA in BEVs between healthy and periodontal saliva was observed. Moreover, the taxonomic compositions of the microbial cells isolated from the same saliva samples were quite similar between healthy people and periodontal patients, and the observed difference in the taxonomic signature between healthy and periodontal samples was more significant in BEVs than microbiome. Those results suggest that DNA in BEVs possessed more discriminatory signatures depending on the health status. Detailed analysis of the DNA cargo in BEVs further revealed that the enriched genomic regions in BEVs from the same bacterial species are different depending on the oral health status (Takano et al. 2024). The presence of enriched loci in the parental bacterial chromosome and distinct taxonomic profiles in DNA cargo depending on the health status demonstrate that a "genetic barcode" in BEV-derived DNA has great potential for a diagnosis target. While droplet sequencing would be one promising approach for mining reliable biomarkers from BEVs, it is costly to apply this method for larger-scale cohort studies. Therefore, the combination with conventional metagenomics or quantitative PCR would be more validatable (e.g., screening the barcodes by detailed analysis with droplet sequencing with a small sample size, then validating that by metagenomics or PCR with a large sample size). Also, it should be tested how much BEV-DNA profiles can increase predictability in diagnostic models when combined with other information that is regarded as a promising target (e.g., miRNA profiles in exosomes) (Fig. 4).

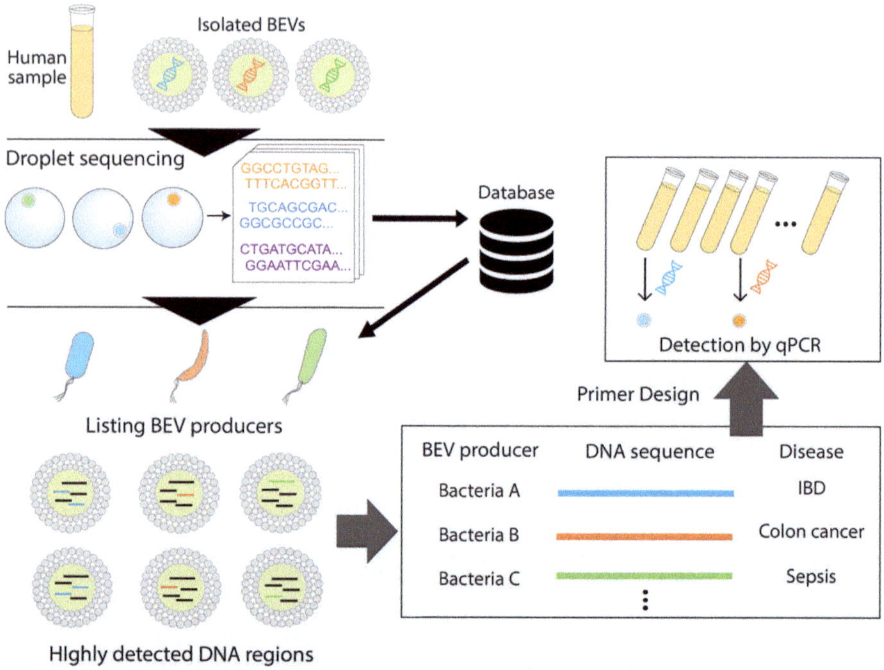

Fig. 4 Application of "DNA barcode" for diagnosis. Droplet sequencing enable comprehensive profiling of internal DNA in BEVs, and it is possible to characterize both BEV producers and highly detected DNA regions in BEVs. Such "DNA barcodes" can be applicable for the simpler and lower-cost detection method such as quantitative PCR in large-scale cohort study

7 Redox-Active BEV for Supporting Energy Acquisition Under Anaerobic Condition

In addition to vesicles, bacteria have been reported to produce nanoparticles possessing physicochemical properties such as metal conductivity, a subject that has been increasingly studied in recent years (Deng et al. 2022, 2020; Murugan et al. 2018; Cao et al. 2021). These entities exhibit capabilities such as aiding metabolism in environments where soluble electron acceptors and donors are depleted. Should BEVs exhibit such functionalities, it would suggest that their formation is less a response to extrinsic stimuli and more a critical outcome driven by intrinsic necessities for the maintenance of biological systems. This perspective lends substantial support to the notion that the genesis of BEVs is an evolutionarily conserved attribute, essential for the sustenance of microbial life under energy-limited conditions. A prominent instance of this phenomenon is observed in electroactive bacteria that utilize BEVs to transfer electrons, effectively turning these vesicles into electron carriers (Liu et al. 2020; Yu et al. 2023).

Electrogenic bacteria are specialized microbes with the capability to exchange electrons with external solid materials like electrodes or minerals (Shi et al. 2016). Unlike traditional electron acceptors like oxygen that can permeate the cell membrane, these bacteria exploit solid materials, which cannot intrude the cell. One might wonder how these bacteria manage to transfer their intracellular reducing power externally. These bacteria usually possess transmembrane electron transport enzymes that efficiently convey electrons from inside the cell to the exterior (Fig. 5A) (Hartshorne et al. 2009; Edwards et al. 2020; Okamoto et al. 2013). This distinctive characteristic enables these bacteria to serve as electrode catalysts, paving the way for innovative bioprocesses like microbial fuel cells, which simultaneously generate electricity and treat wastewater, and microbial electrode synthesis, which produces valuable substances from CO_2 (Rabaey and Rozendal 2010).

The outer-membrane cytochrome complex enzyme of electrogenic bacteria. This complex contains a chain of around twenty heme iron units, forming a "molecular wire" through the outer membrane, enabling electron transfer to extracellular electron acceptor or redox catalysis. BEVs produced by *Shewanella*, therefore endowed with the OMCs, have been reported to concurrently conduct hydrogen oxidation and iron-reduction reactions. Though these MVs do not proliferate, they mimic live bacteria in driving metabolic reactions, possibly accelerating iron dissolution reactions in the environment.

Recent studies highlight microbes' ability to produce redox-active nanoparticles pivotal for anaerobic respiration (Liu et al. 2020; Yu et al. 2023). Some bacteria like *Shewanella oneidensis* MR-1, with membrane-bound redox-active enzymes, outer-membrane cytochromes (OMCs), release BEVs that facilitate extracellular electron transport. Electroactive bacteria also produce nano wires, chain-like structures of MVs, as revealed by cryo-electron microscopy in *Shewanella* strains (Subramanian et al. 2018) (Fig. 5B). These nano wires, even in desiccated conditions, have been found to exhibit electrical conductivity. Moreover, they have been associated with

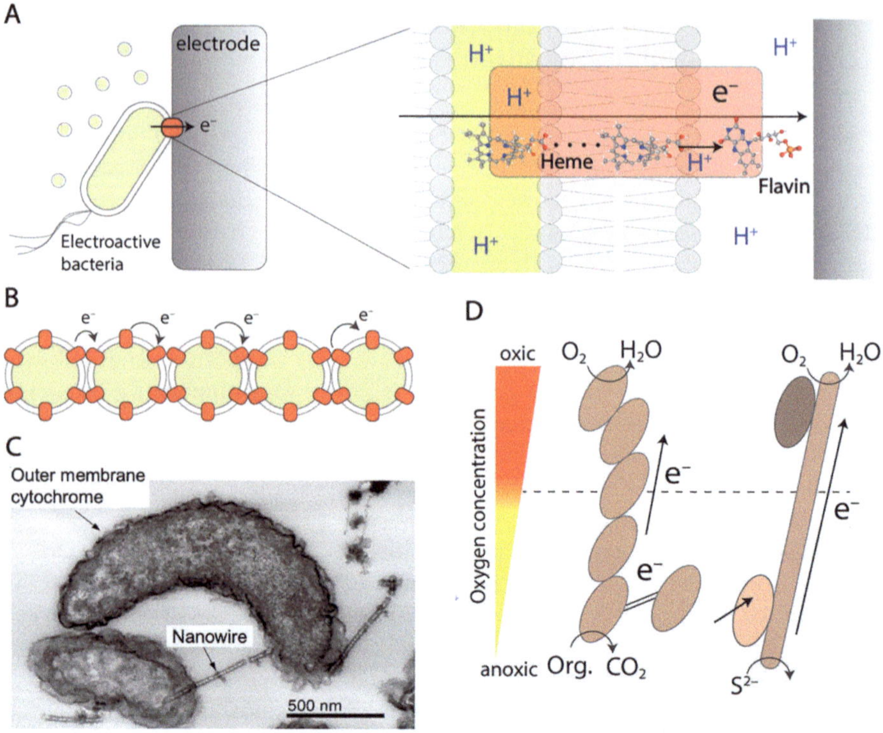

Fig. 5 BEVs with Outer Membrane Cytochromes (OMCs) from Electrogenic Bacteria. **A** The model electrogenic bacterium *Shewanella oneidensis* MR-1 facilitates extracellular electron transfer via OMCs. OMCs receive reducing power from the periplasmic space, transferring electrons from the heme reaction centers to the flavin molecules, and subsequently to the electrode surface. Chemical structures were generated in PubChem (Kim et al. 2023). **B** The formation of wire-like structures extending from the cell surface is observed when BEVs containing OMCs connect. It is believed that electrons are transported through these surface BEVs via OMCs. **C** The transmission electron microscopy image shows the nanowires of iron-oxidizing bacteria, specifically *Desulfovibrio ferrophilus* IS5 strain, which have been selectively stained to visualize redox reactions (Deng et al. 2018). **D** A conceptual diagram depicting electrical symbiosis through intercellular electron transfer mediated by BEVs, nanowires (Wegener et al. 2015) and filamentous bacteria (Pfeffer et al. 2012). Under anaerobic conditions, bacteria can transfer electrons to symbiotic partners or cable bacteria, enabling them to respire using oxygen

bacteria known to corrode iron (Deng et al. 2020) (Fig. 5C). Intercellular electron transfer mediated by nano wires has been suggested as a possible mechanism for electrical symbiosis (Fig. 5D). While the nanowire here is not proposed to be made with BEVs (Wegener et al. 2015), a similar symbiosis may occur with BEVs (Miran et al. 2021). This electron exchange is crucial in certain ecological niches, often involving different species. This electrical symbiosis can extend survival in specific environmental conditions. Thus, electron transport by BEVs is considered to play a significant role in the metabolism of electrogenic bacteria, functioning through active production and regulation by the microbes.

Furthermore, human microbiomes also contain electroactive bacteria. Some, such as *Klebsiella pneumoniae* and *Enterococcus avium*, have been isolated from human intestines (Naradasu et al. 2018). Others, like *Aggregatibacter actinomycetemcomitans* and *Porphyromonas gingivalis*, are frequently found in human oral microbiomes and are known pathogens that form BEVs in the biofilm (Miran et al. 2021). While the significance of BEV-mediated electron transfer in such contexts remains under investigation, it offers a compelling avenue for understanding microbial interactions, pathogenicity and energy conservation in biofilm.

8 Redox BEVs for Photocatalytic Therapy

Redox-active bacterial extracellular vesicles (BEVs) show potential for improving microbial fuel cell efficiency, but their low recovery yield has been a challenge. The Liposome fusion-induced Membrane Exchange (LIME) technique (Ref Wei-Peng Li biorxiv) successfully integrates outer membrane cytochromes (OMCs) from *S. oneidensis* MR-1 into high-yield membrane-integrated liposomes (MILs). This allows even non-electrogenic bacteria like *Escherichia coli* to produce electrical currents (Liu et al. 2020; Yu et al. 2023). Furthering BEV applications, titanium dioxide nanoparticles (NPs) coated with OMC-enriched MILs have demonstrated enhanced photocatalytic activity under X-ray irradiation, increasing superoxide anion production (Chen et al. 2023). This suggests potential uses in radiocatalytic systems for medical treatments, such as orthotopic liver tumor therapy, showcasing BEVs' promise for diverse technological applications.

9 Future Perspective

In this chapter, we have compiled potential evidence supporting the active biological production of BEVs, highlighting research on their DNA and energy contents. These functions are closely linked to microbial physiology and strongly indicate that BEVs are actively produced. However, the mechanisms behind DNA encapsulation and the formation of conductive nanowires remain to be fully elucidated, which would provide additional proof of BEVs' active production. Additionally, we have presented associated applications of BEVs. As our understanding of the function and role of BEVs deepens, the scope of these applications is expected to broaden.

References

Biller SJ, Schubotz F, Roggensack SE, Thompson AW, Summons RE, Chisholm SW. Bacterial vesicles in marine ecosystems. Science. 2014;343(6167):183–6.

Biller SJ, McDaniel LD, Breitbart M, Rogers E, Paul JH, Chisholm SW. Membrane vesicles in sea water: heterogeneous DNA content and implications for viral abundance estimates. ISME J. 2017;11(2):394–404.

Bitto NJ, Chapman R, Pidot S, Costin A, Lo C, Choi J, et al. Bacterial membrane vesicles transport their DNA cargo into host cells. Sci Rep. 2017;7(1):7072.

Bonnington KE, Kuehn MJ. Outer Membrane Vesicle Production Facilitates LPS Remodeling and Outer Membrane Maintenance in Salmonella during Environmental Transitions. mBio. 2016;7(5).

Cao B, Zhao Z, Peng L, Shiu HY, Ding M, Song F, et al. Silver nanoparticles boost charge-extraction efficiency in Shewanella microbial fuel cells. Science. 2021;373(6561):1336–40.

Carvalho AL, Fonseca S, Miquel-Clopes A, Cross K, Kok KS, Wegmann U, et al. Bioengineering commensal bacteria-derived outer membrane vesicles for delivery of biologics to the gastrointestinal and respiratory tract. J Extracell Vesicles. 2019;8(1):1632100.

Chen YC, Li YT, Lee CL, Kuo YT, Ho CL, Lin WC, et al. Electroactive membrane fusion-liposome for increased electron transfer to enhance radiodynamic therapy. Nat Nanotechnol. 2023.

Chiura HX, Kogure K, Hagemann S, Ellinger A, Velimirov B. Evidence for particle-induced horizontal gene transfer and serial transduction between bacteria. Fems Microbiology Ecology. 2011;76(3):576–91.

Choi HI, Choi JP, Seo J, Kim BJ, Rho M, Han JK, et al. Helicobacter pylori-derived extracellular vesicles increased in the gastric juices of gastric adenocarcinoma patients and induced inflammation mainly via specific targeting of gastric epithelial cells. Exp Mol Med. 2017;49(5):e330.

Chronopoulos A, Kalluri R. Emerging role of bacterial extracellular vesicles in cancer. Oncogene. 2020;39(46):6951–60.

Cuesta CM, Guerri C, Urena J, Pascual M. Role of Microbiota-Derived Extracellular Vesicles in Gut-Brain Communication. Int J Mol Sci. 2021;22(8).

Deng X, Dohmae N, Nealson KH, Hashimoto K, Okamoto A. Multi-heme cytochromes provide a pathway for survival in energy-limited environments. Sci Adv. 2018;4(2).

Deng X, Dohmae N, Kaksonen AH, Okamoto A. Biogenic Iron Sulfide Nanoparticles to Enable Extracellular Electron Uptake in Sulfate-Reducing Bacteria. Angew Chem Int Edit. 2020;59(15):5995–9.

Deng X, Luo D, Okamoto A. Defined and unknown roles of conductive nanoparticles for the enhancement of microbial current generation: A review. Bioresour Technol. 2022;350:126844.

Dorward DW, Garon CF, Judd RC. Export and intercellular transfer of DNA via membrane blebs of Neisseria gonorrhoeae. J Bacteriol. 1989;171(5):2499–505.

Dubey GP, Ben-Yehuda S. Intercellular nanotubes mediate bacterial communication. Cell. 2011;144(4):590–600.

Edwards MJ, White GF, Butt JN, Richardson DJ, Clarke TA. The Crystal Structure of a Biological Insulated Transmembrane Molecular Wire. Cell. 2020;181(3):665-+.

Engevik MA, Danhof HA, Ruan W, Engevik AC, Chang-Graham AL, Engevik KA, et al. Fusobacterium nucleatum Secretes Outer Membrane Vesicles and Promotes Intestinal Inflammation. mBio. 2021;12(2).

Gilbert JA, Blaser MJ, Caporaso JG, Jansson JK, Lynch SV, Knight R. Current understanding of the human microbiome. Nat Med. 2018;24(4):392–400.

Gole J, Gore A, Richards A, Chiu YJ, Fung HL, Bushman D, et al. Massively parallel polymerase cloning and genome sequencing of single cells using nanoliter microwells. Nat Biotechnol. 2013;31(12):1126–32.

Grenier D, Mayrand D. Functional characterization of extracellular vesicles produced by Bacteroides gingivalis. Infect Immun. 1987;55(1):111–7.

Hacker G, Redecke V, Hacker H. Activation of the immune system by bacterial CpG-DNA. Immunology. 2002;105(3):245–51.
Hartshorne RS, Reardon CL, Ross D, Nuester J, Clarke TA, Gates AJ, et al. Characterization of an electron conduit between bacteria and the extracellular environment. P Natl Acad Sci USA. 2009;106(52):22169–74.
Herrmann IK, Wood MJA, Fuhrmann G. Extracellular vesicles as a next-generation drug delivery platform. Nat Nanotechnol. 2021;16(7):748–59.
Hosokawa M, Nishikawa Y, Kogawa M, Takeyama H. Massively parallel whole genome amplification for single-cell sequencing using droplet microfluidics. Sci Rep. 2017;7(1):5199.
Jang SC, Kim SR, Yoon YJ, Park KS, Kim JH, Lee J, et al. In vivo kinetic biodistribution of nano-sized outer membrane vesicles derived from bacteria. Small. 2015;11(4):456–61.
Johnsborg O, Eldholm V, Havarstein LS. Natural genetic transformation: prevalence, mechanisms and function. Res Microbiol. 2007;158(10):767–78.
Jones EJ, Booth C, Fonseca S, Parker A, Cross K, Miquel-Clopes A, et al. The Uptake, Trafficking, and Biodistribution of Bacteroides thetaiotaomicron Generated Outer Membrane Vesicles. Front Microbiol. 2020;11:57.
Jones E, Stentz R, Telatin A, Savva GM, Booth C, Baker D, et al. The Origin of Plasma-Derived Bacterial Extracellular Vesicles in Healthy Individuals and Patients with Inflammatory Bowel Disease: A Pilot Study. Genes (Basel). 2021;12(10).
Kadurugamuwa JL, Beveridge TJ. Virulence factors are released from Pseudomonas aeruginosa in association with membrane vesicles during normal growth and exposure to gentamicin: a novel mechanism of enzyme secretion. J Bacteriol. 1995;177(14):3998–4008.
Kahn ME, Barany F, Smith HO. Transformasomes: specialized membranous structures that protect DNA during Haemophilus transformation. Proc Natl Acad Sci U S A. 1983;80(22):6927–31.
Kim OY, Park HT, Dinh NTH, Choi SJ, Lee J, Kim JH, et al. Bacterial outer membrane vesicles suppress tumor by interferon-gamma-mediated antitumor response. Nat Commun. 2017;8(1):626.
Kim SI, Kang N, Leem S, Yang J, Jo H, Lee M, et al. Metagenomic Analysis of Serum Microbe-Derived Extracellular Vesicles and Diagnostic Models to Differentiate Ovarian Cancer and Benign Ovarian Tumor. Cancers (Basel). 2020;12(5).
Kim S, Chen J, Cheng T, Gindulyte A, He J, He S, et al. PubChem 2023 update. Nucleic Acids Res. 2023;51(D1):D1373–D80.
Klieve AV, Yokoyama MT, Forster RJ, Ouwerkerk D, Bain PA, Mawhinney EL. Naturally occurring DNA transfer system associated with membrane vesicles in cellulolytic Ruminococcus spp. of ruminal origin. Appl Environ Microbiol. 2005;71(8):4248–53.
Kolling GL, Matthews KR. Export of virulence genes and Shiga toxin by membrane vesicles of Escherichia coli O157:H7. Appl Environ Microbiol. 1999;65(5):1843–8.
Koskella B, Meaden S. Understanding bacteriophage specificity in natural microbial communities. Viruses. 2013;5(3):806–23.
Lee H, Lee HK, Min SK, Lee WH. 16S rDNA microbiome composition pattern analysis as a diagnostic biomarker for biliary tract cancer. World J Surg Oncol. 2020;18(1):19.
Lee JH, Choi JP, Yang J, Won HK, Park CS, Song WJ, et al. Metagenome analysis using serum extracellular vesicles identified distinct microbiota in asthmatics. Sci Rep. 2020;10(1):15125.
Liao S, Klein MI, Heim KP, Fan Y, Bitoun JP, Ahn SJ, et al. Streptococcus mutans extracellular DNA is upregulated during growth in biofilms, actively released via membrane vesicles, and influenced by components of the protein secretion machinery. J Bacteriol. 2014;196(13):2355–66.
Liu X, Jing X, Ye Y, Zhan J, Ye J, Zhou S. Bacterial Vesicles Mediate Extracellular Electron Transfer. Environmental Science & Technology Letters. 2020;7(1):27–34.
Macosko EZ, Basu A, Satija R, Nemesh J, Shekhar K, Goldman M, et al. Highly Parallel Genome-wide Expression Profiling of Individual Cells Using Nanoliter Droplets. Cell. 2015;161(5):1202–14.
Mell JC, Redfield RJ. Natural competence and the evolution of DNA uptake specificity. J Bacteriol. 2014;196(8):1471–83.

Miran W, Naradasu D, Okamoto A. Pathogens electrogenicity as a tool for in-situ metabolic activity monitoring and drug assessment in biofilms. Iscience. 2021;24(2).

Murugan M, Miran W, Masuda T, Lee DS, Okamoto A. Biosynthesized Iron Sulfide Nanocluster Enhanced Anodic Current Generation by Sulfate Reducing Bacteria in Microbial Fuel Cells. Chemelectrochem. 2018;5(24):4015–20.

Naradasu D, Miran W, Sakamoto M, Okamoto A. Isolation and Characterization of Human Gut Bacteria Capable of Extracellular Electron Transport by Electrochemical Techniques. Front Microbiol. 2018;9.

Naradasu D, Miran W, Sharma S, Takenawa S, Soma T, Nomura N, et al. Biogenesis of Outer Membrane Vesicles Concentrates the Unsaturated Fatty Acid of Phosphatidylinositol in Capnocytophaga ochracea. Front Microbiol. 2021;12:682685

Naradasu D, Miran W, Sharma S, Takenawa S, Soma T, Nomura N, et al. Biogenesis of Outer Membrane Vesicles Concentrates the Unsaturated Fatty Acid of Phosphatidylinositol in. Front Microbiol. 2021;12.

O'Donoghue EJ, Krachler AM. Mechanisms of outer membrane vesicle entry into host cells. Cell Microbiol. 2016;18(11):1508–17.

Okamoto A, Hashimoto K, Nealson KH, Nakamura R. Rate enhancement of bacterial extracellular electron transport involves bound flavin semiquinones. P Natl Acad Sci USA. 2013;110(19):7856–61.

Orench-Rivera N, Kuehn MJ. Differential Packaging Into Outer Membrane Vesicles Upon Oxidative Stress Reveals a General Mechanism for Cargo Selectivity. Front Microbiol. 2021;12:561863.

Paludan SR, Reinert LS, Hornung V. DNA-stimulated cell death: implications for host defence, inflammatory diseases and cancer. Nat Rev Immunol. 2019;19(3):141–53.

Pfeffer C, Larsen S, Song J, Dong M, Besenbacher F, Meyer RL, et al. Filamentous bacteria transport electrons over centimetre distances. Nature. 2012;491(7423):218–21.

Philley JV, Kannan A, Olusola P, McGaha P, Singh KP, Samten B, et al. Microbiome Diversity in Sputum of Nontuberculous Mycobacteria Infected Women with a History of Breast Cancer. Cell Physiol Biochem. 2019;52(2):263–79.

Poore GD, Kopylova E, Zhu Q, Carpenter C, Fraraccio S, Wandro S, et al. Microbiome analyses of blood and tissues suggest cancer diagnostic approach. Nature. 2020;579(7800):567–74.

Rabaey K, Rozendal RA. Microbial electrosynthesis — revisiting the electrical route for microbial production. 2010:1–11.

Renelli M, Matias V, Lo RY, Beveridge TJ. DNA-containing membrane vesicles of Pseudomonas aeruginosa PAO1 and their genetic transformation potential. Microbiology (Reading). 2004;150(Pt 7):2161–9.

Riley DR, Sieber KB, Robinson KM, White JR, Ganesan A, Nourbakhsh S, et al. Bacteria-human somatic cell lateral gene transfer is enriched in cancer samples. PLoS Comput Biol. 2013;9(6):e1003107.

Rumbo C, Fernandez-Moreira E, Merino M, Poza M, Mendez JA, Soares NC, et al. Horizontal transfer of the OXA-24 carbapenemase gene via outer membrane vesicles: a new mechanism of dissemination of carbapenem resistance genes in Acinetobacter baumannii. Antimicrob Agents Chemother. 2011;55(7):3084–90.

Samra M, Nam SK, Lim DH, Kim DH, Yang J, Kim YK, et al. Urine Bacteria-Derived Extracellular Vesicles and Allergic Airway Diseases in Children. Int Arch Allergy Immunol. 2019;178(2):150–8.

Schwechheimer C, Kuehn MJ. Outer-membrane vesicles from Gram-negative bacteria: biogenesis and functions. Nat Rev Microbiol. 2015;13(10):605–19.

Shen Y, Giardino Torchia ML, Lawson GW, Karp CL, Ashwell JD, Mazmanian SK. Outer membrane vesicles of a human commensal mediate immune regulation and disease protection. Cell Host Microbe. 2012;12(4):509–20.

Shi L, Dong HL, Reguera G, Beyenal H, Lu AH, Liu J, et al. Extracellular electron transfer mechanisms between microorganisms and minerals. Nature Reviews Microbiology. 2016;14(10):651–62.

Subramanian P, Pirbadian S, El-Naggar MY, Jensen GJ. Ultrastructure of MR-1 nanowires revealed by electron cryotomography. P Natl Acad Sci USA. 2018;115(14):E3246–E55.
Surve MV, Anil A, Kamath KG, Bhutda S, Sthanam LK, Pradhan A, et al. Membrane Vesicles of Group B Streptococcus Disrupt Feto-Maternal Barrier Leading to Preterm Birth. PLoS Pathog. 2016;12(9):e1005816.
Takano S, Takenawa S, Divya N, Kangmin Y, Maehara T, Nomura N et al. Droplet Sequencing Reveals Virulence Gene Clusters in Oral Biofilm Extracellular Vesicles. 2024. bioRxiv: 2024.2009.2018.613607.
Tomaru Y, Nagasaki K. Flow cytometric detection and enumeration of DNA and RNA viruses infecting marine eukaryotic microalgae. J Oceanogr. 2007;63(2):215–21.
Toyofuku M, Nomura N, Eberl L. Types and origins of bacterial membrane vesicles. Nat Rev Microbiol. 2019;17(1):13–24.
Tran F, Boedicker JQ. Genetic cargo and bacterial species set the rate of vesicle-mediated horizontal gene transfer. Sci Rep. 2017;7(1):8813.
Tulkens J, Vergauwen G, Van Deun J, Geeurickx E, Dhondt B, Lippens L, et al. Increased levels of systemic LPS-positive bacterial extracellular vesicles in patients with intestinal barrier dysfunction. Gut. 2020;69(1):191–3.
Tulkens J, De Wever O, Hendrix A. Analyzing bacterial extracellular vesicles in human body fluids by orthogonal biophysical separation and biochemical characterization. Nat Protoc. 2020;15(1):40–67.
Wegener G, Krukenberg V, Riedel D, Tegetmeyer HE, Boetius A. Intercellular wiring enables electron transfer between methanotrophic archaea and bacteria. Nature. 2015;526(7574):587-90.
Yaron S, Kolling GL, Simon L, Matthews KR. Vesicle-mediated transfer of virulence genes from Escherichia coli O157:H7 to other enteric bacteria. Appl Environ Microbiol. 2000;66(10):4414–20.
Yasuda M, Yamamoto T, Nagakubo T, Morinaga K, Obana N, Nomura N, et al. Phage Genes Induce Quorum Sensing Signal Release through Membrane Vesicle Formation. Microbes Environ. 2022;37(1).
Yu H, Lu Y, Lan F, Wang Y, Hu C, Mao L, et al. Engineering Outer Membrane Vesicles to Increase Extracellular Electron Transfer of Shewanella oneidensis. ACS Synthetic Biology. 2023;12(6):1645–56.
Zhou BT, Xu KL, Zheng X, Chen T, Wang J, Song YM, et al. Application of exosomes as liquid biopsy in clinical diagnosis. Signal Transduct Tar. 2020;5(1).

Macropinocytosis and the Related Actin-Driven Cellular Uptake Pathways for Extracellular Fine Particles

Shiroh Futaki, Hisaaki Hirose, and Yoshimasa Kawaguchi

Abstract Macropinocytosis is an actin-driven and fluid-phase endocytosis in which actin reorganization induces ruffling of the plasma membrane and engulfment of extracellular solutes into the cell. In this chapter, we will briefly summarize the contribution of macropinocytosis to the cellular uptake of extracellular fine particles. We also present examples of macropinocytosis-related cellular uptake pathways that we have found in our uptake studies of extracellular proteins and fine particles.

1 Introduction

Extracellular fine particles of various sizes from nanometers to micrometers are taken up by the cells and can affect cellular functions. These particulates include artificial materials such as polymer/lipid-based nanomaterials, microplastics, and PM2.5. Particles that originate from within the body, such as extracellular vesicles including microvesicles and exosomes, are also known to be taken up by the cells. Understanding the cellular uptake modes and intracellular dynamics of these fine particles is extremely important for elucidating and controlling biological responses induced by extracellular fine particles. The understanding is also important from the perspective of accelerating delivery of biomacromolecules such as antibodies and intracellular drug delivery using nanoparticles as carriers.

Various forms of endocytosis can be involved in the uptake of extracellular fine particles. Representative forms of endocytosis include clathrin-dependent endocytosis and caveolae-dependent endocytosis, macropinocytosis, and phagocytosis (Fig. 1) (Conner and Schmid 2003; Doherty and McMahon 2009). The size of endosomes formed by clathrin-dependent endocytosis and caveolae-dependent endocytosis is generally in the range of tens to 200 nm. Therefore, the internalization of sub-micrometer to micrometer sized extracellular particles requires a form of endocytosis that produces vesicles of larger size. Phagocytosis and macropinocytosis may be candidates for the uptake pathway.

S. Futaki (✉) · H. Hirose · Y. Kawaguchi
Institute for Chemical Research, Kyoto University, Kyoto, Japan
e-mail: futaki@scl.kyoto-u.ac.jp

Y. Baba et al. (eds.), *Extracellular Fine Particles*,
https://doi.org/10.1007/978-981-97-7067-0_10

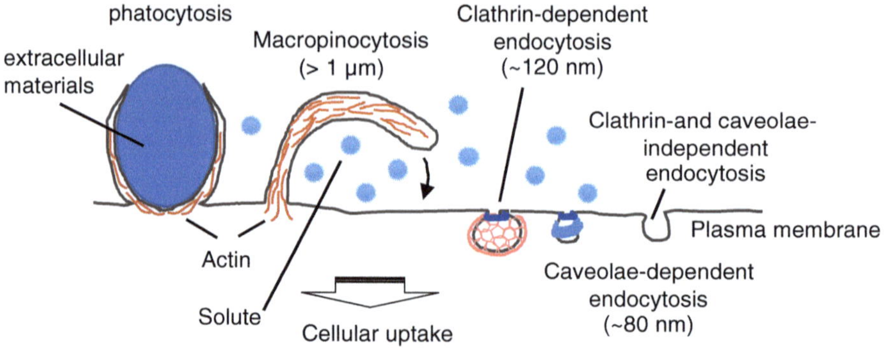

Fig. 1 The representative forms of endocytosis. Adopted with permission from *J. Japanese Biochem. Soc.* 93, 137 (2021) [https://doi.org/10.14952/SEIKAGAKU.2021.930137]. Copyright 2021, Japanese Biochemical Society

Phagocytosis is observed only in specialized cells called phagocytes, including macrophages, neutrophils, monocytes, dendritic cells, and osteoclasts (Hallett 2020). Phagocytosis is the specialized form of endocytosis by which phagocytic cells engulf and ingest large foreign substances, including microbial pathogens and dead cells (e.g., old erythrocytes). In phagocytotic uptake, the plasma membrane tightly adheres to large extracellular substances, which are then internalized into the cell with the exclusion of the extracellular fluid surrounding the particles (Hallett 2020). This adherence of the plasma membrane is accomplished by reorganization of the cytoskeletal actin filaments. Although phagocytosis is similar to macropinocytosis in that it is actin-driven, the modes of uptake are such that the cell membrane adheres tightly around the foreign substance to be taken up, excluding the extracellular fluids surrounding the particles to be taken up into the cell.

Macropinocytosis, like phagocytosis, is an actin-driven endocytosis, but it is a fluid-phase endocytosis in which actin reorganization induces ruffling of the plasma membrane, followed by engulfment of extracellular fluid (and solutes) into the cell (Mooren et al. 2012; Mylvaganam et al. 2021; Jones 2007; Maekawa et al. 2022; Salloum et al. 2023). Macropinocytosis is accompanied by the formation of an endosome called macorpinosome with a diameter of 0.2–10 µm. Whereas phagocytosis is considered specific to phagocytic cells, macropinocytosis is not limited to these cells. Therefore, our research group is interested in macropinocytosis as one of the potential uptake pathways for various extracellular fine particles. Macropinocytosis is also considered as a promising mechanism for intracellular delivery of biopharmaceuticals (Futaki and Nakase 2017). In addition, while studying the cellular uptake modes of peptides developed for intracellular delivery of biomacromolecules and extracellular particles, we found pathways with similarities to macropinocytosis. In this chapter, we will briefly summarize the contribution of macropinocytosis to the cellular uptake of extracellular fine particles. We also present examples of macropinocytosis-related cellular uptake pathways that we have found in our uptake studies of extracellular proteins and fine particles.

2 Macropinocytosis

As described above, macropinocytosis is an actin-driven, fluid-phase endocytosis that allows cells to engulf large amounts of external materials and solutes. Macropinocytosis is considered to be a pathway that enables the delivery of extracellular fine particles with different morphologies and physical properties into the cell (Mooren et al. 2012; Mylvaganam et al. 2021; Jones 2007). In addition, cancer cells take up serum albumin and cellular debris as nutrients for their growth (Ramirez et al. 2019; Commisso et al. 2013; Zhang and Commisso 2019). Although macropinocytosis is known to be an evolutionarily well conserved uptake mechanism, the factors that regulate this process are still unclear compared to other endocytic pathways.

In non-phagocytic cells, macropinocytosis is transiently induced by external stimuli, such as receptor stimulation by growth factors [e.g., epidermal growth factor (EGF)] and chemochines [e.g., stromal cell-derived factor 1 (SDF-1)] (Mercer and Helenius 2009; Tanaka et al. 2012). This leads to activation of intracellular Ras and Rac (conversion from GDP- to GTP-bound form) and phosphatidylinositol phosphorylation [conversion of PI(4,5)P2 to PI(3,4,5)P3], resulting in actin reorganization and cell membrane ruffling. Recently, it has been reported that the canonical Wnt signaling pathway also activates macropinocytosis and lysosomal degradation of extracellular proteins (Tejeda-Muñoz et al. 2019). It has also been reported that macropinocytosis is constantly activated in Ras mutant cancer cells, independent of external stimuli (Ramirez et al. 2019; Commisso et al. 2013). These cancer cells actively take up serum proteins (e.g., albumin) and cellular debris as a source of energy for cancer proliferation through micropinocytosis (Recouvreux and Commisso 2017). This pathway has therefore attracted much attention in relation to nutrient uptake and proliferation in cancer cells, and in relation to anticancer drug delivery and cancer therapy using this pathway. The possibility of its application to cancer therapy based on the difference between the amount of serum proteins taken up by these cells and that of non-cancerous cells has also been proposed (Zhang and Commisso 2019).

In macrophages and other phagocytic cells, macropinocytosis is known to be induced by lipopolysaccharide (LPS) stimulation (Doodnauth et al. 2019). Alternatively, constitutive macropinocytosis, in which macropinocytosis is always activated without specific external stimuli, also occurs and is thought to play a role in biological defense. The involvement of calcium-sensing receptors (CaSRs) in constitutive macropinocytosis of macrophages has been suggested (Doodnauth et al. 2019).

The involvement of macropinocytosis in cellular uptake was determined by structural changes in the plasma membrane and its sensitivity to inhibitors. Induction of macropinocytosis is accompanied by reorganization of actin filaments, formation of lamellipodia and ruffling of the plasma membrane, or cup-like deformation of the plasma membrane and its occlusion, leading to cellular uptake of extracellular fluid (Mooren et al. 2012; Mylvaganam et al. 2021; Jones 2007; Maekawa et al. 2022; Salloum et al. 2023). These are evaluated by observation of cell morphology using electron and light microscopy, staining and visualization of actin filaments, or sensitivity to inhibitors of macropinocytosis. The inhibitors include cytochalasin D (actin polymerization inhibitor), 5-(*N*-ethyl-*N*-isopropyl)amiloride (EIPA; Na^+/H^+ exchanger (NHE) inhibitor), wortmannin (phosphatidylinositol 3-kinase (PI3K) inhibitor, endosomal membrane fusion inhibitor). Macropinocytosis is a type of fluid-phase endocytosis that is not accompanied by specific receptors and marker proteins. Therefore there are very few specific endogenous markers of macropinocytosis. However, 70 kDa dextran, which is not easily internalized by other forms of endocytosis due to its large size, is commonly used as a marker for macropinocytic uptake.

3 Use of Macropinocytosis for Intracellular Delivery

Many attempts have been reported to promote intracellular delivery of nanoparticles by modifying them with peptides or polymers that induce macropinocytosis. Arginine-rich basic peptides have the ability to deliver drugs and proteins into cells. These peptides are collectively referred to as cell-penetrating peptides (CPPs), such as oligoarginine and basic peptides derived from the HIV-1 Tat protein (TAT peptide: GRKKRRQRRRPQ) (Futaki and Nakase 2017). For example, it has been reported that modification of the surface of extracellular vesicles, including liposomes and extracellular vesicles, with oligoarginine or TAT enhances intracellular uptake of these particles. Interaction of arginine-rich peptides with heparan sulfate proteoglycans (HSPGs) on the cell surface, resulting in their concentration on the cell surface and induction of macropinocytosis and promotion of their cellular uptake (Nakase et al. 2017).

Considering that molecules that can induce macropinocytosis are promising in terms of drug delivery promotion, we have developed peptides with macropinocytosis activation capabilities and analyzed the mechanism of intracellular translocation. We found that dodecaarginine (R12) activates macropinocytosis via the chemokine receptor CXCR4, and that stromal cell-derived factor SDF-1α, the original ligand for CXCR4, also induces micropinocytosis (Tanaka et al. 2012). In addition, the N-terminal peptide of SDF-1α (SN21: KPVSLSYRCPCRFFESHVARA-amide) and its truncated analogous peptide (P4A: YRCACRFF-amide) were also found to have the ability to induce macropinocytosis and promote their cellular uptake (Arafiles et al. 2020, 2021). Although its effects are less pronounced in the presence of serum, the N-terminally stearyl-linked intramolecular oxidized form of P4A (C18-oxP4A: stearyl-YRC*AC*RFF-amide, where the "C*" represents the formation of a disulfide bridge between these cysteines) (Arafiles et al. 2021) was shown to have an ability to induce macropinocytosis and promote intracellular translocation even in the presence of serum (Nakagawa et al. 2022). The effective delivery of extracellular fine particles, including extracellular vesicles, has been demonstrated in mRNA delivery using small extracellular vesicles, or exosomes. When the expression level of exosome-encapsulated mRNA in recipient cells was examined in the presence of this peptide, two to three-times higher mRNA expression was observed than in the absence of C18-oxP4A.

When macropinocytosis is used for drug delivery, the drugs of interest must be delivered in the cytosol. Therefore, mechanisms to promote their release from endosomes (macropinosomes) into the cytosol (i.e. endosomal escape) are important. In this context, it should be noted that while the above mentioned peptides are able to promote macropinocytic uptake of extracellular particles, their ability to promote endosomal escape is not necessarily high enough. In addition, the efficacy of larger molecules and particles in promoting endosomal escape is generally lower than that of smaller materials.

Therefore, our laboratory has developed a conjugate of SN21 and LK15 peptides (SN21-LK15: KPVSLSYRCPCRFESHVARAGGKLLLLLLLLLLK-amide) (Fig. 2). LK15 is an amphiphilic peptide composed of basic Lys and hydrophobic Leu and is known to have membrane lytic activity (Arafiles et al. 2020). Co-administration of SN21-LK15 resulted in intracellular delivery of plasmids, siRNA, antibodies and various proteins, leading to the expression of their biological activities.

In addition to arginine-rich peptides, helical peptides (staple peptides) whose structure is constrained by hydrocarbon chains (Sakagami et al. 2018), various nanoparticles (Xiao et al. 2020), viruses (Mercer and Helenius 2009), and extracellular vesicles (Hirose et al. 2022a) can also be taken up by macropinocytosis. Macropinocytosis may thus be involved in the cellular uptake of various exogenous molecules.

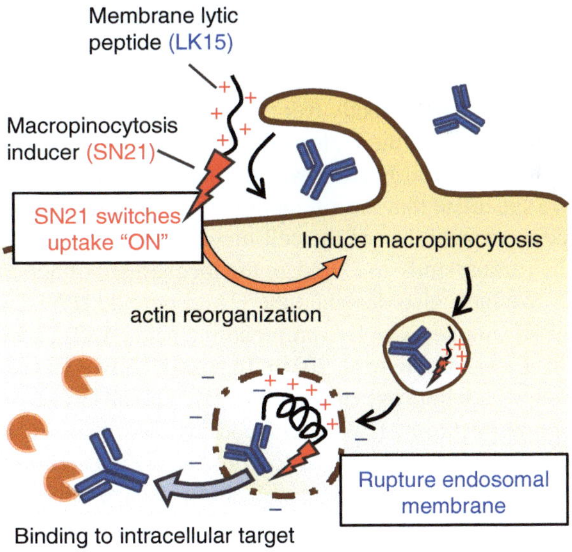

Fig. 2 SN21-LK15: combining a macropinocytosis-inducing peptide (physiological stimulant of cellular uptake) and a membrane-lytic peptide (physicochemical means to disrupt the barriers to cytosolic translocation) for cytosolic delivery of bioactive cargoes. Reprinted with permission from Reference (Arafiles et al. 2020). Copyright 2020, American Chemical Society

4 Actin-Driven Endocytosis Similar to Macropinocytosis

With the aim of promoting the intracellular delivery of antibodies (immunoglobulin Gs, IgGs) and other biofunctional materials, we have developed a series of attenuated cationic amphiphilic lytic (ACAL) peptides, including the L17E peptide (IWLTALKFLGKHAAKHEAKQQLSKL-amide, Fig. 3) (Akishiba et al. 2017). This peptide was derived from spider venom M-lycotoxin, a cationic amphiphilic peptide with hemolytic activity, in which one of the hydrophobic amino acids (Leu at position 17) of M-lycotoxin was replaced with Glu. By attenuating the lytic activity, successful delivery of IgGs and extracellular vesicles into the cell was achieved in the presence of the L17E peptide. The original design concept of L17E was that the side chain carboxy group of Glu would dissociate extracellularly and its negative charge would reduce the interaction of L17E with the cell membrane and the lytic effect on the cell membrane; the overall basic nature of the molecule allows it to adsorb to the cell surface and induce macropinocytosis, facilitating the uptake of L17E and molecules of interest to be delivered into endosomes (Fig. 2b). Inside the endosomes, a decrease in pH induces an increase in the ratio of uncharged molecular forms (non-dissociated forms). This increases the hydrophobicity of the peptide and its affinity with the endosomal membrane, which is expected to effectively disrupt the endosomal membrane (Fig. 3b–(i)). Indeed, effective intracellular (cytosolic) translocation of large proteins such as IgGs and binding to intracellular target molecules were observed when cells were treated with L17E.

L17E treatment induced a cellular response similar to macropinocytosis, such as actin rearrangement and stimulation of extracellular fluid uptake (Akishiba and Futaki 2019). However, unexpectedly, 5 min after L17E and antibodies were added to the cell culture medium, the antibodies had reached the cytosol, suggesting that most antibodies may be transferred to the cytosol at a very early stage of endocytosis or almost directly from the cell membrane, rather than being incorporated into endosomes and then released into the cytosol (Fig. 3b–(ii)). The results suggest that endosomal acidification was not necessary for cytosolic IgG delivery using L17E. We speculate that the interaction of L17E with the plasma membrane leads to a transient permeability of the cell membrane, allowing the cytosolic influx of IgGs, but the further study to elucidate the mechanism of action should be required.

We have also designed the HAad peptide [IWLTALKFLGKAAAKAXAKQQX LSKL-amide; X = L-2-aminoadipic acid (Aad)], which has a modified sequence of L17E (Sakamoto et al. 2020). In HAad, (i) two His residues in L17E are replaced by Ala, which improves the amphipathic helical structure, and (ii) Glu at position 17 and Gln at position 21 were replaced by Aad residues. Ada has higher hydrophobicity and lower p*K*a compared to Glu and is expected to interact more with endosomal membranes. With these modifications, HAad allows cytosolic IgG translocation like L17E (Fig. 3b–(ii)), but also together with endosomal maturation at reduced pH (Fig. 3b–(i)). Therefore, HAad achieves a higher cytosolic IgG delivery capacity than L17E.

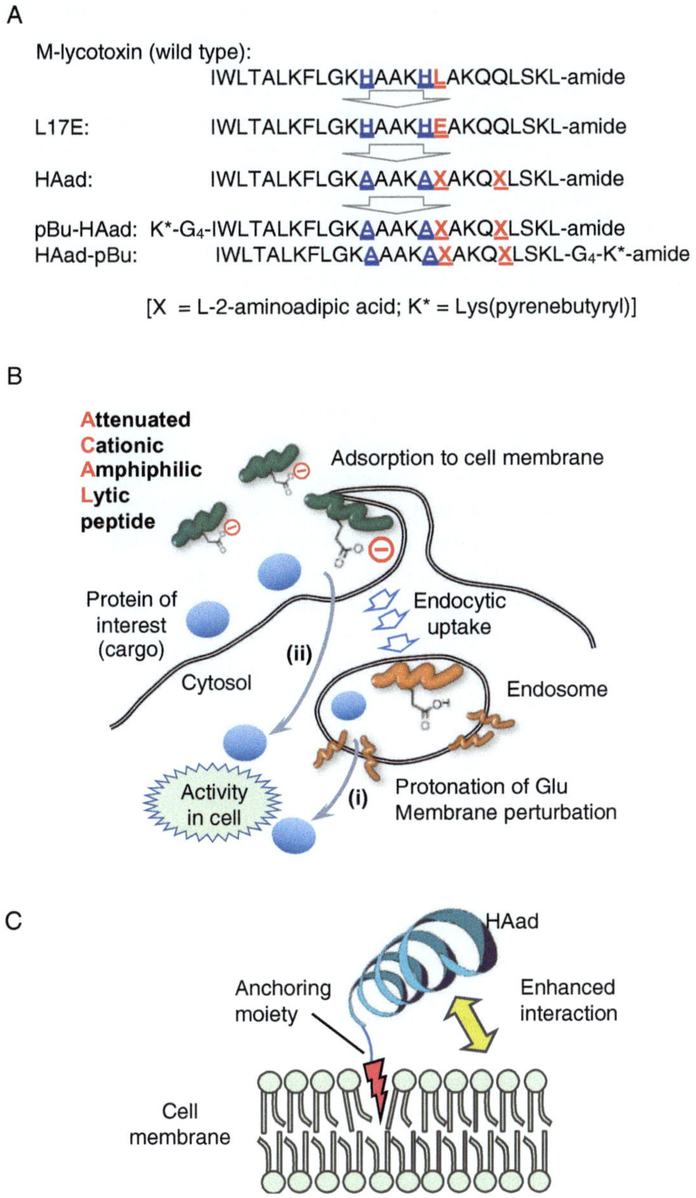
A
M-lycotoxin (wild type):
IWLTALKFLGKHAAKHLAKQQLSKL-amide
L17E: IWLTALKFLGKHAAKHEAKQQLSKL-amide
HAad: IWLTALKFLGKAAAKAXAKQXLSKL-amide
pBu-HAad: K*-G4-IWLTALKFLGKAAAKAXAKQXLSKL-amide
HAad-pBu: IWLTALKFLGKAAAKAXAKQXLSKL-G4-K*-amide
[X = L-2-aminoadipic acid; K* = Lys(pyrenebutyryl)]
B
Attenuated
Cationic
Amphiphilic
Lytic
peptide
Adsorption to cell membrane
Protein of interest (cargo)
Endocytic uptake
(ii)
Cytosol
Endosome
Activity in cell
(i)
Protonation of Glu
Membrane perturbation
C
HAad
Anchoring moiety
Enhanced interaction
Cell membrane

◂**Fig. 3** L17E and the related peptides for intracellular delivery of biomacromolecules, including antibodies. **a** L17E and the analogues. **b** Mode of action of attenuated cationic amphiphilic lytic (ACAL) peptides. Amino acid replacement in hydrophobic face of a cationic amphiphilic peptide to negatively charged amino acid (e.g., Glu) attenuates its lytic activity. Following endocytic uptake and endosomal acidification leads to decrease in the negative charge (protonation of Glu), recovery of its lytic activity and the cytosolic release of the cargo protein (route (i) in original design). Later study elucidated that ACAL peptide may also induce membrane ruffling, which allow transient perturbation of cell membrane and cargo translocation into cells at very early stage of endocytosis (route (ii)). Translocation using route (ii) is not observed under energy deficient condition (e.g. 4 °C), and therefore different from canonical method of pore formation in the membrane. **c** Attachment of a membrane anchoring moiety to HAad (a representative ACAL peptide) may potentiate the membrane interaction of the peptide and the delivery activity. **b, c** Reprinted with permission from Reference Sakamoto et al. (2021b). Copyright 2021, American Chemical Society

As described above, arginine residues are frequently found in the sequences of cell-penetrating peptides, and the guanidino group plays the central role in accelerating membrane permeation. We therefore substituted Lys in HAad with homoarginine to analyze the effect on the delivery activity of HAad (Sakamoto et al. 2021a). The resulting peptide had comparable or better delivery efficacy for Cre recombinase, antibodies, and the Cas9/sgRNA complex at one-fourth the concentration of HAad. By appropriately enhancing the membrane interaction of these peptides, we may be able to obtain more efficient delivery peptides. With this hope in mind, we also prepared a conjugate of HAad with pyrenebutyric acid as the membrane anchoring unit (pBu-HAad) (Sakamoto et al. 2021b). Pyrenebutyric acid is known to have an ability to loosen the packing of lipids (Murayama et al. 2017). pBu-HAad demonstrated protein delivery into cells with only 1/20 the concentration of HAad. Time-course and inhibitor studies suggested that the pyrene moiety attached to HAad enhanced membrane anchoring of HAad, loosening lipid packing to facilitate cytosolic translocation across membranes. Although increasing the membrane interaction of HAad led to an increase in delivery activity, as demonstrated in the case of the homoarginine-substituted HAad and the pyrenebutyric acid conjugate, the cytotoxicity of these peptides also increased. Therefore, approaches to increase delivery activity without increasing cytotoxicity are needed.

The feasibility of using HAad to deliver a protein nanocage (artificial nanocage) was also evaluated (Sakamoto et al. 2022). Nanocarriers that deliver functional proteins to the interior of cells are an attractive platform for intracellular delivery of intact proteins with potential in vivo applicability. The self-assembled β-annulus peptide (24 amino acid peptide) protein nanocage was used as a model. HAad was displayed on the surface of the nanocage to achieve cell permeability, and the nitrilotriacetic acid moiety was displayed on the interior of the nanocage to accept hexahistidine-tagged proteins in the presence of Ni^{2+}. Successful intracellular delivery of cargo proteins and targeting of cytosolic proteins by a nanobody was achieved.

With the aim of more efficient intracellular delivery and in vivo application, we prepared a conjugate of the peptide FcBP, which has been reported to have binding properties to the Fc region of IgG, with a trimer of L17E [FcB(L17E)$_3$]. Effective intracellular (cytosol) translocation of IgG was achieved by treating cells with the mixtures of Alexa Fluor 488-labeled IgG (IgG-Alexa) with FcB(L17E)$_3$ in a 1:2 molar ratio (Fig. 4a). Unexpectedly, we found that liquid droplets or coacervates were formed by mixing IgG-Alexa with FcB(L17E)$_3$ (Fig. 4b) (Iwata et al. 2021). For liquid droplet formation, the adaptation of negative charges on IgG by Alexa488 labeling plays a critical role. A rapid cytosolic influx of IgG-Alexa across the plasma membrane was induced by the interaction of liquid droplets or coacervates formed by mixing IgG-Alexa with FcB(L17E)$_3$. The influx was not due to simple pore formation in the plasma membrane by the L17E peptide, but dynamic change in membrane structure was induced by the process. The intracellular influx of antibody is markedly inhibited under conditions of ATP depletion or in the presence of macropinocytosis inhibitors (cytochalasin D, EIPA, wortmannin). Notably, when the droplet was attached to the cell, a transient accumulation of actin and thus plasma membrane was observed around the droplet, which dissipated upon the internalization of the antibody into the cell. Therefore, a certain cell biological machinery sharing characteristics with macropinocytosis may be involved in this intracellular influx of IgGs.

In addition, we found endocytic uptake of 3 μm polystyrene by non-phagocytic HeLa cells by an actin-, cholesterol-, and membrane-protrusion-dependent noncanonical uptake pathway distinct from the canonical macropinocytic and phagocytic pathways (Hirose et al. 2022b). This result, together with that obtained in the case of L17E-related ACAL peptides, suggests the possible existence of other noncanonical and undiscovered pathways for cellular uptake of extracellular fine particles.

5 Conclusions

In this chapter, we briefly reviewed macropinocytosis and its involvement in the uptake of extracellular particles. We also pointed out the possibility of the existence of actin-dependent uptake pathways that are similar to macropinocytosis but not identical to canonical macropinocytosis, which may be overlooked when identifying uptake pathways when only inhibitor studies are performed. Understanding the process of cellular uptake is important for analyzing their intracellular fate and effects on cell function. Although not discussed in detail in this chapter, the relationship between cancer cell proliferation and macropinocytosis and the possibility of its application to cancer therapy is also awaiting further investigation.

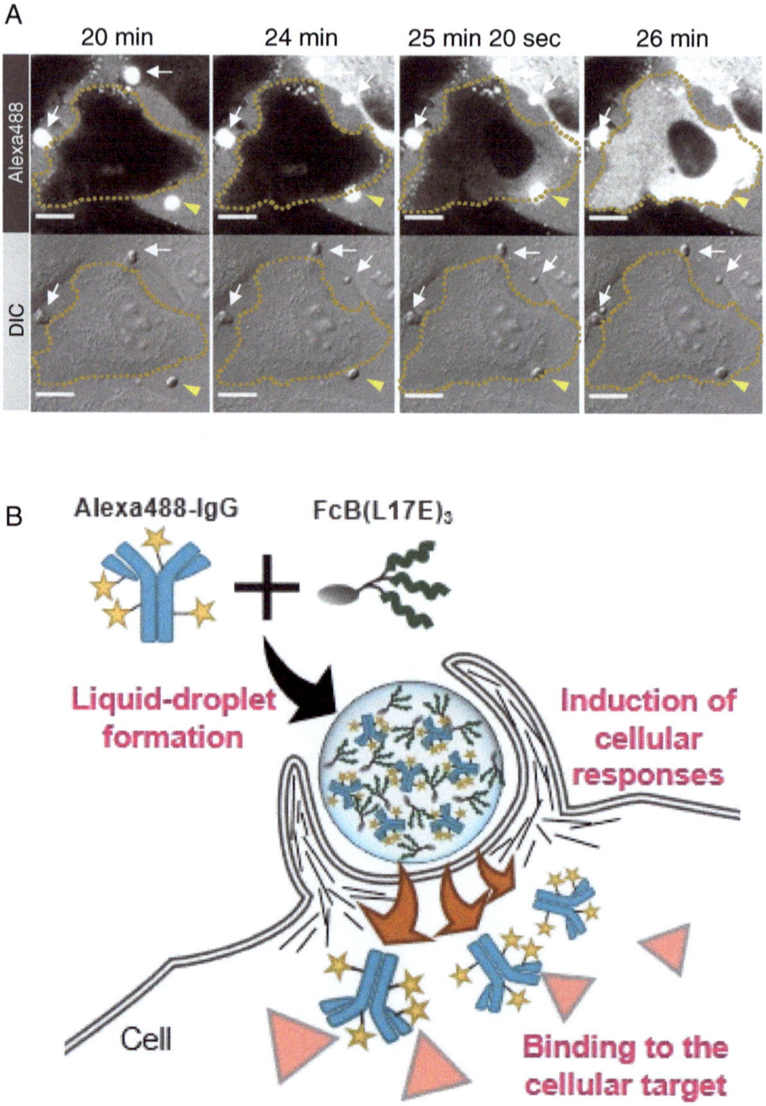

Fig. 4 Liquid droplets mediated intracellular influx of Alexa Fluor 488 labeled IgG (Alexa488-IgG) in the presence of Fc region-binding peptide conjugated to attenuated lytic peptide L17E trimer [FcB(L17E)$_3$]. **a** Time-lapse imaging of cells treated with a mixture of Alexa Fluor 488-labeled IgG (Alexa488-IgG) with FcB(L17E)$_3$. The imaging interval was 20 s. Scale bar denotes 10 μm. Arrows indicate the sedimentation of particle-like structures on culture dishes, which eventually contacted the plasma membrane and led to the distribution of Alexa488-IgG signals inside the cells. Cytosolic spreading of Alexa488-IgG was completed within 1 min. The particle-like structure with yellow arrowhead represents a particle-like structure typically yielding influx of Alexa488-IgG into the cell (the cell boundary is highlighted with a dotted line). **b** Liquid droplets were formed by mixing Alexa488-IgG with FcB(L17E)$_3$. Droplet contact with cell membrane led to quick influx of Alexa488-IgG throughout cells. Successful delivery of functional proteins into cells allowed binding to subcellular targets in the presence of FcB(L17E)$_3$. Reprinted with permission from Reference Sakamoto et al. (2022). Copyright 2021, John Wiley and Sons

Acknowledgements This work was supported by JST CREST (Grant Number JPMJCR18H5) to S.F.

References

Akishiba, M. & Futaki, S. Inducible Membrane Permeabilization by Attenuated Lytic Peptides: A New Concept for Accessing Cell Interiors through Ruffled Membranes. *Mol Pharm* **16**, 2540–2548, https://doi.org/10.1021/acs.molpharmaceut.9b00156 (2019).

Akishiba, M. *et al.* Cytosolic antibody delivery by lipid-sensitive endosomolytic peptide. *Nat Chem* **9**, 751–761, https://doi.org/10.1038/nchem.2779 (2017).

Arafiles, J. V. V. *et al.* Stimulating Macropinocytosis for Intracellular Nucleic Acid and Protein Delivery: A Combined Strategy with Membrane-Lytic Peptides To Facilitate Endosomal Escape. *Bioconjug Chem* **31**, 547–553, https://doi.org/10.1021/acs.bioconjchem.0c00064 (2020).

Arafiles, J. V. V. *et al.* Discovery of a Macropinocytosis-Inducing Peptide Potentiated by Medium-Mediated Intramolecular Disulfide Formation. *Angew Chem Int Ed Engl* **60**, 11928–11936, https://doi.org/10.1002/anie.202016754 (2021).

Commisso, C. *et al.* Macropinocytosis of protein is an amino acid supply route in Ras-transformed cells. *Nature* **497**, 633–637, https://doi.org/10.1038/nature12138 (2013).

Conner, S. D. & Schmid, S. L. Regulated portals of entry into the cell. *Nature* **422**, 37–44, https://doi.org/10.1038/nature01451 (2003).

Doherty, G. J. & McMahon, H. T. Mechanisms of endocytosis. *Annu Rev Biochem* **78**, 857–902, https://doi.org/10.1146/annurev.biochem.78.081307.110540 (2009).

Doodnauth, S. A., Grinstein, S. & Maxson, M. E. Constitutive and stimulated macropinocytosis in macrophages: roles in immunity and in the pathogenesis of atherosclerosis. *Philos Trans R Soc Lond B Biol Sci* **374**, 20180147, https://doi.org/10.1098/rstb.2018.0147 (2019).

Futaki, S. & Nakase, I. Cell-Surface Interactions on Arginine-Rich Cell-Penetrating Peptides Allow for Multiplex Modes of Internalization. *Acc Chem Res* **50**, 2449–2456, https://doi.org/10.1021/acs.accounts.7b00221 (2017).

Hallett, M. B. An Introduction to Phagocytosis. *Adv Exp Med Biol* **1246**, 1–7, https://doi.org/10.1007/978-3-030-40406-2_1 (2020).

Hirose, H., Hirai, Y., Sasaki, M., Sawa, H. & Futaki, S. Quantitative Analysis of Extracellular Vesicle Uptake and Fusion with Recipient Cells. *Bioconjug Chem* **33**, 1852–1859, https://doi.org/10.1021/acs.bioconjchem.2c00307 (2022).

Hirose, H. *et al.* A noncanonical endocytic pathway is involved in the internalization of 3 μm polystyrene beads into HeLa cells. *Biomater Sci* **10**, 7093–7102, https://doi.org/10.1039/d2bm01353c (2022).

Iwata, T. *et al.* Liquid Droplet Formation and Facile Cytosolic Translocation of IgG in the Presence of Attenuated Cationic Amphiphilic Lytic Peptides. *Angew Chem Int Ed Engl* **60**, 19804–19812, https://doi.org/10.1002/anie.202105527 (2021).

Jones, A. T. Macropinocytosis: searching for an endocytic identity and role in the uptake of cell penetrating peptides. *J Cell Mol Med* **11**, 670–684, https://doi.org/10.1111/j.1582-4934.2007.00062.x (2007).

Maekawa, M., Natsume, R. & Arita, M. Functional significance of ion channels during macropinosome resolution in immune cells. *Front Physiol* **13**, 1037758, https://doi.org/10.3389/fphys.2022.1037758 (2022).

Mercer, J. & Helenius, A. Virus entry by macropinocytosis. *Nat Cell Biol* **11**, 510–520, https://doi.org/10.1038/ncb0509-510 (2009).

Mooren, O. L., Galletta, B. J. & Cooper, J. A. Roles for actin assembly in endocytosis. *Annu Rev Biochem* **81**, 661–686, https://doi.org/10.1146/annurev-biochem-060910-094416 (2012).

Murayama, T. *et al.* Loosening of Lipid Packing Promotes Oligoarginine Entry into Cells. *Angew Chem Int Ed Engl* **56**, 7644–7647, https://doi.org/10.1002/anie.201703578 (2017).

Mylvaganam, S., Freeman, S. A. & Grinstein, S. The cytoskeleton in phagocytosis and macropinocytosis. *Curr Biol* **31**, R619–r632, https://doi.org/10.1016/j.cub.2021.01.036 (2021).

Nakagawa, Y. *et al.* Stearylated Macropinocytosis-Inducing Peptides Facilitating the Cellular Uptake of Small Extracellular Vesicles. *Bioconjug Chem* **33**, 869–880, https://doi.org/10.1021/acs.bioconjchem.2c00113 (2022).

Nakase, I. *et al.* Arginine-rich cell-penetrating peptide-modified extracellular vesicles for active macropinocytosis induction and efficient intracellular delivery. *Sci Rep* **7**, 1991, https://doi.org/10.1038/s41598-017-02014-6 (2017).

Ramirez, C., Hauser, A. D., Vucic, E. A. & Bar-Sagi, D. Plasma membrane V-ATPase controls oncogenic RAS-induced macropinocytosis. *Nature* **576**, 477–481, https://doi.org/10.1038/s41586-019-1831-x (2019).

Recouvreux, M. V. & Commisso, C. Macropinocytosis: A Metabolic Adaptation to Nutrient Stress in Cancer. *Front Endocrinol (Lausanne)* **8**, 261, https://doi.org/10.3389/fendo.2017.00261 (2017).

Sakagami, K., Masuda, T., Kawano, K. & Futaki, S. Importance of Net Hydrophobicity in the Cellular Uptake of All-Hydrocarbon Stapled Peptides. *Mol Pharm* **15**, 1332–1340, https://doi.org/10.1021/acs.molpharmaceut.7b01130 (2018).

Sakamoto, K. *et al.* Optimizing charge switching in membrane lytic peptides for endosomal release of biomacromolecules. *Angew Chem Int Ed Engl*, https://doi.org/10.1002/anie.202005887 (2020).

Sakamoto, K. *et al.* Use of homoarginine to obtain attenuated cationic membrane lytic peptides. *Bioorg Med Chem Lett* **40**, 127925, https://doi.org/10.1016/j.bmcl.2021.127925 (2021).

Sakamoto, K. *et al.* Potentiating the Membrane Interaction of an Attenuated Cationic Amphiphilic Lytic Peptide for Intracellular Protein Delivery by Anchoring with Pyrene Moiety. *Bioconjug Chem* **32**, 950–957, https://doi.org/10.1021/acs.bioconjchem.1c00101 (2021).

Sakamoto, K. *et al.* Artificial Nanocage Formed via Self-Assembly of β-Annulus Peptide for Delivering Biofunctional Proteins into Cell Interiors. *Bioconjug Chem* **33**, 311–320, https://doi.org/10.1021/acs.bioconjchem.1c00534 (2022).

Salloum, G., Bresnick, A. R. & Backer, J. M. Macropinocytosis: mechanisms and regulation. *Biochem J* **480**, 335–362, https://doi.org/10.1042/bcj20210584 (2023).

Tanaka, G. *et al.* CXCR4 stimulates macropinocytosis: implications for cellular uptake of arginine-rich cell-penetrating peptides and HIV. *Chem Biol* **19**, 1437–1446, https://doi.org/10.1016/j.chembiol.2012.09.011 (2012).

Tejeda-Muñoz, N., Albrecht, L. V., Bui, M. H. & De Robertis, E. M. Wnt canonical pathway activates macropinocytosis and lysosomal degradation of extracellular proteins. *Proc Natl Acad Sci U S A* **116**, 10402–10411, https://doi.org/10.1073/pnas.1903506116 (2019).

Xiao, Y. *et al.* Effect of Surface Modifications on Cellular Uptake of Gold Nanorods in Human Primary Cells and Established Cell Lines. *ACS Omega* **5**, 32744–32752, https://doi.org/10.1021/acsomega.0c05162 (2020).

Zhang, Y. & Commisso, C. Macropinocytosis in Cancer: A Complex Signaling Network. *Trends Cancer* **5**, 332–334, https://doi.org/10.1016/j.trecan.2019.04.002 (2019).

Relationship Between Bio-Distribution of Environmental Particles and Induction of Biological and Immune Response in the Respiratory System

Tomoya Sagawa, Raga Ishikawa, Sakiko Akaji, Akiko Honda, Takayuki Kameda, and Hirohisa Takano

Abstract Environmental particles, such as PM2.5, diesel exhaust particles, and metal nanoparticles, are known to affect the respiratory and immune systems. Although environmental particles exhibit toxicity owing to their unique physical and chemical properties, little is known about the relationship between the bio-distribution and dynamics of environmental particles and health effects. This review introduces a novel method in which Raman microscopy is combined with histological staining, such as hematoxylin–eosin staining, immunostaining, and Diff-Quik staining, to evaluate the bio-distribution of environmental particles and the biological responses. This hybrid approach enables the visualization of environmental particles and cellular structures in the same field-of-view. Using this integrated method, we successfully visualized diesel exhaust particles and their components in respiratory epithelial cells as well as titanium dioxide particles in alveolar macrophages of mouse lung tissue and the bronchoalveolar lavage fluid. These findings demonstrate the potential of Raman microscopy for elucidating the mechanisms by which

T. Sagawa · R. Ishikawa (✉) · H. Takano
Department of Natural Resources, Graduate School of Environmental Studies, Kyoto University, Kyoto, Japan
e-mail: ishikawa.raga.3t@kyoto-u.ac.jp

S. Akaji · A. Honda
Department of Environmental Engineering, Graduate School of Engineering, Kyoto University, Kyoto, Japan

T. Sagawa
Inflammation and Immunology, Graduate School of Medical Science, Kyoto Prefectural University of Medicine, Kyoto, Japan

T. Kameda
Department of Socio-Environmental Energy Science, Graduate School of Energy Science, Kyoto University, Kyoto, Japan

H. Takano
Institute for International Academic Research, Kyoto University of Advanced Science, Kyoto, Japan

Y. Baba et al. (eds.), *Extracellular Fine Particles*,
https://doi.org/10.1007/978-981-97-7067-0_11

environmental particles affect the respiratory and immune systems, offering new perspectives for the management of environmental particles.

1 Introduction

Environmental particles, such as particulate matter with a diameter of less than 2.5 μm (PM2.5), diesel exhaust particles (DEPs), and metal nanoparticles, are known to enter the body through inhalation, thus contributing to the development and exacerbation of diseases related to the respiratory and immune systems (Ural et al. 2022; Yang et al. 2020; Sonwani et al. 2021). Environmental particles have distinctive physical properties and chemical components derived from their sources and induce cytotoxicity depending on the characteristics. Consequently, environmental particles have been implicated in the initiation and exacerbation of various diseases, including lung cancer (Guo et al. 2020, Wang et al. 2023), chronic obstructive pulmonary disease (COPD) (Duan et al. 2021; Zhao et al. 2019), pulmonary fibrosis (Singh and Singh 2021; Winterbottom et al. 2018; Liu et al. 2022), bronchial asthma (Liu et al. 2022; Aldegunde et al. 2022), and allergic rhinitis (Jia et al. 2022; Fukuoka et al. 2016). These effects occur via various mechanisms depending on the type of particles exposed. Some reports indicate that environmental particles activate the respiratory and immune systems through specific processes, such as oxidative stress, chronic airway inflammation, and immune response induction (Liu et al. 2022; Li and Liu 2021; Thangavel et al. 2022). However, most of the existing studies on environmental particles are focused on specific health impacts associated with particle sources, compositions, or toxicity. Studies addressing the bio-distribution and dynamics of environmental particles in the respiratory and immune systems and their relationship to health effects are notably limited.

The bio-distribution of environmental particles is evaluated with various methods, including direct observation of particles labeled with fluorescent dyes or radiotracers (Li et al. 2019), direct morphological observation using electron microscopy (Naav et al. 2020; Okada et al. 2021), histopathological examination of hematoxylin–eosin (HE) stained specimens (Rattanapinyopituk et al. 2013), and detection of components in tissue samples using inductively coupled plasma mass spectrometry (ICP-MS) (Rattanapinyopituk et al. 2013; Sadauskas et al. 2009a, 2009b). However, environmental particles in real-world settings have broad size distributions, complex origins, heterogeneous morphologies, and diverse elemental compositions (Zhao et al. 2021; Tian et al. 2022). Owing to these complexities, accurate detection of particles and tracking of particle dynamics within cells and tissues have been challenging.

Raman spectroscopy is commonly used to identify unlabeled substances through an analysis of the characteristic Raman scattering of incident laser light. In addition, Raman spectroscopy has recently been applied to image cellular and tissue samples (Krafft et al. 2017). Laser-based Raman microscopy, a combination of Raman spectroscopy and microscopy, is a powerful observational and analytical tool for the precise identification of specific constituents and molecular structures of biological

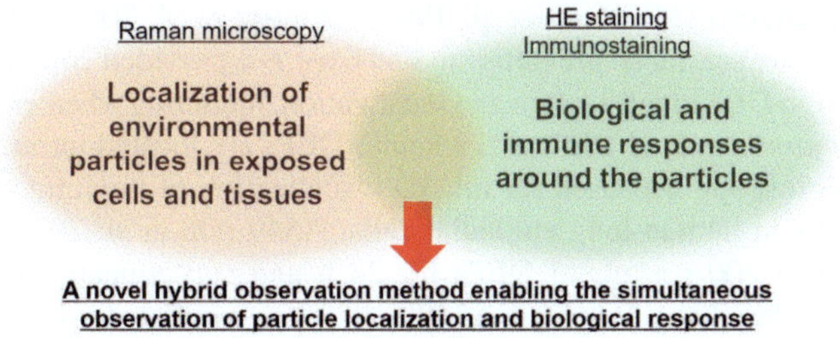

Fig. 1 Concept of the hybrid method for the simultaneous visualization of environmental particle distribution and biological responses

samples (Li et al. 2021; Ember et al. 2017). In addition, three-dimensional (3D) analysis offers the potential to visualize phenomena such as particle internalization and organelle localization within cells, which may not be discernible through two-dimensional (2D) analysis. Thus, using Raman microscopy to evaluate the bio-distribution of environmental particles may provide information for the elucidation of previously elusive relationships between environmental particle localization and associated health effects. Although tissue staining methods, such as HE staining and immunostaining, enable the observation of tissue structures for the assessment of inflammatory responses to environmental particles, these techniques do not allow for the accurate identification of environmental particles. Therefore, we developed a new platform to enable the simultaneous evaluation of the bio-distribution of environmental particles using Raman microscopy and the assessment of biological and immune responses by tissue staining.

In this review, we present our recent works on the intracellular localization of environmental particles and their components using Raman microscopy and on the correlation of particle localization with biological response using a novel hybrid technique combining Raman microscopy and conventional histopathological staining methods (Fig. 1).

2 3D Visualization of the Intracellular Distribution of Environmental Particles and Their Components

DEPs are particulate matter generated through the incomplete combustion of diesel (Wang et al. 2019). DEPs consist of an elemental carbon core and adhered chemical substances, such as metals, inorganic ions, and polycyclic aromatic hydrocarbons (PAHs) and their derivatives (Wang et al. 2019; Sharma et al. 2005; Dandajeh et al. 2022). PAHs and their derivatives are known to induce mutagenicity and carcinogenicity in living organisms through oxidative stress and DNA damage (Kwon et al. 2023; Cao et al. 2020). Furthermore, DEPs have been shown to exacerbate allergic diseases, including asthma, by inducing eosinophilic airway inflammation in the presence of allergens (Miyabara et al. 1998; Ichinose et al. 2002; Alvarez-Simón et al. 2017). However, these previous studies have not provided sufficient information for thorough DEP identification and localization, including whether the observed particulate matter is indeed DEPs and whether DEPs are located intracellularly.

Raman microscopy was used for the 3D visualization of the intracellular distribution of DEPs in human lung epithelial cells, A549 (Ou et al. 2022). A549 cells were exposed to 200 μg/mL of DEPs for 24 h and subsequently fixed with 4% paraformaldehyde (PFA). Prepared A549 cells were irradiated with 532 nm laser light. Scanning was performed at intervals of 1 μm in the z-axis direction to obtain 3D Raman spectra. Detection of carbon-derived signals (red; 1340, 1600 cm^{-1}), nucleus-derived signals (blue; 1080, 1220 cm^{-1}), and cytoplasm-derived signals (green; 1450, 1650, 2800–2900 cm^{-1}) allowed for the 3D visualization of DEPs within A549 cells (Fig. 2a).

Raman microscopy was also used for the 3D visualization of the intracellular distribution of 1-hydroxypyrene (1-OHPy), a type of PAH (Louro et al. 2022; Casjens et al. 2023). A549 cells were exposed to 30 μM of 1-OHPy for 24 h and subsequently fixed with 4% paraformaldehyde (PFA). Prepared A549 cells were irradiated with 785 nm laser light. For 3D image analysis, scanning was conducted at intervals of 1 μm in the z-axis direction. Analysis of 2D images revealed that 1-OHPy (610,

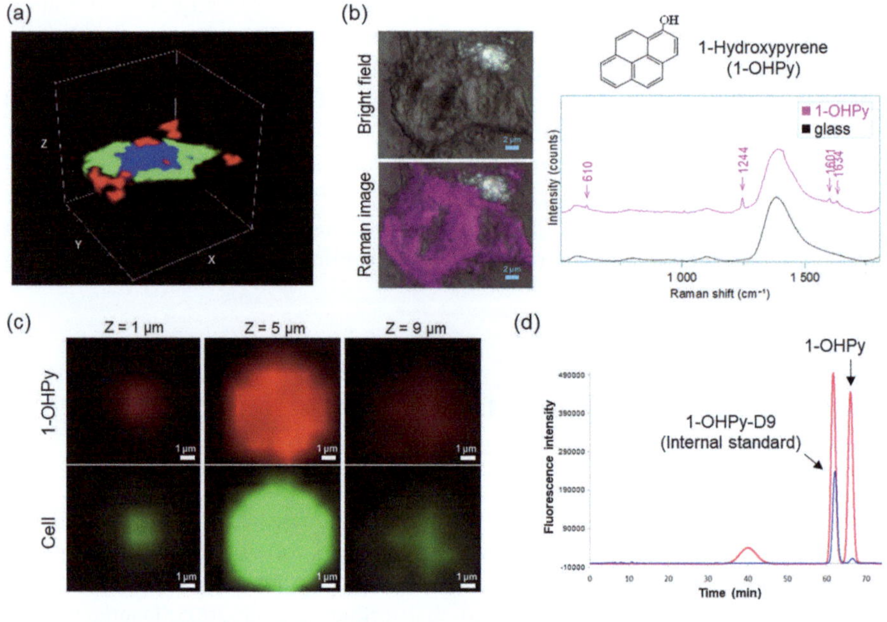

Fig. 2 **a** Representative 3D Raman mapping of A549 cells exposed to DEPs (Red: DEPs, Green: Cytoplasm, Blue: Nucleus). Adapted with permission (Sagawa et al. 2023). **b** 2D Raman mapping image and bright field image of A549 cells exposed to 1-OHPy, with Raman spectra (Pink: 1-OHPy). **c** 3D Raman mapping of A549 cells exposed to 1-OHPy (Red: 1-OHPy, Green: Cytoplasm). **d** HPLC chromatograms of A549 cell extract (blue) and standard sample (pink)

1244 cm^{-1}) was present in the cells (Fig. 2b), while analysis of 3D images revealed that 1-OHPy was uniformly distributed throughout the cell rather than localized within specific regions of the cell (Fig. 2c). In addition, intracellular 1-OHPy was quantified using high-performance liquid chromatography (HPLC). 1-OHPy was detected in A549 extract samples, and the concentration of 1-OHPy was quantified using the peak area (Fig. 2d).

The above studies demonstrated the potential utility of Raman microscopy for the detection and visualization of environmental particles or their constituents in cells. While we successfully visualized the intracellular distribution of individual components, such as carbon particles and 1-OHPy molecules, it is important to note that real-world environmental particles are complex mixtures containing various constituents. Therefore, visualizing the distribution of multiple components simultaneously will be crucial in future evaluations. If differences in the intracellular localization and dynamics of each component become apparent, it would pave the way toward a deeper understanding of biological immune responses triggered by environmental particles.

3 Visualization of Environmental Particles and Cellular Structures in HE-Stained Specimens in the Same Field-Of-View

HE staining of tissues is used to distinguish the cellular structures of tissues with optical microscopy, which makes it an indispensable method for histopathological diagnosis. In tissues exposed to environmental particles, foreign materials resembling particles are often in proximity to areas with inflammatory cell infiltration. However, it is challenging to definitively confirm whether these foreign materials are indeed environmental particles using HE-stained specimens alone. Therefore, to determine the effect of environmental particles on the host's immune system, we developed a hybrid technique by combining Raman microscopy with HE staining for the visualization of environmental particles and tissue structures in the same field-of-view (Akaji et al. 2022). To demonstrate this approach, we used commercially available rutile-type titanium dioxide (TiO_2) particles as model particles. Eosin strongly fluoresces when it is excited with 530 nm light, which could interfere with Raman signals when 532 nm excitation laser is used for Raman microscopy. Therefore, we focused on the decolorization of HE dyes with citrate buffer (Ozawa and Sakaue 2020), and postulated that removing the dyes from HE-stained specimens would enable the visualization and identification of environmental particles and cellular structures in the same field-of-view.

Tissue sections of mouse lung intratracheally administrated with TiO_2 particles were stained with HE dyes and examined using optical microscopy. Some cells contained features attributed to TiO_2 particles (Fig. 3, left). Subsequently, the HE stains were decolorized, and these lung tissue sections were examined using Raman microscopy. Areas corresponding to lung tissue contained peaks attributed to nucleic acids, lipids, proteins (1000, 1450, 1650 cm^{-1}), and cytochrome c (750 cm^{-1}), suggesting that these peaks were derived from cellular components. Furthermore, in areas believed to contain TiO_2 particles, we captured the signals of rutile-type TiO_2 (230, 446, and 608 cm^{-1}), in addition to the signals of cellular components. A classical least squares analysis of the acquired Raman spectra revealed that TiO_2 was incorporated within the cells (Fig. 3, right). Tissue damage or structural changes attributed to laser irradiation were not observed. Therefore, it was possible to acquire Raman spectra of environmental particles and cellular components in the same field-of-view by decolorizing HE-stained specimens.

Although this method was applied to investigate lung tissues exposed to TiO_2, an environmental particle, it can also be applied to study other organs exposed to various environmental particles. It is expected that future research will leverage this method to understand the entry of environmental particles into the body and to examine the ensuing biological responses.

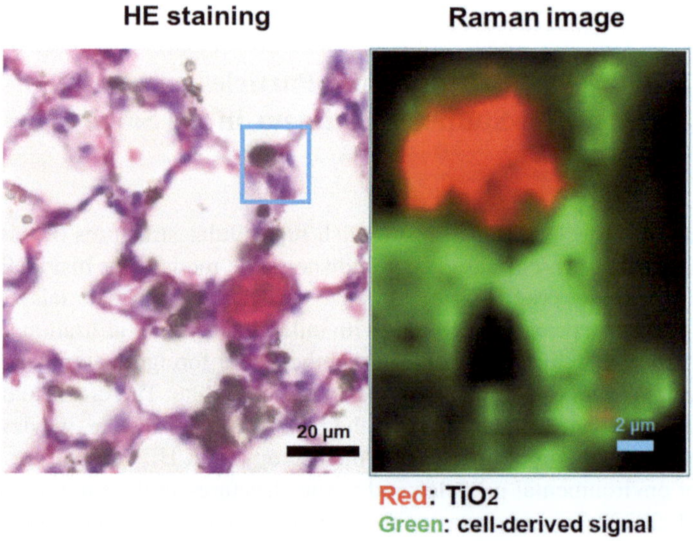

Fig. 3 Optical microscopy (left) and Raman microscopy (right) images of HE-stained mouse lung tissue exposed to TiO_2 (intratracheal administration). The Raman image was obtained from the area encompassed by the blue square on the left panel. Red represents TiO_2 and green represents cellular components

4 Visualization of Environmental Particles and Cell Types in Immunohistochemical Specimens in the Same Field-Of-View

Traditionally, TiO_2 nanoparticles are considered non-toxic, and thus they are found in cosmetics, sunscreens, and food additives (Baan et al. 2006; Park et al. 2009; Weir et al. 2012). In recent years, there has been growing recognition of the health risks associated with the inhalation of TiO_2 (Johnston et al. 2009; Zhu et al. 2019; Zhao et al. 2018). Animal studies have found that acute exposure to TiO_2 particles leads to pulmonary inflammation and TiO_2 phagocytosis by alveolar macrophages (AMs) (Ichinose et al. 2008; Warheit et al. 1997; Husain et al. 2013). However, the specific responses of TiO_2-phagocytized AMs and how these responses lead to inflammation have not been elucidated to date.

Immunostaining is a standard tissue staining technique used to detect proteins and identify cell types in tissue sections. It is crucial to identify cell types that have phagocytized particles for the accurate evaluation of the distribution of environmental particles in the respiratory system. Therefore, we applied decolorization to immunostained sections in an attempt to observe the distribution of cell types and the localization of environmental particles in the same field-of-view using Raman microscopy (Sagawa et al. 2021). Mice were intratracheally administered rutile-type TiO_2, and lung tissues were collected 24 h later. HE staining was performed on these lung tissue sections, followed by decolorization with citrate buffer. Subsequently, immunostaining for macrophage marker Mac-3 was carried out to visualize macrophages in the lung tissue. After another round of decolorization and antibody stripping with citrate buffer, the sections were observed using Raman microscopy. As a result, signals derived from rutile-type TiO_2 were observed within AMs, indicating that TiO_2 particles were phagocytized by AMs (Fig. 4).

Bronchoalveolar lavage fluid (BALF) was collected from mice intratracheally administered with TiO_2. BALF cells were processed using the cytospin protocol and then stained with Diff-Quik. Counting of stained cells revealed that the number of inflammatory cells, such as neutrophils and macrophages, increased following TiO_2 exposure. Furthermore, the concentrations of chemokines in the BALF supernatant were measured using the enzyme-linked immunosorbent assay (ELISA). The results indicated that the concentration of CCL3, involved in the migration of neutrophils, increased following TiO_2 exposure. Raman microscopy of decolorized Diff-Quik stained samples confirmed that rutile-type TiO_2 was localized within AMs (Fig. 5a).

BALF cells were stained with Annexin V and 7-AAD to examine the apoptosis/necrosis of cells. The results showed that TiO_2-exposed cells, especially TiO_2-phagocytized AMs, had a significantly higher rate of 7-AAD positivity than unexposed cells. In addition, the majority of TiO_2-phagocytized cells were necrotic, as characterized by 7-AAD+Annexin V-. Previous reports have indicated that TiO_2-phagocytized cells undergo apoptosis (Xu et al. 2020; Chen et al. 2006). Various types of regulated cell death have been reported, and TiO_2 is known to induce necroptosis in vitro (Honarpisheh et al. 2017; Mohammadalipour et al. 2020; Zhou et al.

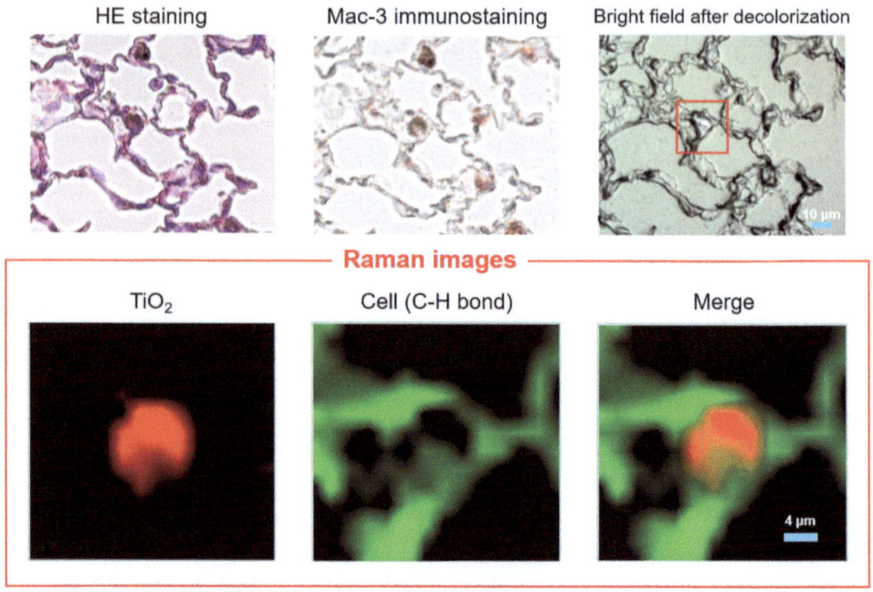

Fig. 4 HE staining, Mac-3 immunostaining, and Raman images of the same tissue section derived from murine lung intratracheally administered with TiO_2 particles. The Raman images was obtained from the area encompassed by the red square in the bright field image obtained after decolorization. Red represents TiO_2 and green represents cellular components. Adapted with permission (Sagawa et al. 2023)

2020). Therefore, we examined the expression of phosphorylated MLKL (pMLKL), a trigger molecule for necroptosis, in primary cultured AMs. The results indicated that pMLKL was expressed by AMs that had phagocytized TiO_2 (Fig. 5b). In addition, pMLKL was expressed by Mac-3 positive cells (macrophages) in the lung tissue of mice intratracheally administered with TiO_2. Furthermore, intraperitoneal administration of necroptosis inhibitor Necrostatin-1s (Nec-1s) before intratracheal administration of TiO_2 significantly suppressed the appearance of 7-AAD positive AMs, neutrophils, and CCL3 in BALF cells. The addition of Nec-1s to primary cultured AMs also significantly suppressed the fluorescence intensity of ethidium homodimer III (Eth-D-III), a marker of dead cells, and the secretion of CCL3. Overall, these results indicated that the phagocytosis of TiO_2 induces the necroptosis of AMs, which leads to the secretion of CCL3 and the migration of neutrophils, resulting in acute inflammation.

Within the living body, cell death occurs via various mechanisms involving diverse signals, which changes over time. Whether cell death invariably worsens inflammation and which signals are involved remain unclear. Clarifying these issues is one of the challenges that should be addressed in the future.

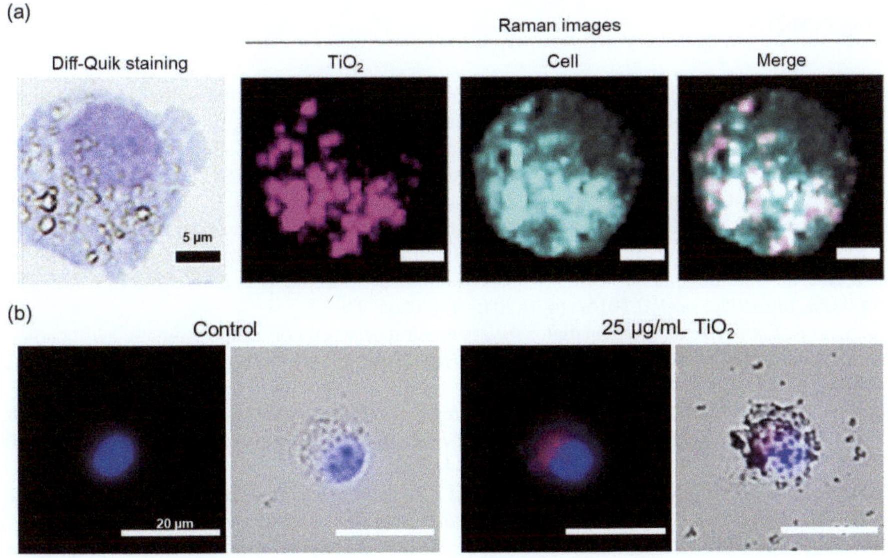

Fig. 5 **a** Representative Raman images of Diff-Quik stained BALF cells of a mouse intratracheally injected with TiO_2. Pink represents TiO_2 and cyan represents cellular components. **b** Representative immunofluorescence stained images of primary cultured alveolar macrophages for pMLKL detection. The two images on the left show control AMs and the two images on the right show AMs exposed to 25 μg/mL TiO_2. Red represents fluorescence derived from pMLKL immunostaining, and blue represents fluorescence derived from nuclear staining using 4′,6-diamidino-2-phenylindole (DAPI)

5 Conclusion

We developed a hybrid method by combining Raman microscopy with histological staining for evaluating the bio-distribution of environmental particles and their components. In this review, we primarily presented the results of Raman microscopy applied to HE-stained, immunostained, and Diff-Quik-stained samples. However, it is expected that this method will be effective for visualizing samples exhibiting more complex biological and immune responses to environmental particles. Further development of this method will enable investigations of the mechanisms by which environmental particles affect the respiratory and immune systems, which will provide insights for controlling environmental particles.

Acknowledgements This study was supported by the Japan Science and Technology Agency (JST) (CREST Grant Number JPMJCR19H3). We thank Edanz (https://jp.edanz.com/ac) for editing a draft of this manuscript.

References

Akaji, S. *et al.* Post-staining Raman analysis of histological sections following decolorization. *Analyst* **147**, 4473–4479 (2022). https://doi.org/10.1039/d2an01138g

Aldegunde, J. A., Sánchez, A. F., Saba, M., Bolaños, E. Q. & Caraballo, L. R. Spatiotemporal Analysis of PM2.5 Concentrations on the Incidence of Childhood Asthma in Developing Countries: Case Study of Cartagena de Indias, Colombia. *Atmosphere* **13** (2022).

Alvarez-Simón, D. *et al.* Effects of diesel exhaust particle exposure on a murine model of asthma due to soybean. *PLOS ONE* **12**, e0179569 (2017). https://doi.org/10.1371/journal.pone.0179569

Baan, R. *et al.* Carcinogenicity of carbon black, titanium dioxide, and talc. *Lancet Oncol* **7**, 295–296 (2006). https://doi.org/10.1016/s1470-2045(06)70651-9

Cao, L. *et al.* Oxidative damage mediates the association between polycyclic aromatic hydrocarbon exposure and lung function. *Environmental Health* **19**, 75 (2020). https://doi.org/10.1186/s12940-020-00621-x

Casjens, S. *et al.* Impact of diesel exhaust exposure on urinary 1-hydroxypyrene in underground salt and potash workers. *International Journal of Hygiene and Environmental Health* **251**, 114190 (2023). https://doi.org/10.1016/j.ijheh.2023.114190

Chen, H.-W. *et al.* Titanium dioxide nanoparticles induce emphysema-like lung injury in mice. *The FASEB Journal* **20**, 2393–2395 (2006). https://doi.org/10.1096/fj.06-6485fje

Dandajeh, H. A., Talibi, M., Ladommatos, N. & Hellier, P. Polycyclic aromatic hydrocarbon and soot emissions in a diesel engine and from a tube reactor. *Journal of King Saud University - Engineering Sciences* **34**, 435–444 (2022). https://doi.org/10.1016/j.jksues.2020.12.007

Duan, R. *et al.* Adverse effects of short-term personal exposure to fine particulate matter on the lung function of patients with chronic obstructive pulmonary disease and asthma: a longitudinal panel study in Beijing, China. *Environ Sci Pollut Res Int* **28**, 47463–47473 (2021). https://doi.org/10.1007/s11356-021-13811-y

Ember, K. J. I. *et al.* Raman spectroscopy and regenerative medicine: a review. *npj Regenerative Medicine* **2**, 12 (2017). https://doi.org/10.1038/s41536-017-0014-3

Fukuoka, A., Matsushita, K., Morikawa, T., Takano, H. & Yoshimoto, T. Diesel exhaust particles exacerbate allergic rhinitis in mice by disrupting the nasal epithelial barrier. *Clinical & Experimental Allergy* **46**, 142–152 (2016). https://doi.org/10.1111/cea.12597

Guo, H., Li, W. & Wu, J. Ambient PM2.5 and Annual Lung Cancer Incidence: A Nationwide Study in 295 Chinese Counties. *Int J Environ Res Public Health* **17** (2020). https://doi.org/10.3390/ijerph17051481

Honarpisheh, M. *et al.* Phagocytosis of environmental or metabolic crystalline particles induces cytotoxicity by triggering necroptosis across a broad range of particle size and shape. *Scientific Reports* **7**, 15523 (2017). https://doi.org/10.1038/s41598-017-15804-9

Husain, M. *et al.* Pulmonary instillation of low doses of titanium dioxide nanoparticles in mice leads to particle retention and gene expression changes in the absence of inflammation. *Toxicology and Applied Pharmacology* **269**, 250–262 (2013). https://doi.org/10.1016/j.taap.2013.03.018

Ichinose, T. *et al.* Enhancement of antigen-induced eosinophilic inflammation in the airways of mast-cell deficient mice by diesel exhaust particles. *Toxicology* **180**, 293–301 (2002). https://doi.org/10.1016/S0300-483X(02)00420-1

Ichinose, T. *et al.* Effects of Asian Sand Dust, Arizona Sand Dust, Amorphous Silica and Aluminum Oxide on Allergic Inflammation in the Murine Lung. *Inhalation Toxicology* **20**, 685–694 (2008). https://doi.org/10.1080/08958370801935133

Jia, X. *et al.* Association of fine particulate matter to allergic rhinitis: A systematic review and meta-analysis. *European Journal of Inflammation* **20**, 1721727X221089839 (2022). https://doi.org/10.1177/1721727X221089839

Johnston, H. J. *et al.* Identification of the mechanisms that drive the toxicity of TiO(2)particulates: the contribution of physicochemical characteristics. *Part Fibre Toxicol* **6**, 33 (2009). https://doi.org/10.1186/1743-8977-6-33

Krafft, C. *et al.* Label-Free Molecular Imaging of Biological Cells and Tissues by Linear and Nonlinear Raman Spectroscopic Approaches. *Angew Chem Int Ed Engl* **56**, 4392–4430 (2017). https://doi.org/10.1002/anie.201607604

Kwon, M. *et al.* Diesel exhaust particle exposure accelerates oxidative DNA damage and cytotoxicity in normal human bronchial epithelial cells through PD-L1. *Environmental Pollution* **317**, 120705 (2023). https://doi.org/10.1016/j.envpol.2022.120705

Li, Y. *et al.* Review of Stimulated Raman Scattering Microscopy Techniques and Applications in the Biosciences. *Advanced Biology* **5**, 2000184 (2021). https://doi.org/10.1002/adbi.202000184

Li, X. & Liu, X. Effects of PM2.5 on Chronic Airway Diseases: A Review of Research Progress. *Atmosphere* **12** (2021).

Li, D. *et al.* Fluorescent reconstitution on deposition of PM(2.5) in lung and extrapulmonary organs. *Proc Natl Acad Sci U S A* **116**, 2488–2493 (2019). https://doi.org/10.1073/pnas.1818134116

Liu, H. *et al.* Different exposure modes of PM2.5 induces bronchial asthma and fibrosis in male rats through macrophage activation and immune imbalance induced by TIPE2 methylation. *Ecotoxicology and Environmental Safety* **247**, 114200 (2022). https://doi.org/10.1016/j.ecoenv.2022.114200

Louro, H. *et al.* The Use of Human Biomonitoring to Assess Occupational Exposure to PAHs in Europe: A Comprehensive Review. *Toxics* **10** (2022).

Miyabara, Y., Ichinose, T., Takano, H., Lim, H.-B. & Sagai, M. Effects of diesel exhaust on allergic airway inflammation in mice. *Journal of Allergy and Clinical Immunology* **102**, 805–812 (1998). https://doi.org/10.1016/S0091-6749(98)70021-1

Mohammadalipour, Z., Rahmati, M., Khataee, A. & Moosavi, M. A. Differential effects of N-TiO2 nanoparticle and its photo-activated form on autophagy and necroptosis in human melanoma A375 cells. *Journal of Cellular Physiology* **235**, 8246–8259 (2020). https://doi.org/10.1002/jcp.29479

Naav, A. *et al.* Urban PM2.5 Induces Cellular Toxicity, Hormone Dysregulation, Oxidative Damage, Inflammation, and Mitochondrial Interference in the HRT8 Trophoblast Cell Line. *Front Endocrinol (Lausanne)* **11**, 75 (2020). https://doi.org/10.3389/fendo.2020.00075

Okada, T., Iwayama, T., Murakami, S., Torimura, M. & Ogura, T. Nanoscale observation of PM2.5 incorporated into mammalian cells using scanning electron-assisted dielectric microscope. *Sci Rep* **11**, 228 (2021). https://doi.org/10.1038/s41598-020-80546-0

Ou, L. *et al.* Application of three-dimensional Raman imaging to determination of the relationship between cellular localization of diesel exhaust particles and the toxicity. *Toxicol Mech Methods* **32**, 333–340 (2022). https://doi.org/10.1080/15376516.2021.2008569

Ozawa, A. & Sakaue, M. New decolorization method produces more information from tissue sections stained with hematoxylin and eosin stain and masson-trichrome stain. *Ann Anat* **227**, 151431 (2020). https://doi.org/10.1016/j.aanat.2019.151431

Park, E. J., Yoon, J., Choi, K., Yi, J. & Park, K. Induction of chronic inflammation in mice treated with titanium dioxide nanoparticles by intratracheal instillation. *Toxicology* **260**, 37–46 (2009). https://doi.org/10.1016/j.tox.2009.03.005

Rattanapinyopituk, K. *et al.* Ultrastructural changes in the air-blood barrier in mice after intratracheal instillations of Asian sand dust and gold nanoparticles. *Exp Toxicol Pathol* **65**, 1043–1051 (2013). https://doi.org/10.1016/j.etp.2013.03.003

Sadauskas, E. *et al.* Protracted elimination of gold nanoparticles from mouse liver. *Nanomedicine* **5**, 162–169 (2009). https://doi.org/10.1016/j.nano.2008.11.002

Sadauskas, E. *et al.* Biodistribution of gold nanoparticles in mouse lung following intratracheal instillation. *Chem Cent J* **3**, 16 (2009). https://doi.org/10.1186/1752-153X-3-16

Sagawa, T. *et al.* Role of necroptosis of alveolar macrophages in acute lung inflammation of mice exposed to titanium dioxide nanoparticles. *Nanotoxicology* **15**, 1312–1330 (2021). https://doi.org/10.1080/17435390.2021.2022231

Sagawa, T., Honda, A. & Takano, H. Elucidation of the dynamics/kinetics of environmental particles and the underlying mechanisms by which they cause biological and immune responses, and identification of the exogenous/endogenous factors and molecules contributing to their health effects. *J Jpn Biochem Soc* **95**, 169–176 (2023). https://doi.org/10.14952/seikagaku.2023.950169

Sharma, M., Agarwal, A. K. & Bharathi, K. V. L. Characterization of exhaust particulates from diesel engine. *Atmospheric Environment* **39**, 3023–3028 (2005). https://doi.org/10.1016/j.atmosenv.2004.12.047

Singh, N. & Singh, S. Interstitial Lung Diseases and Air Pollution: Narrative Review of Literature. *Pulm Ther* **7**, 89–100 (2021). https://doi.org/10.1007/s41030-021-00148-7

Sonwani, S. *et al.* Inhalation Exposure to Atmospheric Nanoparticles and Its Associated Impacts on Human Health: A Review. *Frontiers in Sustainable Cities* **3** (2021). https://doi.org/10.3389/frsc.2021.690444

Thangavel, P., Park, D. & Lee, Y.-C. Recent Insights into Particulate Matter (PM2.5)-Mediated Toxicity in Humans: An Overview. *International Journal of Environmental Research and Public Health* **19** (2022).

Tian, Y. *et al.* Size distribution, meteorological influence and uncertainty for source-specific risks: PM2.5 and PM10-bound PAHs and heavy metals in a Chinese megacity during 2011–2021. *Environmental Pollution* **312**, 120004 (2022). https://doi.org/10.1016/j.envpol.2022.120004

Ural, B. B. *et al.* Inhaled particulate accumulation with age impairs immune function and architecture in human lung lymph nodes. *Nature Medicine* **28**, 2622–2632 (2022). https://doi.org/10.1038/s41591-022-02073-x

Wang, X., Wang, Y., Bai, Y., Wang, P. & Zhao, Y. An overview of physical and chemical features of diesel exhaust particles. *Journal of the Energy Institute* **92**, 1864–1888 (2019). https://doi.org/10.1016/j.joei.2018.11.006

Wang, T. H. *et al.* PM2.5 promotes lung cancer progression through activation of the AhR-TMPRSS2-IL18 pathway. *EMBO Mol Med* **15**, e17014 (2023). https://doi.org/10.15252/emmm.202217014

Warheit, D. B. *et al.* Inhalation of High Concentrations of Low Toxicity Dusts in Rats Results in Impaired Pulmonary Clearance Mechanisms and Persistent Inflammation. *Toxicology and Applied Pharmacology* **145**, 10–22 (1997). https://doi.org/10.1006/taap.1997.8102

Weir, A., Westerhoff, P., Fabricius, L., Hristovski, K. & von Goetz, N. Titanium dioxide nanoparticles in food and personal care products. *Environ Sci Technol* **46**, 2242–2250 (2012). https://doi.org/10.1021/es204168d

Winterbottom, C. J. *et al.* Exposure to Ambient Particulate Matter Is Associated With Accelerated Functional Decline in Idiopathic Pulmonary Fibrosis. *Chest* **153**, 1221–1228 (2018). https://doi.org/10.1016/j.chest.2017.07.034

Xu, S. *et al.* Titanium dioxide nanoparticles induced the apoptosis of RAW264.7 macrophages through miR-29b-3p/NFAT5 pathway. *Environmental Science and Pollution Research* **27**, 26153–26162 (2020). https://doi.org/10.1007/s11356-020-08952-5

Yang, L., Li, C. & Tang, X. The Impact of PM2.5 on the Host Defense of Respiratory System. *Frontiers in Cell and Developmental Biology* **8** (2020). https://doi.org/10.3389/fcell.2020.00091

Zhao, L. *et al.* Cardiopulmonary effects induced by occupational exposure to titanium dioxide nanoparticles. *Nanotoxicology* **12**, 169–184 (2018). https://doi.org/10.1080/17435390.2018.1425502

Zhao, X., Zhou, W. & Han, L. The spatial and seasonal complexity of PM2.5 pollution in cities from a social-ecological perspective. *Journal of Cleaner Production* **309** (2021). https://doi.org/10.1016/j.jclepro.2021.127476

Zhao, J. *et al.* Role of PM(2.5) in the development and progression of COPD and its mechanisms. *Respir Res* **20**, 120 (2019). https://doi.org/10.1186/s12931-019-1081-3

Zhou, T. *et al.* Nec-1 Attenuates Neurotoxicity Induced by Titanium Dioxide Nanomaterials on Sh-Sy5y Cells Through RIP1. *Nanoscale Research Letters* **15**, 65 (2020). https://doi.org/10.1186/s11671-020-03300-5

Zhu, Y. *et al.* Recent advances in the biotoxicity of metal oxide nanoparticles: Impacts on plants, animals and microorganisms. *Chemosphere* **237**, 124403 (2019). https://doi.org/10.1016/j.chemosphere.2019.124403

Direct Observation of Biological Fine Particles in Water by Scanning Electron Assisted Dielectric Microscopy

Toshihiko Ogura and Tomoko Okada

Abstract Investigation of the effects of nanoparticles and extracellular vesicles on cells requires direct observation and analysis at nano-level resolution, both in vitro and in living cells. Recently, we have developed a scanning electron assisted dielectric microscope (SE-ADM) that enables the direct observation of living cells and organic materials without staining or fixation. Here, we present an overview of SE-ADM and scanning electron impedance microscopy including our recent results analyzing the effects of extracellular particles on biological specimens. Our system can be easily used to examine unstained biological specimens including bacteria, viruses and protein complexes. Furthermore, it can be used for diverse liquid samples across a broad range of scientific fields, e.g. nanoparticles, nanotubes and organic and catalytic materials.

1 Introduction

Direct observation at nano-level resolution is important for analyzing the effects of endogenous and exogenous extracellular particles on cells. Furthermore, establishing such effects requires direct observation of intact living cells. We have developed a scanning electron assisted dielectric microscope (SE-ADM) that allows direct observation of living cells and organic materials in solution without staining or fixation. Here, we present an overview of SE-ADM and the recently developed scanning electron impedance microscope (IP-SEM), and report on recent results using these techniques, including the investigation of extracellular fine particles.

Many cells secrete extracellular vesicles such as exosomes and microvesicles. The diameters of these particles range widely from 30 nm to a few micrometers. Since the resolution of ordinary optical microscopes is limited to around 200 nm due to the diffraction limit of light, it is difficult to directly observe extracellular particles smaller than this. Therefore, it is necessary to use a higher resolution observation

T. Ogura (✉) · T. Okada
Health and Medical Research Institute, National Institute of Advanced Industrial Science and Technology (AIST), Central 6, Higashi, Tsukuba 305-8566, Ibaraki, Japan
e-mail: t-ogura@aist.go.jp

Y. Baba et al. (eds.), *Extracellular Fine Particles*,
https://doi.org/10.1007/978-981-97-7067-0_12

device such as an electron microscope. Since electron microscopes use electron beams (EB) as observation probes, they can perform observations with an extremely high resolution of 1 nm or less. However, in a typical electron microscope, it is necessary to maintain a vacuum inside the microscope, so biological samples, organic materials, etc. in solution cannot be directly studied. If a solution sample is directly introduced into the microscope, the water in the sample will be evaporated and the sample will be completely dried. Therefore, a capsule-shaped sample holder has been developed to maintain atmospheric pressure and has been introduced in an electron microscope to observe liquid samples (Nagata and Ishikawa 1972; Parsons 1974; Thiberge et al. 2004; Jonge et al. 2009). The inside of the sample holder is hermetically sealed and is maintained at atmospheric pressure even under vacuum. The development of such sample holders has been going on for over 50 years, and various types of holders are still being developed (Nagata and Ishikawa 1972; Parsons 1974).

2 Overview of Conventional Capsule-Type Sample Holders

A typical capsule-type sample holder has an observation window made of a thin film that transmits electron beams well. The sample is introduced into this area and the capsule is sealed. The thin film is often made of silicon nitride (SiN) or silicon oxide, which can be produced in large quantities with uniform thickness using a semiconductor process. Furthermore, SiN thin films have high mechanical strength and can be made with a thickness of less than 50 nm. There are two types of electron microscopes: transmission electron microscopes (TEM) and scanning electron microscopes (SEM). The atmospheric pressure sample holder of the TEM has two observation windows, between which the sample is sealed and observed by an electron beam (EB) transmitted through the windows (Nagata and Ishikawa 1972; Parsons 1974; Jonge et al. 2009). On the other hand, an SEM has an observation window on the top through which the electron beam (EB) passes, and observation is made via the scattered electrons reflected from the sample (Thiberge et al. 2004). Although observation in liquid using such electron microscopes has been performed for more than 50 years, the contrast of the samples is low and damage of the samples by the EB is large. Furthermore, when the thickness of the samples is more than 1 μm, the transmission of the EB is reduced, resulting in reduced resolution and contrast. To solve these problems, a new method of observation without directly irradiating the sample with an EB is needed.

3 Newly Developed Scanning Electron Assisted Dielectric Microscopy (SE-ADM)

To observe extracellular particles, cultured cells, and organic materials in aqueous solution with high contrast and low damage using electron microscopy, it is important to avoid direct irradiation of the sample. However, detection of transmitted electrons and scattered electrons using an EB as a probe requires direct irradiation of the sample with the EB. To resolve this problem, we have developed a method to irradiate a thin film with an EB, converting it into an alternative physical form that produces less damage and higher contrast, thus indirectly irradiating the sample. Our SE-ADM visualizes differences in the dielectric constants of samples in aqueous solution by irradiating and absorbing a low-voltage EB on a thin film in the observation window (Ogura 2014a, 2015). In this method, the tungsten layer on the SiN thin film in the sample holder is irradiated by an EB, creating a local potential change within the thin film (Fig. 1a) (Ogura 2014a, 2015). This potential change is transmitted through the sample and detected by the measurement terminal underneath (Fig. 1a). Since the permeability of such potential changes is determined by the dielectric constant of the sample, high contrast images can be obtained in the intact state without any staining. Water has a high dielectric constant of 80, which makes it highly permeable to potential changes (Ogura 2014a, 2015). On the other hand, the dielectric constant of biological samples and organic materials is as low as 2–3, which suppresses the permeability to potential change. By detecting the difference in the transmission of potential change due to the dielectric constant, samples in aqueous solution can thus be observed with high contrast (Fig. 1a). In addition, since most of the incident electrons are scattered and absorbed by the tungsten layer, the EB does not irradiate the sample directly, resulting in low damage (Ogura 2014a, 2015).

4 High-Resolution SE-ADM System

To increase image resolution, it is necessary to improve the detection sensitivity of potential change. We have found that the detection sensitivity can be greatly improved by applying a bias voltage to the tungsten metal layer on SiN film (Ogura 2014b). Moreover, the introduction of a dielectric constant observation system to the high-resolution field emission SEM (FE-SEM) has made it possible to observe biological samples in solution with higher resolution and clarity (Ogura 2015). The FE-SEM can focus EBs down to 3 nm in diameter at an acceleration voltage of 3–4 kV, which significantly improves resolution at low acceleration voltages compared to a general-purpose thermal electron gun-type SEM. When unstained and unfixed bacteria and protein complexes in solution were observed using this system, clear images were obtained (Ogura 2015).

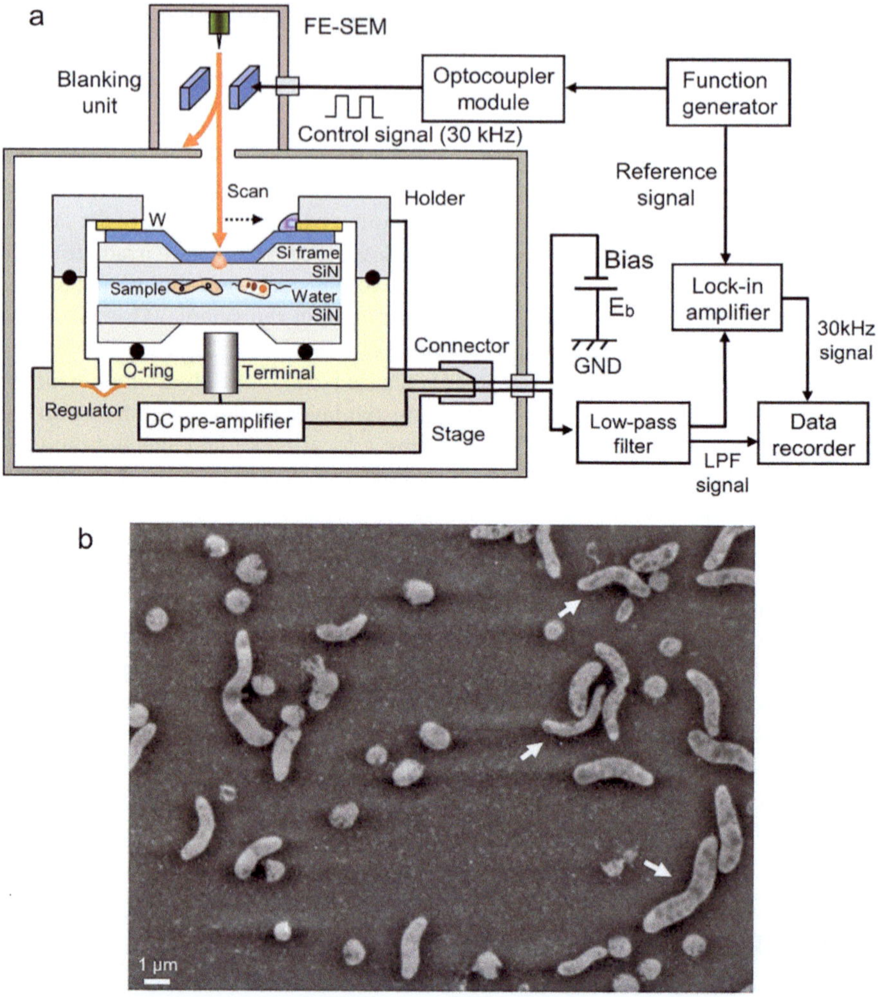

Fig. 1 Experimental set-up of the high-resolution SE-ADM system (Ogura 2015). **a** A schematic of the SE-ADM system based on FE-SEM. Biological specimens in water were injected into the space between two SiN films in the sample holder. **b** Observed image of bacteria in solution by the SE-ADM system. White arrows indicate areas where flagella and each bacteria connected

5 Direct Observation of Biological Samples Using SE-ADM in the Liquid Phase

We have been directly observing biological samples such as bacteria and mammalian cells in aqueous solution using high-resolution SE-ADM (Fig. 1a) (Ogura 2015, 2014b; Okada and Ogura 2016, 2017). Formerly, observation of bacteria and cells by electron microscopy requires fixation, dehydration, and staining with heavy metals.

SE-ADM, on the other hand, allows direct observation of biological samples in liquid solution without these steps. Figure 1b shows the results of direct observation of photosynthetic bacteria in an aqueous solution by enclosing them in a sample holder (Ogura 2015). This photosynthetic bacterium has an elongated cylindrical shape and a flagellum. Furthermore, inside the bacteria, there are many spherical structures that appear to be vacuoles and areas where genes and proteins are densely packed (Fig. 1b). Further, by using SE-ADM, we succeeded in directly observing mouse IgM antibodies, a type of protein complex, in an aqueous solution (Fig. 2a, d). The structure of an IgM antibody is 45–50 nm in diameter, with a dome-shaped bulge in the center and five IgG-like structures arranged in a star shape around the periphery (Czajkowsky and Shao 2009). The IgM antibodies in aqueous solution observed by SE-ADM are particles with a diameter of 45–50 nm, which corresponds to the size of IgM previously reported (Ogura 2015).

In addition, we have directly observed raw milk using SE-ADM. Ordinary milk is a mixture of milk fat and protein-rich casein micelles dispersed in an aqueous solution (Gallier et al. 2010; McMahon and Oommen 2008). Previous observations using fluorescence microscopy and cryo-TEM have reported that the shape of milk fat is spherical with a diameter of several micrometers (Gallier et al. 2010; McMahon and Oommen 2008). The casein micelles have an aggregated structure of protein and calcium phosphate with a diameter of approximately 100 nm (Gallier et al. 2010; McMahon and Oommen 2008). Therefore, milk can be regarded as a natural emulsion. By using our SE-ADM system, it is possible to directly observe emulsion samples in which lipid and protein particles are suspended. Figure 2e, f show images of milk that was sealed in an aqueous state in a sample holder and observed using an SE-ADM system (Ogura and Okada 2017). Figure 2e is an SE-ADM image of whole milk, where many white spherical particles of several μm size are seen; these are thought to be milk fat (Ogura and Okada 2017). Furthermore, many black particles with a diameter of 100–300 nm are also observed, which we believe are casein micelles (Ogura and Okada 2017). When the milk fat was removed by centrifugation, the white spheres had disappeared, confirming that these were indeed milk fat (Fig. 2f). When the milk fat area was enlarged, it was observed with good contrast that the fat was surrounded by black casein micelles (Ogura and Okada 2017). Furthermore, when we compared milk fat images observed using SE-ADM with an optical microscope and a conventional atmospheric pressure capsule, we found that with the latter method the contrast was low and the resolution was extremely poor (Ogura and Okada 2017). Moreover, the particle structure of casein micelles was not evident. On the other hand, it was confirmed that the images observed using the SE-ADM had high resolution and high contrast (Ogura and Okada 2017).

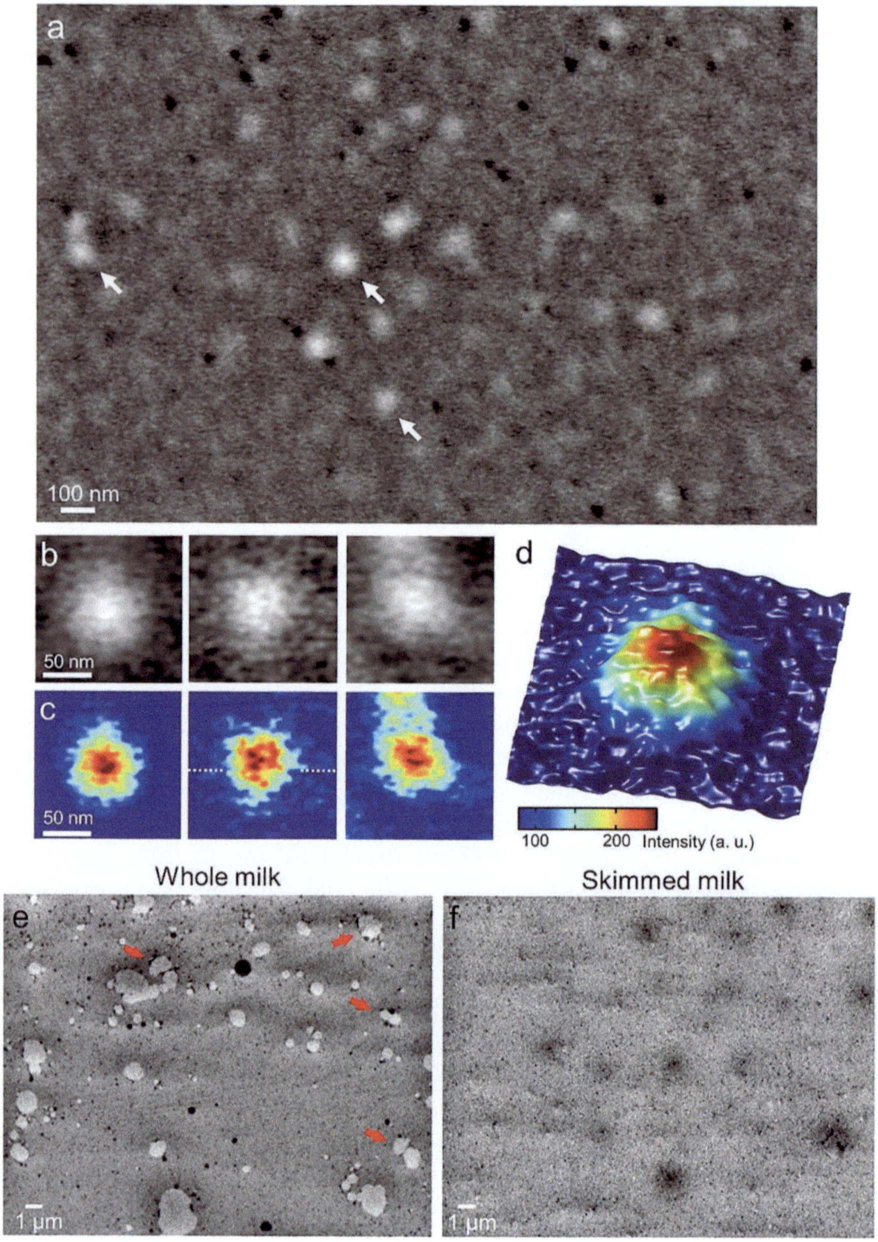

Fig. 2 Direct observation of IgM antibodies and whole milk in solution. **a** Image of unstained mouse IgM particles in water observed using the SE-ADM system (Ogura 2015). **b** Three individual IgM molecules are indicated with white arrows in **a** (Ogura 2015). **c** Pseudo-color maps of **b** (Ogura 2015). **d** A 3D color map of the IgM molecules on the left side of **c** (Ogura 2015). **e** A dielectric-impedance image of untreated liquid whole-milk. Red arrows indicate milk fat (Ogura and Okada 2017). **f** Liquid skimmed-milk image (Ogura and Okada 2017)

6 Direct Observation of Cultured Cells by SE-ADM

Direct observation of cells in culture medium at high resolution would allow analysis of the effects of extracellular particles on cells. We have developed an original culture dish for observation of cultured cells using SE-ADM and succeeded in directly observing living cells at high resolution (Okada and Ogura 2016, 2017). Figure 3 shows an overview of the system for cultured cell observation (Okada and Ogura 2016, 2017). A square is cut from the center of the bottom of a commercially available culture dish, and an SiN film is fitted into it. The cells are then seeded onto the SiN film and cultured to form a monolayer (Fig. 3a). To observe the cells, the aluminum holder with the SiN thin film is removed, inverted, and placed with culture medium on another bottom holder including another SiN film (Fig. 3b, c). It was then sealed and placed inside an SEM for observation by SE-ADM (Fig. 3d). Figure 3e, f show SE-ADM images of mouse breast cancer cells at low magnification (Okada and Ogura 2016). Cell nuclei are elliptical in shape with a diameter of approximately 10 μm, and complex internal cellular structures can be observed around them (Fig. 3e). Furthermore, at a higher magnification of 10,000×, inner membrane structures were observed in the cytoplasm, and elongated filament structures were also observed in some areas (Fig. 3f, g). White intracellular vesicles are attached to these filament structures, suggesting that the filaments are involved in vesicle migration.

We have also observed and analyzed cell surface molecules such as CD44 (Okada and Ogura 2017) and integrin β1 (Okada and Ogura 2018). Further, we have succeeded in visualizing mineralizing nanovesicles in intact osteoblasts and clarifying their biogenesis and trafficking (Iwayama et. al. 2019). We also analyzed melanosome distribution in melanoma cells (Mastrangelo et. al. 2023).

These results indicate that SE-ADM can be used for high-resolution transmission observation of relatively thick biological samples such as mammalian cells (Okada and Ogura 2016, 2017, 2018; Iwayama et. al. 2019).

7 Analysis of PM2.5 Uptake to Cultured Cells

Next, we analyzed the effects of atmospheric fine particle matter, PM2.5 on cells. PM2.5 is thought to affect health because it is composed of particulate matter such as soot and oxidized metals emitted into the air and attached to organic matter and nitrogen oxides (Canagaratna et al. 2007; Abbas et al. 2009). We used SE-ADM to analyze the state of PM2.5 when it is taken up by cells (Okada et al. 2021). First, we checked whether PM2.5 in water could be observed by SE-ADM. The PM2.5 in solution consisted of small particles that aggregated into a cluster-like structure with a diameter of several micrometers (Okada et al. 2021). The PM2.5 was added to cultured cells and the cells were directly observed by SE-ADM after 3–24 h (Fig. 4). Three hours after the addition of PM2.5, a small amount of PM2.5 is taken up by cells (Okada et al. 2021). On the other hand, after 5–9 h, a large amount of PM2.5 is

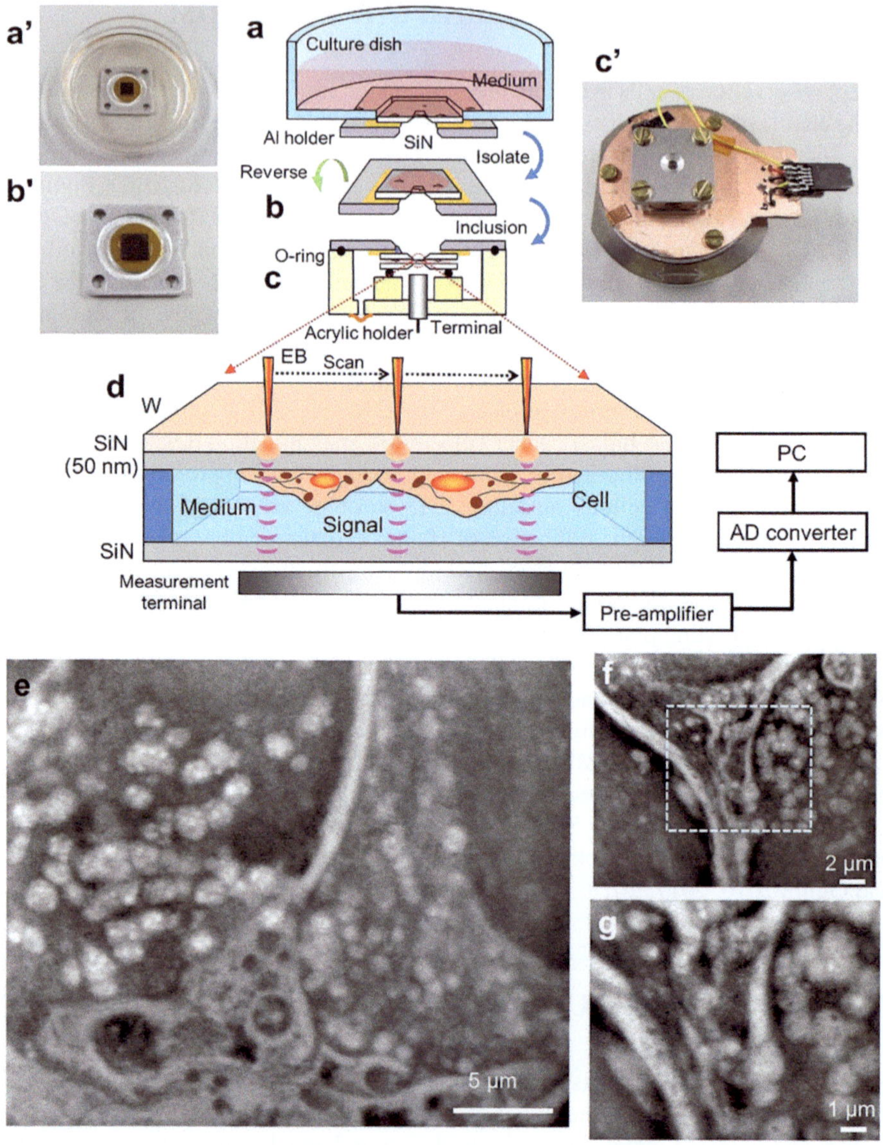

Fig. 3 Overview of the method for observing cultured cells using the SE-ADM system (Okada and Ogura 2016). **a**, **a′** Al holder with W-coated SiN film is attached to the bottom of a medium-filled culture dish. **b**, **b′** The holder containing culture cells was separated from the plastic culture dish after removing most of the medium. **c**, **c′** The Al holder was attached upside down onto another SiN film on a square acrylic plate. **d** Schematic figure of our high-contrast imaging method producing low-level radiation damage to untreated cells in the medium under W-coated SiN film. **e** SE-ADM image near the nuclear regions of cultured cell in medium. **f** Dielectric image of a vesicle rich region. **g** High-magnification image of the framed area in **f** at 10,000× magnification

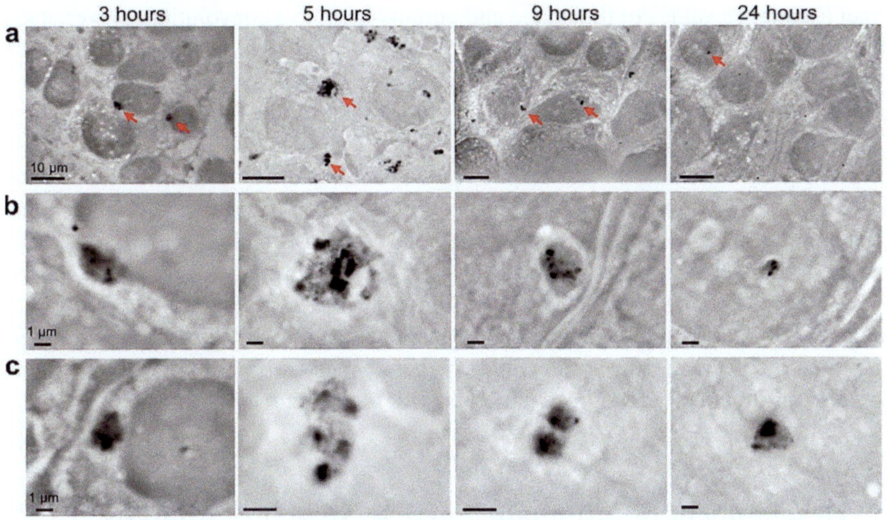

Fig. 4 SE-ADM images of OBA9 cells after the addition of PM2.5 (Okada et al. 2021). **a** Low magnification (1500–2500×) images of OBA9 cells at 3, 5, 9 and 24 h after the addition of PM2.5. **b**, **c** High-magnification (**b**, 10,000×; **c**, 10,000–20,000×) images of the areas indicated by red arrows in **a** show the PM2.5 aggregates

taken up (Fig. 4, 5 and 9 h). PM2.5 in the cells is surrounded by the inner membrane. This suggests that the cells detoxify the captured particles by surrounding them with an inner membrane structure, similar to autophagy (Okada et al. 2021). The amount of PM2.5 taken up by the cells after 24 h was significantly reduced, suggesting that PM2.5, once taken up intracellularly, was extruded from the cells (Okada et al. 2021).

8 Development of an Automatic Recognition System for Intracellular Particles Using a Deep Neural Network

To examine the structure and dispersion of biological and material particles in solution, it is necessary to locate the particles in the image and analyze distances, particle diameters, etc. Image processing by visual inspection requires a lot of time and effort. Therefore, it is necessary to use an automatic recognition system to analyze the particle images.

Recently, we have developed a technique for automatic recognition and analysis of melanin pigment vesicles inside cultured melanoma cells (Okada et al. 2023). In this method, a deep neural network (DNN) with a convolutional layer is used to recognize particles. The melanin pigment vesicles in the image can be recognized and their locations identified (Okada et al. 2023). Figure 5a–d is an overview of this image processing system. On the left (Fig. 5c), there is an input layer that takes the pixel luminance values of the image and outputs them to a convolutional layer, itself consisting of four layers (Okada et al. 2023). In the convolutional layer, the size is gradually reduced toward the output layer on the right side; at the fourth layer, the size is 3 × 3 pixels (Fig. 5c). After the output of the fourth layer, there is a free-connecting layer, and finally, the outputs are aggregated into a single output layer (Okada et al. 2023). The output value is 2 for melanin pigment vesicles and 1 for background images without particles (Fig. 5a, b). 200 particle images and the same number of background images were used for training, and almost 100% recognition accuracy was achieved (Fig. 5d). When the image of a melanoma cell was input into this trained DNN, the system was able to recognize the position of melanin pigment vesicles (melanosomes) inside the cell with high accuracy (Fig. 5e, f). When using images of extracellular melanosomes as training data, this system can also automatically recognize particles with high accuracy.

9 Development of Automatic Particle Structure Analysis System

Next, we developed an image processing system that automatically recognizes and extracts structural information of particles from particle images (Okada et al. 2023). To analyze the structure of a particle image, the contours of the particle must first be masked. By using this mask data, the area, long axis, short axis, and roundness of the particles are calculated. For this masking process, too, we developed a system based on a DNN with a convolutional computational layer (Okada et al. 2023). As learning data, 200 melanosome images and manually masked images were used (Okada et al. 2023). First, a 50 × 50 pixel area within the particle image was randomly selected and its intensity values were entered into the input layer. As the output value, the value of the central portion of the mask-processed image corresponding to this cut-out range was used. At this time, when an image inside the particle was input, the mask value was designated as 2, and when an image outside the particle is input, the mask value was 1. By repeating this learning process, it became possible to perform automatic mask processing with an accuracy close to that of visual mask processing (Okada et al. 2023). This enables automatic recognition of particles in the SE-ADM image, and in conjunction with mask processing, the long axis, short axis, and circularity distribution of particles can be analyzed in a short time (Okada et al. 2023).

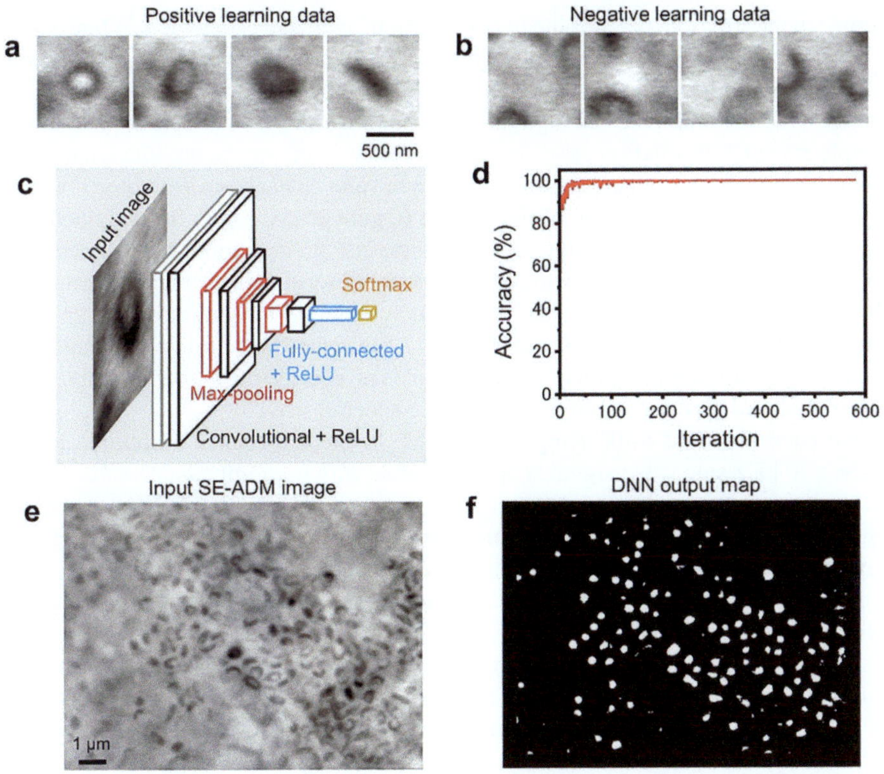

Fig. 5 Detection and analysis of melanosomes using the DNN system (Okada et al. 2023). **a** Typical melanosome images in MNT-1 cells observed by SE-ADM. We manually selected 239 typical images. These images were presented to the DNN system as positive learning data. **b** Typical non-melanosome images observed by SE-ADM. We manually selected 307 non-melanosome images in the cell. **c** A schematic diagram of the DNN system. **d** Recognition accuracy for each learning iteration calculated from unlearned 1000 images. **e** Typical SE-ADM image of MNT-1 cells (10,000×) inputs to the DNN system. **f** DNN output map for the input image in **e**. The output value at the particle position is 2 (white area) at other positions, 1 (dark area)

10 Development of Scanning Electron Impedance Microscope and Observation of Biological Samples

Biological samples, extracellular particles, and emulsions in aqueous solution can be directly observed using our SE-ADM without staining or fixation. However, it is not possible to analyze the composition of the observed samples. We are therefore developing a new impedance microscope (scanning electron impedance microscope, IP-SEM) that can analyze the impedance values of samples (Ogura 2019, 2021). In this microscope, a sinusoidal input signal is applied to the metal terminal at the bottom of the sample holder, and the output signal is detected from the tungsten layer on the upper SiN film (Ogura 2019, 2021). The applied potential signal passes through the sample in the solution, reaches the upper tungsten layer, and is detected. When an EB is applied to the upper tungsten layer in this state, the impedance of the SiN film at the EB irradiation position changes. As a result, the detected impedance signal also changes, but the amount of change differs depending on the impedance value of the sample at the EB irradiation position. Therefore, by detecting the output signal while scanning the EB and calculating the impedance value at each position, it is possible to obtain an impedance image of the sample (Ogura 2019, 2021). In the impedance signal, both the amplitude value and the phase can be determined, and images of each can be obtained. Furthermore, by repeatedly observing the same location while changing the frequency of the AC signal, it is possible to determine the spectral components of impedance (Ogura 2019, 2021). Since these spectral components vary depending on the composition of the sample, it is possible to analyze the composition of the observed structure from the impedance spectral waveform.

We have observed milk using an IP-SEM. In the impedance amplitude image when a 20 kHz sine wave signal is applied, areas with high impedance values of 20 MΩ or more are spread throughout the image, whereas in the 250 kHz image, areas with high impedance values decrease. This suggests that the impedance changes in a frequency-dependent manner depending on the composition of the sample. This shows that it is possible to analyze the composition of a sample from such spectral information (Ogura 2019, 2021).

Recently, we have been developing an impedance microscope that can observe multiple wavelengths simultaneously by mixing multiple wavelengths and inputting them to the input terminal and separating the frequency components using multiple lock-in amplifiers (Ogura 2019, 2021). In this system, eight wavelengths of sinusoidal signals from 20 to 500 kHz are mixed and applied to the input terminal, making it possible to observe eight wavelengths simultaneously. Observations of sunscreen samples have confirmed that the impedance frequency components differ depending on oil, water, and metal oxides (Ogura 2019, 2021). In the future, we expect that nano-level impedance analysis of extracellular particles and biological samples will become possible by increasing the number of wavelengths. However, the resolution of the observed images is still low at a 30 nm level, since this system is installed in a thermionic gun type scanning electron microscope, which has relatively low resolution. In the future, we plan to achieve a resolution of 10 nm or less by equipping this system with a higher-resolution FE-SEM (Ogura 2019, 2021).

11 Conclusion

Here, we have reported an overview and the results of analysis using SE-ADM and IP-SEM, which are new observation methods for analyzing the structure of intracellular and extracellular particles in solution and their effects on cells. These systems enable direct observation of unstained and unfixed cultured cells and extracellular particles in solution with nano-level resolution. Radiation damage is minimized since SE-ADM does not directly irradiate the samples with the EB. Furthermore, since this method detects the difference of dielectric constant between water and the sample, biological samples in solution can be observed with high contrast. Thus, exogenous and endogenous extracellular particles can be observed in solution, as well as their effects on cultured cells. Furthermore, in an IP-SEM, a sinusoidal potential signal is applied as an input signal, and the potential signal transmitted through the sample is detected from the tungsten layer on the upper SiN thin film. By scanning irradiation of the upper SiN film by an EB, impedance images of the inside of the sample holder can be acquired. This method allows the frequency spectrum of impedance to be determined by changing the frequency of the input signal and is expected to allow compositional analysis. In the future, we expect that this method will be applied to the observation and composition analysis of a wide variety of organic materials, nanoparticles, biological samples, emulsions, etc. in solution.

References

F. Nagata, I. Ishikawa: "Observation of wet biological materials in a high voltage electron microscope.": Jpn. J. Appl. Phys. 11: p.1239–1244. (1972)

D.F. Parsons.: "Structure of wet specimens in electron microscopy": Science 186: p.407–414. (1974).

S. Thiberge, A. Nechushtan, D. Sprinzak, O. Gileadi, V. Behar, O. Zik, Y. Chowers, S. Michaeli, J. Schlessinger, W. Moses: "Scanning electron microscopy of cells and tissues under fully hydrated conditions.": Proc. Natl. Acad. Sci. USA 101: p.3346–3351 (2004)

N. de Jonge, D.B. Peckys, G.J. Kremers, D.W. Piston: "Elecetron microscopy of whole cells in liquid with nanometer resolution": Proc. Natl. Acad. Sci. USA 106: p.2159–2164. (2009)

T. Ogura: "Direct observation of unstained biological specimens in water by the frequency transmission electric-field method using SEM.": PLOS ONE. 9: e92780 (2014a)

T. Ogura: "Non-destructive observation of intact bacteria and viruses in water by the highly sensitive frequency transmission electric-field method based on SEM": Biochem. Biophys. Res. Commun. 450: p.1684–1689. (2014b)

T. Ogura: "Nanoscale analysis of unstained biological specimens in water without radiation damage using high-resolution frequency transmission electric-field system based on FE-SEM": Biochem. Biophys. Res. Commun. 459: p.521–528. (2015)

T. Okada, T. Ogura: "Nanoscale imaging of untreated mammalian cells in a medium with low radiation damage using scanning electron-assisted dielectric microscopy": Sci. Rep. 6: 29169 (2016)

T. Okada, T. Ogura: "High-resolution imaging of living mammalian cells bound by nanobeads-connected antibodies in a medium using scanning electron assisted dielectric microscopy": Sci. Rep. 7: 43025 (2017)

D.M. Czajkowsky, Z. Shao: "The human IgM pentamer is a mushroom-shaped molecule with a flexural bias": Proc. Natl. Acad. Sci. USA 106: p.14960–14965 (2009)

S. Gallier, D. Gragson, R Jimenz-Flores, D. Everett: "Using confocal laser scanning microscopy to probe the milk fat globule membrane and associated proteins": J. Dairy Sci. 58: p.4350–4257 (2010)

D. J. McMahon, B. S. Oommen: "Supramolecular structure of the casein micelle": J. Agric. Food Chem. 91: p.1709–1721 (2008)

T. Ogura, T. Okada: "Nanoscale observation of the natural structure of milk-fat globules and casein micelles in the liquid condition using a scanning electron assisted dielectric microscopy": Biochem. Biophys. Res. Commun. 491: p.1021–1025. (2017)

T. Okada, T. Ogura: "Nanoscale imaging of the adhesion core including integrin b1 on intact living cells using scanning electron assisted dielectric-impedance microscopy": PLOS ONE. 13: e0204133 (2018)

T. Iwayama *et. al.*: "Osteoblastic lysosome plays a central role in mineralization": Sci. Adv. eaax0672 (2019)

M.R. Canagaratna *et al.*: "Chemical and microphysical characterization of ambient aerosols with the aerodyne aerosol mass spectrometer": Mass Spectrom. Rev. 26, 185–222 (2007)

I. Abbas et al.: "Air pollution particulate matter (PM2.5)-induced gene expression of volatile organic compound and/or polycyclicaromatic hydrocarbon-metabolizing enzymes in an in vitro coculture lung model" Toxicol. Vitro 23, 37–46 (2009).

T. Okada, T. Iwayama, S. Murakami, M. Torimura, T. Ogura: "Nanoscale observation of PM2.5 incorporated into mammalian cells using scanning electron-assisted dielectric microscope": Sci. Rep. 11: 228 (2021)

T. Okada, T. Iwayama, T. Ogura, S. Murakami, T. Ogura: "Structural analysis of melanosomes in living mammalian cells using scanning electron-assisted dielectric microscopy with deep neural network": Comp. Struct. Biotech. J: 21, 506–518 (2023)

T. Ogura: "Direct observation of unstained biological samples in water using newly developed impedance scanning electron microscopy": PLOS ONE. 14: e0221296 (2019)

T. Ogura: "Development of multi-frequency impedance scanning electron microscopy": PLOS ONE. 17: e0263098 (2021)

R. Mastrangelo, T. Okada, T. Ogura, T. Ogura, P. Baglioni: "Direct observation of the effects of chemical fixation in MNT-1 cells: A SE-ADM and Raman study": Proc. Natl. Acad. Sci. USA 120(51): e2308088120 (2023)

Pathways to Repair or Remove Lysosomes Damaged by Extracellular Fine Particles

Akiko Kuma and Tamotsu Yoshimori

Abstract Exogenous and endogenous fine particles such as environmental materials (e.g., silica, asbestos, alum), toxic protein aggregates (e.g., α-synuclein, amyloid-β), and endogenous crystals (e.g., cholesterol crystals, uric acid crystals) are internalized into the cell by the endocytic pathway or phagocytosis. Because lysosomes are the terminal compartments of these pathways, lysosomes are known to be damaged by exocytosed extracellular fine particles. Lysosomal membrane damage allows the leakage of the lysosomal contents such as cathepsins, H^+, Ca^{2+}, and iron into the cytosol, which is harmful to the cell. Numerous studies have suggested that lysosomal damage is tightly associated with toxicity of exogenous particles, inflammatory responses, and diseases including those involving neurodegeneration. To preserve lysosomal integrity, cells have several mechanisms for the repair or elimination of compromised lysosomes collectively called the "lysosomal damage response". This review summarizes recent findings on the responses to lysosomal damage, focusing on extracellular fine particles.

1 Introduction

The living body is continuously exposed to a large number of exogenous fine particulates such as silica, asbestos, alum, and carbon nanotubes from the environment. Understanding the potential risk of these environmental materials and their mechanisms of toxicity is important. For example, crystalline silica (SiO_2) is contained in dust and air pollutants (Pozzi et al. 2003). It is known that prolonged inhalation of large amounts of crystalline silica dust can cause lung fibrosis (Leung et al. 2012). It has also been demonstrated that crystalline silica has toxic effects on tissues

A. Kuma · T. Yoshimori (✉)
Division of Health Sciences, Osaka University Graduate School of Medicine, Suita, Osaka, Japan
e-mail: tamyoshi@sahs.med.osaka-u.ac.jp

T. Yoshimori
Institute for Open and Transdisciplinary Research Initiatives (OTRI), Osaka University, Suita, Osaka, Japan

Y. Baba et al. (eds.), *Extracellular Fine Particles*,
https://doi.org/10.1007/978-981-97-7067-0_13

in mice, and indicated that silica crystals can induce inflammation through activation of the NLRP3 inflammasome (Cassel et al. 2008; Dostert et al. 2008; Hornung et al. 2008; Yang et al. 2016). Besides fine particles from the environment, cells are also exposed to endogenous extracellular particles such as cholesterol crystals or monosodium urate crystals, and toxic protein aggregates, which are associated with arteriosclerosis, gout, and neurodegeneration, respectively (Tall and Yvan-Charvet 2015; Freeman et al. 2013; So and Martinon 2017).Therefore, cells are exposed to a variety of exogenous and endogenous extracellular particles, and lysosomes are thought to play an important role in the toxicity that they exert. As lysosomes are the terminal compartments of the endocytic pathway, lysosomes are damaged by exocytosed extracellular fine particles. Once the lysosomal membrane is damaged, the lysosomal contents including cathepsins, H^+, Ca^{2+}, and iron leak into the cytosol, which can harm the cell. Numerous studies have suggested that lysosomal damage is tightly associated with toxicity of exogenous particles, inflammatory responses, and diseases (Wang et al. 2018; Bonam et al. 2019; Flavin et al. 2017). To cope with these threats, cells have several mechanisms that repair or eliminate compromised lysosomes collectively called the lysosomal damage response (LDR). This review summarizes recent findings on the responses to lysosomal damage, focusing on extracellular fine particles.

2 Lysosomal Damage by Extracellular Fine Particles

Lysosomes are acidic organelles with an intraluminal pH of 4.5-5 containing over 60 different hydrolases such as proteases, nucleases, lipases, and glycosidases (Ballabio and Bonifacino 2020; Xu and Ren 2015). They play a central role in the degradation of materials delivered from either outside and inside the cell via the autophagic and endocytic pathways. As lysosomes are the terminal components of the endocytic pathway, diverse extracellular materials are delivered to them for degradation through phagocytosis and endocytosis. Exogenous fine particulates such as silica and asbestos are known to be phagocytosed by macrophages. Numerous studies have reported that these particles can induce lysosomal membrane rupture, which can lead to oxidative stress, inflammation, indirect DNA damage, and apoptosis or necrosis (Bhabra et al. 2009; Deng et al. 2011; Nel et al. 2009; Xia et al. 2006) (Fig. 1). In addition to the exogenous particles from the environment, endogenous insoluble particles such as monosodium urate crystals and cholesterol crystals can be delivered to lysosomes and are known to induce lysosomal damage (Tall and Yvan-Charvet 2015; So and Martinon 2017) (Fig. 1). These endogenous crystals are associated with pathologies including gout, atherosclerosis, and the development of hyperuricemic nephropathy (Tall and Yvan-Charvet 2015; Freeman et al. 2013; Emmerson et al. 1990; Maejima et al. 2013). Recent reports also indicate that toxic protein aggregates associated with neurodegenerative diseases such as α-synuclein, amyloid-beta, mutant huntingtin, and tau fibrils are endocytosed from the extracellular space and impair endo-lysosomal membrane integrity (Freeman et al. 2013;

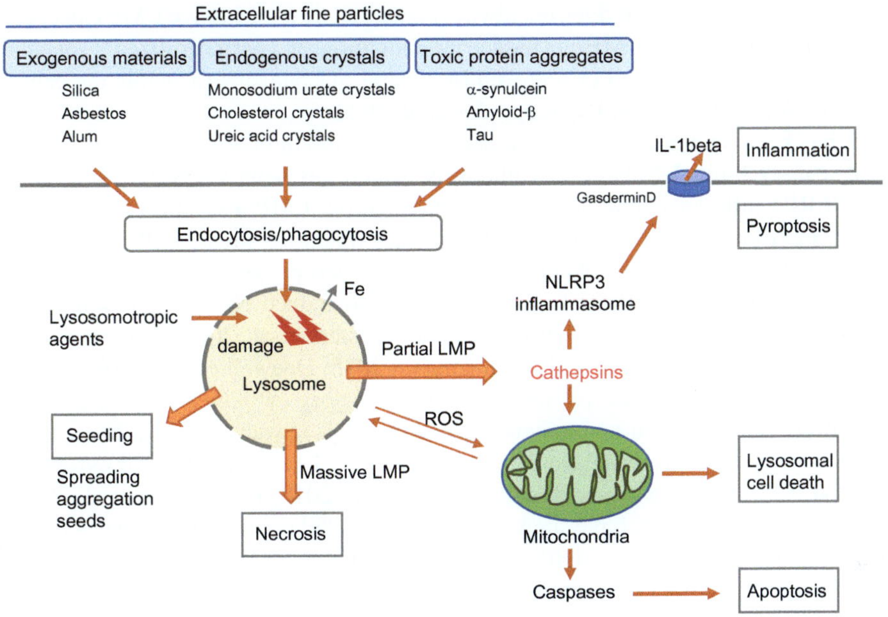

Fig. 1 Lysosomal membrane permeabilizations (LMP) by extracellular fine particles. Numerous extracellular fine particles including exogenous particles, endogenous crystals, and toxic protein aggregates can cause lysosomal damage and LMP. During partial LMP, specific cathepsins are released from the permeabilized lysosomes, resulting in activation of NLRP3 inflammasome, mitochondrial damage, and cell death such as apoptosis and pyroptosis. Massive lysosomal rupture results in uncontrolled release of cathepsins and other lysosomal contents and necrosis. Endocytosed protein aggregates transmit aggregation through the lysosomal rupture

Flavin et al. 2017), which potentially allows to release aggregation-prone proteins into the cytosol (Kakuda et al. 2024). Lysosomes are also known to be damaged by invading bacteria (Huang and Brumell 2014). Therefore, lysosomes are constantly at risk of damage, and when they are damaged, it is harmful and stressful for cells and tightly associated with diseases.

3 Lysosome Membrane Permeabilization (LMP)

One common form of lysosomal damage by various extracellular fine particles is lysosomal membrane permeabilization (LMP) (Fig. 1). LMP refers to the loss of membrane integrity, resulting in the release of luminal contents such as cathepsins into the cytosol (Wang et al. 2018; Aits and Jäättelä 2013). Different forms of stress including extracellular particles, lysosomotropic agents, pathogens, and lipids, reactive oxygen species are known to induce LMP (Wang et al. 2018; Aits and Jäättelä 2013). Although LMP has been described as a mechanism of apoptosis or necrosis, it has become evident that it does not always cause cell death. Massive LMP can cause cell death, while partial LMP can induce inflammation, and permeabilized lysosomal membrane can be repaired or removed by several mechanisms to enable the cell to survive (Wang et al. 2018; Papadopoulos et al. 2020).

It has been shown that LMP is associated with inflammatory responses. Partial LMP causes limited leakage of cathepsins, which has been suggested to activate the NLRP3 inflammasome, leading to inflammation. LMP also allows the release of H^+ and iron into the cytosol, which provokes oxidative stress, ROS production, and mitochondrial damage (Wang et al. 2018). In macrophages, phagocytosed silica, alum, or cholesterol crystals have been shown to induce NLRP3 inflammasome activation through LMP (Cassel et al. 2008; Dostert et al. 2008; Hornung et al. 2008; Yang et al. 2016; Tall and Yvan-Charvet 2015; Duewell et al. 2010). NLRP3 activation drives the robust release of proinflammatory cytokines including IL-1β, promoting inflammation and enhancing pathogenesis (Swanson et al. 2019; Place and Kanneganti 2018). LMP can also induce a type of cell death termed pyroptosis with NLRP3 inflammasome activation (Wang et al. 2018; Bergsbaken et al. 2009). Accumulating evidences have suggested that extracellular particles could trigger the onset and progression of various types of inflammatory diseases such as silicosis, asbestos, gout, pseudogout, atherosclerosis, and metabolic diseases through NLRP3 inflammasome activation following LMP. LMP-mediated cell death is also associated with diseases (Gómez-Sintes et al. 2016).

4 Lysosomal Damage Response

To prevent the leakage of lysosomal contents from damaged lysosomes and restore lysosomal function after damage, cells have several mechanisms for the repair or elimination of compromised lysosomes collectively called the lysosomal damage response (Papadopoulos and Meyer 2017). Lysosomal damage is induced by numerous factors including extracellular particles, lysosome damaging drugs, pathogens, changes in membrane lipid composition, and apoptotic regulators. Experimentally, lysosomotropic agents such as L-leucyl-L-leucine methyl ester (LLOMe) and Gly-Phe-β-naphthylamide (GPN) are often used for studying the lysosomal damage response. LLOMe condenses in lysosomes and forms membranolytic peptides via the action of cathepsin C. These peptides acutely permeabilize membranes. The effects are reversible upon washout of the reagent, which is useful for monitoring the process of lysosomal membrane repair (Papadopoulos and Meyer 2017; Thiele and Lipsky 1990).

Recently, at least six pathways involved in the response to lysosomal damage have been reported; (1) lysophagy, (2) ESCRT-mediated repair, (3) TFEB-mediated repair, (4) PITT-mediated repair, (5) Alix-mediated repair, and (6) stress granule-mediated repair. Lysophagy is a degradation process, while the other mechanisms are involved in the repair of injured membranes. In fact, the repair mechanisms have been shown to act in parallel. Since the discovery of lysophagy in 2013 (Maejima et al. 2013; Hung et al. 2013), accumulating data have led to the suggestion that small-scale lysosomal damage can be repaired by sealing the holes in the membrane. This acute response has been shown to be initiated mainly by Ca^{2+} release from damaged lysosomes. When repair fails, or damage is more profound, the organelle will be cleared by lysophagy initiated by the exposure of glycoproteins in the lysosomal rumen, to avoid further damage from leaked lysosomal contents. Each of these processes is described below.

5 Autophagy

Autophagy is defined as an evolutionarily conserved catabolic process through which intracellular materials are delivered to lysosomes for degradation (Morishita and Mizushima 2019; Yamamoto et al. 2023). Autophagy is induced by various cellular stresses such as nutrient starvation, hypoxia, organelle damage, and invading bacteria, and plays diverse physiological and pathological roles as a prosurvival mechanism of cells. At present, three major forms of autophagy have been described: macroautophagy, microautophagy, and chaperone-mediated autophagy (CMA) (Fig. 2). During macroautophagy (hereinafter referred to as autophagy), cytoplasmic materials are sequestered within a double-membraned vesicle, termed an autophagosome. The autophagosome fuses with the lysosome containing many kinds of hydrolases, which enables degradation of the contents of the autophagosome. Autophagy was initially described as a non-selective, bulk degradation system, but it is now known that autophagy can also selectively degrade intracellular materials including organelles, protein aggregates and infecting bacteria, depending on the inducing stresses.

By contrast, neither CMA nor microautophagy involves autophagosomes. CMA is a process by which individual cytosolic proteins with a KFERQ-like motif are transported into the lysosome for degradation by binding to chaperone proteins (Kaushik and Cuervo 2018). Microautophagy is characterized by invagination of the lysosomal membrane. During microautophagy, cytoplasmic materials are directly sequestered via lysosomal membrane invagination, and the autophagic cargos is then degraded in the lysosomal lumen (Wang et al. 2023). Recent studies have shown that macroautophagy and microautophagy are involved in the lysosomal damage response (discussed later).

In macroautophagy, an ubiquitin-like conjugation system that mediates the conjugation of ATG8 proteins (LC3A/B/C and GABARAP/L1/L2) to phosphatidylethanolamine (PE) plays an important role in autophagosome formation (Fig. 3). ATG8 is conjugated to the double-membrane of the autophagosome. However, it has been reported that the same ATG8 conjugation system plays non-canonical roles independent of autophagy, which is termed the conjugation of ATG8 to single membranes (CASM) of the endo-lysosomal compartments (Durgan and Florey 2022) (Fig. 3). CASM is implicated in a wide range of processes such as phagocytosis, endocytosis, and response to infection, and considered to be involved in immunity, cancer, and neurodegeneration. However, the precise function is not clear

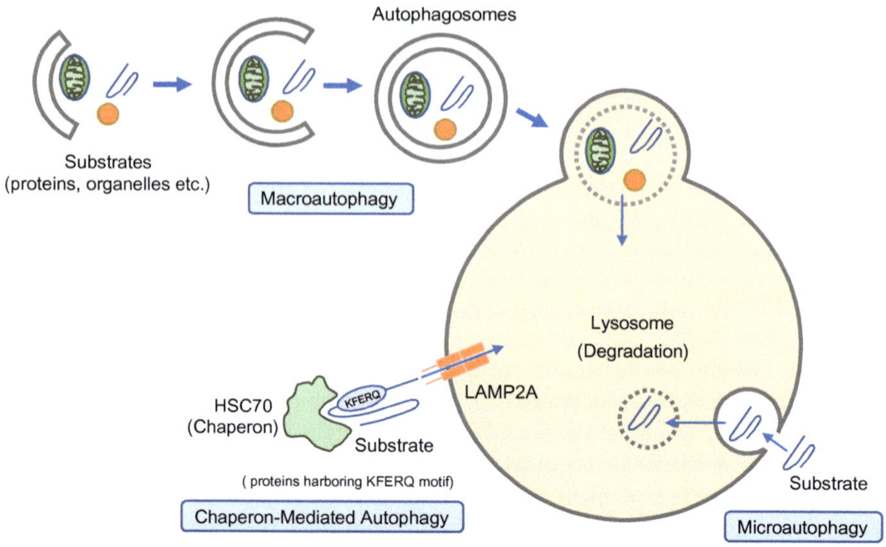

Fig. 2 Three major forms of autophagy. Autophagy is an evolutionarily conserved catabolic process through which intracellular materials are delivered to lysosomes for degradation. Three major forms of autophagy have been described; macroautophagy, microautophagy, and chaperon-mediated autophagy (CMA)

yet. Recent studies have demonstrated that both canonical autophagy and CASM are involved in the lysosomal damage response (Durgan and Florey 2022; Cross et al. 2023).

6 Lysophagy

Lysophagy is a type of macroautophagy in which damaged lysosomes are selectively sequestered by autophagosomes for degradation by other intact lysosomes (Maejima et al. 2013; Hung et al. 2013) (Fig. 4). Diverse factors such as LLOME, silica crystals, urate crystals, and endocytosed proteins aggregates induce lysophagy (Maejima et al. 2013; Kakuda et al. 2024; Papadopoulos et al. 2017). When the lysosomal membrane is damaged, lysosomal membrane become ubiquitinated to initiate lysophagy (Kakuda et al. 2024; Fujita et al. 2013). It is well known that ubiquitin is covalently conjugated to substrate proteins by a cascade of enzymes called E1 ubiquitin-activating enzyme, E2 ubiquitin-conjugating enzyme, and E3 ubiquitin ligase. In lysophagy, the E2 ubiquitin-conjugate enzymes UBE2QL1 and potentially UBE2N have been shown to be involved in the ubiquitination of damaged lysosomal membranes (Koerver et al. 2019; Shima et al. 2023). However, it is not known how E2 enzymes are recruited to damaged lysosomes. Meanwhile, several studies have shown that E3 ligases are recruited to damaged lysosomes in a manner depend on

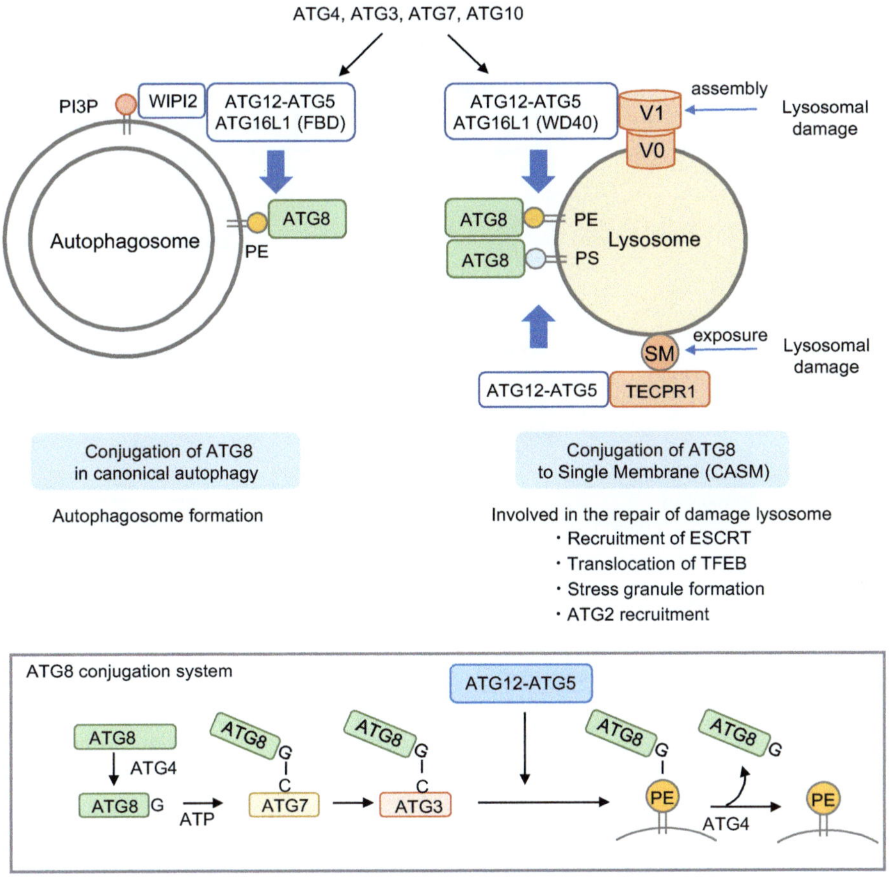

Fig. 3 Features of Atg8 lipidation in canonical autophagy and CASM upon lysosomal damage. Schematic shows canonical and non-canonical roles of ATG8 lipidation in the lysosomal damage response

exposure of glycans that are normally localized in the luminal side of lysosomes. So far, TRIM16, SCF^{FBXO27}, SCF^{FBXO2}, and CUL4A have been reported as E3 ubiquitin ligases in lysophagy (Chauhan et al. 2016; Yoshida et al. 2017; Teranishi et al. 2022; Liu et al. 2020). Exposed glycans are sensed by cytosolic lectins such as galectin-1, -3, -8, and -9, that bind β-galactosides (Chauhan et al. 2016; Thurston et al. 2012; Jia et al. 2018, 2020). Among these galectins, galectin-3 has been shown to recruit TRIM16 (Chauhan et al. 2016). Exposed glycans are also directly sensed by FBXO27 or FBXO2 in SCF E3 ligase (Yoshida et al. 2017). Membrane rupture also allows the recruitment of CUL4A E3 ligase complex to damaged lysosomes and ubiquitination of the luminal side of LAMP2, a major lysosomal membrane-resident glycoprotein (Teranishi et al. 2022).

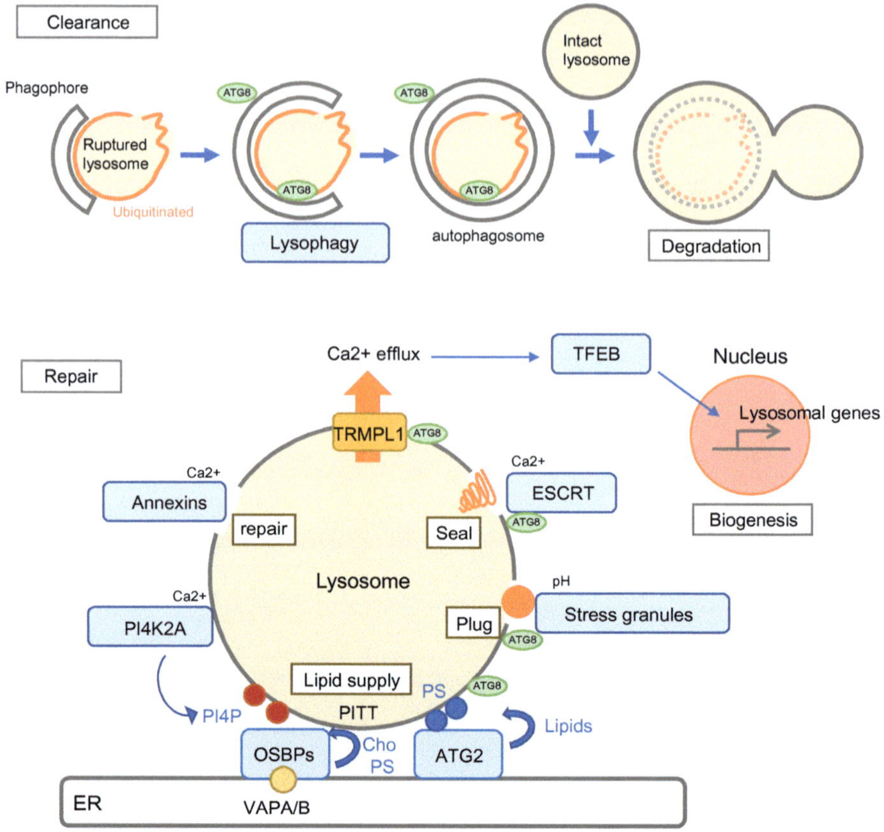

Fig. 4 Pathways in the lysosomal damage response. Small-scale lysosomal damage can be repaired by sealing the holes in the membrane by several early-acting responses including ESCRT-mediated pathway, TFEB-mediated pathway, Annexin-mediated pathway, PITT-mediated pathway, and stress granule-mediated pathway. When damage is more profound, the damaged organelle is cleared by lysophagy

Ubiquitin on the damaged lysosomes then recruits autophagy machinery and autophagy receptors. ATG16L1 and the ULK1 complex are recruited to damaged lysosomes through direct binding with ubiquitin for autophagosome formation (Fujita et al. 2013). Ubiquitin in also recruits autophagy receptors, which links the damaged lysosome to the phagophore. Upon lysosomal damage by LLOMe, all five receptors p62, Nbr1, TAX1BP1, NDP52 and OPTN were observed to be recruited to damaged lysosomes (Shima et al. 2023; Eapen et al. 2021). Among them, TAX1BP1, NDP52, and OPTN have been shown to be important for efficient lysophagy. In addition, Tank binding kinase 1 (TBK1) is recruited to damaged lysosomes in a calcium-dependent manner that is required for the recruitment of OPTN (Shima et al. 2023; Eapen et al. 2021). Moreover, ubiquitination triggers the recruitment of the AAA-ATPase VCP/p97 (Papadopoulos et al. 2017). Given the well-known function of p97 in removing ubiquitinated proteins from cellular structures, it has been predicted that p97 would promote lysophagy by removing ubiquitinated proteins from damaged lysosomes. A recent study has shown that ubiquitinated calponin-2, which functions to assist phagophore formation by regulating actin, is removed by p97. The timely removal of calponin-2 from damaged lysosomes is considered to be important for efficient lysophagy (Kravić et al. 2022).

The physiological importance of lysophagy have also been described. Various groups including ours own have shown that lysosomal damage by endocytosed neurotoxic aggregates induces lysophagy, and lysophagy prevents seeding of aggregation in the affected cells (Kakuda et al. 2024; Papadopoulos et al. 2017; Bussi et al. 2018; Falcon et al. 2018). Lysophagy is also reported to prevent exacerbation of crystalline nephropathy with lysosomal damage by urate crystals in vivo (Maejima et al. 2013).

7 ESCRT-Mediated Repair

If damage is limited, cells can repair the injury to the lysosomal membrane, rather than remove the damaged lysosome (Fig. 4). One mechanism for this is endosomal sorting complex required for transport (ESCRT)-mediated lysosomal membrane repair. ESCRT machinery consists of the functionally distinct complexes ESCRT-0, ESCRT-I, ESCRT-II, and ESCRT-III, and the AAA + ATPase VPS4, and plays essential roles in processes such as endosomal trafficking, viral budding, multi vesicular body formation (MVB), and plasma membrane sealing (Vietri et al. 2020). Upon lysosomal damage, ESCRT-I, ESCRT-III, and ALIX are rapidly recruited to holes in the lysosomal membrane by Ca^{2+} release from damaged lysosomes (Skowyra et al. 2018; Radulovic et al. 2018). Galectin-3 is also shown to recruit ALIX to damaged lysosomes (Jia et al. 2020). Diverse factors such as LLOME, silica crystals, and endocytosed protein aggregates induce the recruitment of ESCRTs. Notably, ESCRT proteins were found to be recruited within only minutes after LLOMe treatment; by contrast, ubiquitination and lysophagy was observed ~30 min after LLOMe treatment (Skowyra et al. 2018; Radulovic et al. 2018). It was also shown that ESCRTs preferentially sensed small ruptures in the lysosomal membrane. Regarding the timing and degree of damage, ESCRT-mediated repair is considered as an acute response to repair relatively small ruptures; by contrast, lysophagy is a late-acting response to remove damaged lysosomes for degradation. Although it remains unclear how ESCRT seals the injured lysosomal membrane, the ability of ESCRT proteins to promote inward constriction of membrane edges suggests that this function is responsible for resealing holes on the lysosomal membrane.

Recently, our group demonstrated that ESCRT-mediated lysosomal repair is part of microautophagy (Ogura et al. 2023). As mentioned above, microautophagy is a type of autophagy in which the lysosomal membrane enwraps and transports cytoplasmic components into the lumen of the lysosomes for degradation, and has been reported to occur in an ESCRT-dependent or ESCRT-independent manner (Wang et al. 2023). Upon lysosomal damage, microautophagy is initiated by CASM on the lysosomal membrane, which is required for ESCRT assembly on damaged lysosomes via interaction with ALIX. The AGC kinase STK38 recruits VPS4 for the disassembly of ESCRTs in the late step of membrane invagination (Ogura et al. 2023).

8 TFEB-Mediated Repair

Another branch of the lysosomal damage response is the activation of transcription factor EB (TFEB). TFEB, one of four members of the MiT/TFE family of basic helix-loop-helix leucine zipper transcription factors, is a master regulator of gene expression involved in lysosome biogenesis and autophagy. The mechanistic target of rapamycin complex 1 (mTORC1) is a major kinase known to phosphorylate TFEB and regulate its intracellular localization. Under normal conditions, TFEB is retained and inactivated in the cytoplasm by phosphorylation by mTORC1. Upon starvation, mTORC1 is inactivated which leads to the dephosphorylation and activation of TFEB. Activated TFEB translocates from the cytosol to the nucleus, where it upregulates the transcription of genes related to lysosomal biogenesis and autophagy (Puertollano et al. 2018). TFEB is also known to be activated in response to lysosomal stress or damage to promote lysosomal biogenesis (Chauhan et al. 2016; Jia et al. 2018).

Our recent study showed that lysosomal damage-induced TFEB activation depends on the unconventional lipidation of LC3 (Nakamura et al. 2020). The leakage of Ca^{2+} from damaged lysosomes induces the conjugation of LC3 to damaged lysosomes. Subsequently, lipidated LC3 interacts with the calcium channel MCOLN1/TRPML1 on the lysosome, which is assumed to induce larger amounts of calcium efflux, leading to TFEB activation. GABARAPs are also shown to be required for TRPML1-mediated TFEB activation (Goodwin et al. 2021). The exact mechanism by which TFEB is involved in lysosomal repair is not known. Presumably it is needed for the restoration of lysosomal function after damage, or the replacement of irreversibly damaged lysosomes with new ones through the activation of lysosomal biogenesis. The importance of TFEB-mediated lysosomal repair has also been shown in vivo. It was demonstrated that kidney injury induced by CaOx crystals was exacerbated in proximal tubule-specific TFEB knockout mice compared with that in a control (Nakamura et al. 2020). Moreover, human crystal nephropathy specimens show clear lysosomal damage and reduction in TFEB expression, highlighting the correlation between TFEB expression and progression of crystal nephropathy (Nakamura et al. 2020).

9 PITT-Mediated Repair

Two recent studies uncovered a mechanism of lysosomal membrane repair that involves lipid transfer to damaged lysosomes from the ER via a membrane contact site (Tan and Finkel 2022; Radulovic et al. 2022). This pathway is termed phosphoinositide-initiated membrane tethering and lipid transport (PITT) (Tan and Finkel 2022). During PITT, phosphatidylinositol 4-kinase type 2α (PI4K2A) is rapidly recruited to damaged lysosomes in a Ca^{2+}-dependent manner and generates high levels of phosphatidylinositol 4-phosphate (PtdIns4P) on the lysosomes. Accumulated PtdIns4P recruits OSBP and members of the ORP family, which interact with the ER-resident proteins VAPA and VAPB to form an ER-lysosome contact site to mediate the transfer of phosphatidylserine (PS) and cholesterol. It is thought that cholesterol supplied from the ER protects lysosomes from damage, and PS regulates the membrane binding of the lipid transporter ATG2, known to deliver phospholipids to phagophores, at the contact site to reseal the damaged membrane. Although the rapid accumulation of PtdIns4P on damaged lysosomes occurs with kinetics similar to that for ESCRT recruitment, these pathways have been shown to act independently. Interestingly, a recent study showed that lipidated LC3A/B on the damaged lysosome interacts with ATG2, suggesting that the recruitment of ATG2 to the damaged lysosomes by ATG8 lipidation may occurs in parallel with PITT-mediated repair (Cross et al. 2023). How the delivery of PS, cholesterol, and other lipids from the ER repairs lysosomal injury remains to be elucidated. Presumably, an ER-mediated lipid supply might be necessary for sealing the hole in the membrane or restoring the integrity of the lysosome after damage.

10 Annexin-Mediated Repair

Annexins are a superfamily of calcium and phospholipid binding proteins connected to various membrane-related events and cellular functions (Gerke et al. 2005). Annexins are known to be involved in plasma membrane repair, when the plasma membrane is disrupted (Koerdt et al. 2019; Boye and Nylandsted 2016). Plasma membrane repair is triggered by Ca^{2+} influx into the cytosol through the membrane pore. Numerous studies have suggested that annexins play roles in vesicle movement, fusion, and patch formation during membrane repair. A recent study showed that Annexin 1 and Annexin 2 are also important for lysosomal membrane repair (Yim et al. 2022), although the precise mechanism involved remains unclear. Upon lysosomal damage, Annexin 1 and Annexin 2 are recruited to damaged lysosomes in a calcium-dependent manner. Lysosomal membrane repair by Annexin 1 and Annexin 2 is independent of ESCRTs. ESCRTs are considered to repair small membrane pores, by contrast, annexins preferentially localize to lysosomes with larger holes (>4.6 nm in diameter) to prevent excessive lysosomal leakage (Yim et al. 2022).

11 Stress Granule-Mediated Repair

Stress granules are membrane-less organelles formed in response to different stress stimuli (Alberti et al. 2019; Marcelo et al. 2021). Stress granules are condensates of RNA-binding proteins, translation factors, mRNAs, and other proteins. Stress granule assembly is essential for the cellular response to stress, and implicated in the regulation of translation, mRNA storage and other processes during stress (Marcelo et al. 2021). A recent study showed that lysosomal damage by LLOMe, GPN, and silica crystals triggers stress granule formation on the surface of lysosome, which contributes to mTOR inactivation (Jia et al. 2022). Of note, non-conventional GABARAP lipidation was needed for the recruitment of stress granule components to damaged lysosomes (Jia et al. 2022). More recently, another group observed that stress granules form rapidly and specifically at sites of membrane damage, by tracking the formation of stress granules using super-resolution microscopy (Bussi et al. 2023). This observation indicates that stress granules plug the holes in the lysosomal membrane. This process is required for lysosomal acidification, implying that local changes in milieu conditions, for example decrease in pH, trigger stress granule formation. Stress granule formation and membrane stabilization were required for efficient lysosomal membrane repair through ESCRT-mediated, Annexin-mediated, and PITT-mediated mechanisms. The study proposed a mechanism that stress granules stabilize the damaged membranes, plug the holes to prevent further leakage of the lysosomal contents, and then enable efficient repair damaged membranes (Bussi et al. 2023).

12 ATG8 Lipidation During the Lysosomal Damage Response

It is remarkable that autophagy machinery, especially ATG8 lipidation (both canonical autophagy and CASM), is involved in multiple pathways in the lysosomal damage response. (1) both macroautophagy and microautophagy occur in response to lysosomal damage. (2) CASM has multiple roles for lysosomal membrane repair (3) ATG2 functions in both lysophagy and PITT-mediated repair for lipid transport. How is ATG8 lipidation activated on the damaged lysosome membrane? So far, two mechanisms have been reported, mediated by ATG16L1 or TECPR1. In response to elevated lysosomal pH following damage, the V1 and V0 regions of V-ATPase are assembled on the lysosome to drive re-acidification, which enhances the recruitment of ATG16L (Cross et al. 2023). ATG16L1, in complex with ATG5 and ATG12, is recruited to the lysosomal membrane via binding of its C-terminal WD40 domain to the V-ATPase on the lysosome (Fletcher et al. 2018; Hooper et al. 2022). In parallel with this, the exposure of sphingomyelin on the cytosolic face of damaged membrane recruits TECPR1, which can bind to ATG12-ATG5 (Boyle et al. 2023;

Corkery et al. 2023; Kaur et al. 2023; Niekamp et al. 2022). Therefore, the ATG12–ATG5–ATG16L1 complex or ATG12–ATG5–TECPR1 complex catalyzes the lipid conjugation of LC3 on damaged lysosomes. Although no comprehensive overview has yet been obtained, it would be particularly interesting to understand how lipidated ATG8 acts on damaged lysosomes, whether ATG8 lipidation coordinates all pathways of the lysosomal damage response, and what are the precise downstream effects of ATG8 lipidation on damaged lysosomes. These will be clarified in future studies.

Finally, it should be noted that, from an experimental perspective, caution is required when interpreting the data related to ATG8 lipidation. The detection of the lipidated form of LC3-II is a widely used method to analyze autophagy. The amount of LC3-II or the number of LC3 puncta is relative to the amount of autophagosomes. However, this is not the case for lysophagy, because it is now known that the robust ATG8 lipidation occurs via the CASM pathway upon lysosomal damage. Knockout of ATG genes involved in the conjugation systems also affects both canonical autophagy and CASM upon lysosomal damage.

13 Conclusions

We have summarized the mechanisms by which cells maintain lysosomal integrity during membrane damage by extracellular fine particles. As mentioned above, most of the experiments have been performed using lysosome damaging agents such as LLOME or GPN to mimic physiological lysosomal damage, however, the mechanisms have also been shown to be shared with the responses to extracellular fine particles including environmental materials, endogenous crystals, and toxic protein aggregates. As elucidation of the lysosomal damage response is critical to understanding the toxicity of extracellular microparticles and their relationship to disease, future studies are warranted.

References

Aits, S. and M. Jäättelä, *Lysosomal cell death at a glance.* J Cell Sci, 2013. **126**(Pt 9): p. 1905–12.

Alberti, S., A. Gladfelter, and T. Mittag, *Considerations and Challenges in Studying Liquid-Liquid Phase Separation and Biomolecular Condensates.* Cell, 2019. **176**(3): p. 419–434.

Ballabio, A. and J.S. Bonifacino, *Lysosomes as dynamic regulators of cell and organismal homeostasis.* Nat Rev Mol Cell Biol, 2020. **21**(2): p. 101–118.

Bergsbaken, T., S.L. Fink, and B.T. Cookson, *Pyroptosis: host cell death and inflammation.* Nat Rev Microbiol, 2009. **7**(2): p. 99–109.

Bhabra, G., et al., *Nanoparticles can cause DNA damage across a cellular barrier.* Nat Nanotechnol, 2009. **4**(12): p. 876–83.

Bonam, S.R., F. Wang, and S. Muller, *Lysosomes as a therapeutic target.* Nat Rev Drug Discov, 2019. **18**(12): p. 923–948.

Boye, T.L. and J. Nylandsted, *Annexins in plasma membrane repair.* Biol Chem, 2016. **397**(10): p. 961–9.
Boyle, K.B., et al., *TECPR1 conjugates LC3 to damaged endomembranes upon detection of sphingomyelin exposure.* Embo j, 2023. **42**(17): p. e113012.
Bussi, C., et al., *Stress granules plug and stabilize damaged endolysosomal membranes.* Nature, 2023. **623**(7989): p. 1062–1069.
Bussi, C., et al., *Alpha-synuclein fibrils recruit TBK1 and OPTN to lysosomal damage sites and induce autophagy in microglial cells.* J Cell Sci, 2018. **131**(23).
Cassel, S.L., et al., *The Nalp3 inflammasome is essential for the development of silicosis.* Proc Natl Acad Sci U S A, 2008. **105**(26): p. 9035–40.
Chauhan, S., et al., *TRIMs and Galectins Globally Cooperate and TRIM16 and Galectin-3 Co-direct Autophagy in Endomembrane Damage Homeostasis.* Dev Cell, 2016. **39**(1): p. 13–27.
Corkery, D.P., et al., *An ATG12-ATG5-TECPR1 E3-like complex regulates unconventional LC3 lipidation at damaged lysosomes.* EMBO Rep, 2023. **24**(9): p. e56841.
Cross, J., et al., *Lysosome damage triggers direct ATG8 conjugation and ATG2 engagement via non-canonical autophagy.* J Cell Biol, 2023. **222**(12).
Deng, Z.J., et al., *Nanoparticle-induced unfolding of fibrinogen promotes Mac-1 receptor activation and inflammation.* Nat Nanotechnol, 2011. **6**(1): p. 39–44.
Dostert, C., et al., *Innate immune activation through Nalp3 inflammasome sensing of asbestos and silica.* Science, 2008. **320**(5876): p. 674–7.
Duewell, P., et al., *NLRP3 inflammasomes are required for atherogenesis and activated by cholesterol crystals.* Nature, 2010. **464**(7293): p. 1357–61.
Durgan, J. and O. Florey, *Many roads lead to CASM: Diverse stimuli of noncanonical autophagy share a unifying molecular mechanism.* Sci Adv, 2022. **8**(43): p. eabo1274.
Eapen, V.V., et al., *Quantitative proteomics reveals the selectivity of ubiquitin-binding autophagy receptors in the turnover of damaged lysosomes by lysophagy.* Elife, 2021. **10**.
Emmerson, B.T., et al., *Reaction of MDCK cells to crystals of monosodium urate monohydrate and uric acid.* Kidney Int, 1990. **37**(1): p. 36–43.
Falcon, B., et al., *Galectin-8-mediated selective autophagy protects against seeded tau aggregation.* J Biol Chem, 2018. **293**(7): p. 2438–2451.
Flavin, W.P., et al., *Endocytic vesicle rupture is a conserved mechanism of cellular invasion by amyloid proteins.* Acta Neuropathol, 2017. **134**(4): p. 629–653.
Fletcher, K., et al., *The WD40 domain of ATG16L1 is required for its non-canonical role in lipidation of LC3 at single membranes.* Embo j, 2018. **37**(4).
Freeman, D., et al., *Alpha-synuclein induces lysosomal rupture and cathepsin dependent reactive oxygen species following endocytosis.* PLoS One, 2013. **8**(4): p. e62143.
Fujita, N., et al., *Recruitment of the autophagic machinery to endosomes during infection is mediated by ubiquitin.* J Cell Biol, 2013. **203**(1): p. 115–28.
Gerke, V., C.E. Creutz, and S.E. Moss, *Annexins: linking Ca2+ signalling to membrane dynamics.* Nat Rev Mol Cell Biol, 2005. **6**(6): p. 449–61.
Gómez-Sintes, R., M.D. Ledesma, and P. Boya, *Lysosomal cell death mechanisms in aging.* Ageing Res Rev, 2016. **32**: p. 150–168.
Goodwin, J.M., et al., *GABARAP sequesters the FLCN-FNIP tumor suppressor complex to couple autophagy with lysosomal biogenesis.* Sci Adv, 2021. **7**(40): p. eabj2485.
Hooper, K.M., et al., *V-ATPase is a universal regulator of LC3-associated phagocytosis and non-canonical autophagy.* J Cell Biol, 2022. **221**(6).
Hornung, V., et al., *Silica crystals and aluminum salts activate the NALP3 inflammasome through phagosomal destabilization.* Nat Immunol, 2008. **9**(8): p. 847–56.
Huang, J. and J.H. Brumell, *Bacteria-autophagy interplay: a battle for survival.* Nat Rev Microbiol, 2014. **12**(2): p. 101–14.
Hung, Y.H., et al., *Spatiotemporally controlled induction of autophagy-mediated lysosome turnover.* Nat Commun, 2013. **4**: p. 2111.

Jia, J., et al., *Galectins Control mTOR in Response to Endomembrane Damage.* Mol Cell, 2018. **70**(1): p. 120–135.e8.

Jia, J., et al., *Galectin-3 Coordinates a Cellular System for Lysosomal Repair and Removal.* Dev Cell, 2020. **52**(1): p. 69–87.e8.

Jia, J., et al., *Stress granules and mTOR are regulated by membrane atg8ylation during lysosomal damage.* J Cell Biol, 2022. **221**(11).

Kakuda, K., et al., *Lysophagy protects against propagation of α-synuclein aggregation through ruptured lysosomal vesicles.* Proc Natl Acad Sci U S A, 2024. **121**(1): p. e2312306120.

Kaur, N., et al., *TECPR1 is activated by damage-induced sphingomyelin exposure to mediate noncanonical autophagy.* Embo j, 2023. **42**(17): p. e113105.

Kaushik, S. and A.M. Cuervo, *The coming of age of chaperone-mediated autophagy.* Nat Rev Mol Cell Biol, 2018. **19**(6): p. 365–381.

Koerdt, S.N., A.P.K. Ashraf, and V. Gerke, *Annexins and plasma membrane repair.* Curr Top Membr, 2019. **84**: p. 43–65.

Koerver, L., et al., *The ubiquitin-conjugating enzyme UBE2QL1 coordinates lysophagy in response to endolysosomal damage.* EMBO Rep, 2019. **20**(10): p. e48014.

Kravić, B., et al., *Ubiquitin profiling of lysophagy identifies actin stabilizer CNN2 as a target of VCP/p97 and uncovers a link to HSPB1.* Mol Cell, 2022. **82**(14): p. 2633–2649.e7.

Leung, C.C., I.T. Yu, and W. Chen, *Silicosis.* Lancet, 2012. **379**(9830): p. 2008–18.

Liu, E.A., et al., *Fbxo2 mediates clearance of damaged lysosomes and modifies neurodegeneration in the Niemann-Pick C brain.* JCI Insight, 2020. **5**(20).

Maejima, I., et al., *Autophagy sequesters damaged lysosomes to control lysosomal biogenesis and kidney injury.* Embo j, 2013. **32**(17): p. 2336–47.

Marcelo, A., et al., *Stress granules, RNA-binding proteins and polyglutamine diseases: too much aggregation?* Cell Death Dis, 2021. **12**(6): p. 592.

Morishita, H. and N. Mizushima, *Diverse Cellular Roles of Autophagy.* Annu Rev Cell Dev Biol, 2019. **35**: p. 453–475.

Nakamura, S., et al., *LC3 lipidation is essential for TFEB activation during the lysosomal damage response to kidney injury.* Nat Cell Biol, 2020. **22**(10): p. 1252–1263.

Nel, A.E., et al., *Understanding biophysicochemical interactions at the nano-bio interface.* Nat Mater, 2009. **8**(7): p. 543–57.

Niekamp, P., et al., *Ca(2+)-activated sphingomyelin scrambling and turnover mediate ESCRT-independent lysosomal repair.* Nat Commun, 2022. **13**(1): p. 1875.

Ogura, M., et al., *Microautophagy regulated by STK38 and GABARAPs is essential to repair lysosomes and prevent aging.* EMBO Rep, 2023. **24**(12): p. e57300.

Papadopoulos, C. and H. Meyer, *Detection and Clearance of Damaged Lysosomes by the Endo-Lysosomal Damage Response and Lysophagy.* Curr Biol, 2017. **27**(24): p. R1330–r1341.

Papadopoulos, C., et al., *VCP/p97 cooperates with YOD1, UBXD1 and PLAA to drive clearance of ruptured lysosomes by autophagy.* Embo j, 2017. **36**(2): p. 135–150.

Papadopoulos, C., B. Kravic, and H. Meyer, *Repair or Lysophagy: Dealing with Damaged Lysosomes.* J Mol Biol, 2020. **432**(1): p. 231–239.

Place, D.E. and T.D. Kanneganti, *Recent advances in inflammasome biology.* Curr Opin Immunol, 2018. **50**: p. 32–38.

Pozzi, R., et al., *Inflammatory mediators induced by coarse (PM2.5–10) and fine (PM2.5) urban air particles in RAW 264.7 cells.* Toxicology, 2003. **183**(1–3): p. 243–54.

Puertollano, R., et al., *The complex relationship between TFEB transcription factor phosphorylation and subcellular localization.* Embo j, 2018. **37**(11).

Radulovic, M., et al., *Cholesterol transfer via endoplasmic reticulum contacts mediates lysosome damage repair.* Embo j, 2022. **41**(24): p. e112677.

Radulovic, M., et al., *ESCRT-mediated lysosome repair precedes lysophagy and promotes cell survival.* Embo j, 2018. **37**(21).

Shima, T., et al., *The TMEM192-mKeima probe specifically assays lysophagy and reveals its initial steps.* J Cell Biol, 2023. **222**(12).

Skowyra, M.L., et al., *Triggered recruitment of ESCRT machinery promotes endolysosomal repair.* Science, 2018. **360**(6384).

So, A.K. and F. Martinon, *Inflammation in gout: mechanisms and therapeutic targets.* Nat Rev Rheumatol, 2017. **13**(11): p. 639–647.

Swanson, K.V., M. Deng, and J.P. Ting, *The NLRP3 inflammasome: molecular activation and regulation to therapeutics.* Nat Rev Immunol, 2019. **19**(8): p. 477–489.

Tall, A.R. and L. Yvan-Charvet, *Cholesterol, inflammation and innate immunity.* Nat Rev Immunol, 2015. **15**(2): p. 104–16.

Tan, J.X. and T. Finkel, *A phosphoinositide signalling pathway mediates rapid lysosomal repair.* Nature, 2022. **609**(7928): p. 815–821.

Teranishi, H., et al., *Identification of CUL4A-DDB1-WDFY1 as an E3 ubiquitin ligase complex involved in initiation of lysophagy.* Cell Rep, 2022. **40**(11): p. 111349.

Thiele, D.L. and P.E. Lipsky, *Mechanism of L-leucyl-L-leucine methyl ester-mediated killing of cytotoxic lymphocytes: dependence on a lysosomal thiol protease, dipeptidyl peptidase I, that is enriched in these cells.* Proc Natl Acad Sci U S A, 1990. **87**(1): p. 83–7.

Thurston, T.L., et al., *Galectin 8 targets damaged vesicles for autophagy to defend cells against bacterial invasion.* Nature, 2012. **482**(7385): p. 414–8.

Vietri, M., M. Radulovic, and H. Stenmark, *The many functions of ESCRTs.* Nat Rev Mol Cell Biol, 2020. **21**(1): p. 25–42.

Wang, F., R. Gómez-Sintes, and P. Boya, *Lysosomal membrane permeabilization and cell death.* Traffic, 2018. **19**(12): p. 918–931.

Wang, L., D.J. Klionsky, and H.M. Shen, *The emerging mechanisms and functions of microautophagy.* Nat Rev Mol Cell Biol, 2023. **24**(3): p. 186–203.

Xia, T., et al., *Comparison of the abilities of ambient and manufactured nanoparticles to induce cellular toxicity according to an oxidative stress paradigm.* Nano Lett, 2006. **6**(8): p. 1794–807.

Xu, H. and D. Ren, *Lysosomal physiology.* Annu Rev Physiol, 2015. **77**: p. 57–80.

Yamamoto, H., S. Zhang, and N. Mizushima, *Autophagy genes in biology and disease.* Nat Rev Genet, 2023. **24**(6): p. 382–400.

Yang, M., et al., *Macrophages participate in local and systemic inflammation induced by amorphous silica nanoparticles through intratracheal instillation.* Int J Nanomedicine, 2016. **11**: p. 6217–6228.

Yim, W.W., H. Yamamoto, and N. Mizushima, *Annexins A1 and A2 are recruited to larger lysosomal injuries independently of ESCRTs to promote repair.* FEBS Lett, 2022. **596**(8): p. 991–1003.

Yoshida, Y., et al., *Ubiquitination of exposed glycoproteins by SCF(FBXO27) directs damaged lysosomes for autophagy.* Proc Natl Acad Sci U S A, 2017. **114**(32): p. 8574–8579.

Neurodegenerative Disorder and Fine Particulate Matter

Hironori Kawahara and Rikinari Hanayama

Abstract Extracellular vesicles (EVs), diverse membranous vesicles secreted by cells, include microvesicles, exosomes, and other cell-specific types. They efficiently deliver proteins and nucleic acids to distal parts and are implicated in the pathology of neurodegenerative disorders. Additionally, long-term exposure to extracellular microparticles, notably particulate matter (PM) 2.5, is suspected to induce neuroinflammation via oxidative stress mechanisms. Production of macrovesicles relies on the ARRDC1 and ARF6/RhoA pathways, whereas exosome production involves both ESCRT-dependent and ESCRT-independent pathways. In neurodegenerative disorders, EVs play various roles: microglia-derived EVs activate endothelial cells and neurons in stroke models, EVs accelerate α-synuclein aggregation and hinder autophagy in Parkinson's disease, patient-derived muscle cell small Extracellular vesicles (sEVs) worsen motor neuron death in Amyotrophic lateral sclerosis (ALS), and microglia-derived EVs influence neuronal transmission through the hippocampus, leading to synaptic spine reduction in Alzheimer's disease. However, the precise mechanisms underlying the involvement of EVs in disease onset remain largely unknown, emphasizing the need for further investigations.

Keywords Extracellular vesicles · Fine particulate matter · Neurological diseases

1 Introduction

Extracellular vesicles (EVs) are membrane-bound vesicles secreted by various cells. They exhibit diverse compositions, with membrane proteins and lipids on their surfaces, and cargo that includes proteins and nucleic acids within their lumens. These EVs play a pivotal role in intercellular communication by being taken up

H. Kawahara · R. Hanayama (✉)
Department of Immunology, Graduate School of Medical Sciences, Kanazawa University, Kanazawa 920-8640, Ishikawa, Japan
e-mail: hanayama@med.kanazawa-u.ac.jp

WPI Nano Life Science Institute (NanoLSI), Kanazawa University, Kakuma, Kanazawa 920-1192, Ishikawa, Japan

Y. Baba et al. (eds.), *Extracellular Fine Particles*,
https://doi.org/10.1007/978-981-97-7067-0_14

by recipient cells. Classification of EVs is an ongoing process; however, they are broadly categorized into two main types: microvesicles and exosomes, with the existence of cell-specific vesicles such as apoptotic bodies. Research targeting EVs has gained momentum, not only on their potential as biomarkers related to diseases of the nervous system but also on their role in disease pathogenesis. Furthermore, chronic exposure to ambient fine particulate matter has been reported as a potential etiological factor in neurodegenerative diseases. Here, we provide an overview of the production pathways of extracellular vesicles, their relationship with neurological diseases, and the impact of chronic exposure to extracellular microparticles in the context of these diseases.

2 EVs and Neural EVs and Ambient Fine Particulate Matter

EVs are broadly categorized into approximately two types; however, their experimental isolation is challenging, and specific molecular markers for precise subclassification are lacking. Consequently, in scientific papers, EVs are often described based on particle size, with small EVs (sEVs) of approximately 100 nm and large EVs encompassing all other particle sizes. For instance, it is common to refer to a sedimentation product following centrifugation of cell culture supernatants at approximately 10,000 × *g* and after impurities have been removed as large EVs, while a sediment obtained by ultracentrifugation at approximately 100,000 × *g* of the same supernatant is termed sEVs (Kowal et al. 2016). When EVs are isolated from cultured cells, the presence of a few dead cells results in the production of large EVs, which are often referred to as microvesicles.

Microvesicles, also known as ectosomes, are produced through the budding and pinching of the cell membrane and encapsulate various cell membrane components (Doyle and Wang 2019) (Fig. 1). Microvesicle production is primarily driven by two known pathways: the ARRDC1-dependent and ARF6/RhoA-dependent pathways. The ARRDC1-dependent pathway involves the ESCRT (Endosomal Sorting Complex Required for Transport) complexes, which functions in viral budding and exosome production (Nabhan et al. 2012). ARRDC1 also facilitates recruitment of Tsg101, another exosome marker and member of the ESCRT-I complex, to the cell membrane surface. Subsequently, Tsg101 recruits the ESCRT-III complex to form filamentous structures that encircle the budding site of the vesicle. Vsp4 then cleaves the vesicle from the cell membrane and releases it into the microvesicle. In the ARF6/RhoA-dependent pathway, a phosphorylation cascade of small GTPases (e.g., ARF6 and RhoA) is initiated, leading to rearrangement of actin and myosin cytoskeletal proteins. Myosin, which is involved in muscle contraction, is a motor protein present on actin filaments. Inhibitors of myosin or actin negatively affect microvesicle release. ARF6 activates the ERK signaling pathway via phospholipase D (PLD), eventually stimulating myosin contraction through Myosin Light Chain

Kinase (MLCK) to facilitate microvesicle budding (Muralidharan-Chari et al. 2009). Conversely, the phosphorylation signal of RhoA is transmitted to Rho-associated protein kinase (ROCK) and LIM kinase (LIMK), inducing inactivation (phosphorylation) of cofilin, a protein involved in actin polymerization (Schlienger et al. 2014). This, in turn, promotes actin rearrangement and facilitates release of microvesicles (Li et al. 2012). ROCK also functions in a crosstalk that promotes phosphorylation of MLCK. In summary, reorganization and contraction of the cytoskeleton play crucial roles in microvesicle formation and release.

Exosomes are membrane vesicles produced via endosomes (Fig. 1) that incorporatc various molecules, including nucleic acids (DNA and RNA) and proteins, suggesting their role in long-distance interorgan communication (Fig. 2) (Doyle and Wang 2019). Two primary pathways for exosome biogenesis are known: ESCRT-dependent and ESCRT-independent pathways. The ESCRT complexes play a crucial role in the formation of multivesicular bodies (MVBs). These MVBs are generated from the endosomes when the endosomal membrane undergoes inward budding, leading to the formation of small vesicles inside larger endosomes. When MVBs fuse with lysosomes, their vesicles are degraded. However, when they fuse with the cell membrane, vesicles are released into the extracellular space and are termed as exosomes. Within the ESCRT-dependent pathway, exosome markers such as Syntenin-1 and Alix coordinate with ESCRT-III complexes, promoting inward budding of the endosome membrane to form MVBs. Notably, the absence of Tsg-101, a subunit of ESCRT, reduces secretion of sEVs (Colombo et al. 2013). The ESCRT-independent pathway encompasses tetraspanin proteins and sphingolipids, with the tetraspanin proteins such as CD9 and CD81, which are prominent exosome markers, spanning the membrane four times. These tetraspanins form large complexes by associating with other membrane proteins such as integrins, creating tetraspanin-enriched microdomains (TEMs). A TEM assembly model was proposed to play a role in MVB formation (Hemler 2005). Exosomal membranes tend to concentrate sphingolipids and inhibit n-SMase 2, which synthesizes ceramides from sphingomyelin and reduces exosome production (Trajkovic et al. 2008). Additionally, a low-molecular-weight G protein rab31 resides in lipid rafts, interacts with exosome markers such as flotillin, and associates with ceramides. This interaction with ceramides inhibits MVB and lysosomal fusion, thereby influencing MVB formation (Wei et al. 2021).

In contrast, fusion of MVBs with the autophagic vesicles, autophagosomes, creates amphisomes. These amphisomes merge with the cell membrane, enabling their contents to be secreted as exosomes. This process has been demonstrated through analysis of α-synuclein, the main causative molecule in Parkinson's disease (Minakaki et al. 2018). In recent years, in addition to microvesicles and exosomes, the existence of vesicles isolated using asymmetric flow field-flow fractionation-based methods (Zhang and Lyden 2019), such as exomeres or supermeres obtained through repeated ultracentrifugation (Zhang et al. 2021a), has been proposed (Fig. 1). However, numerous uncertainties remain, including their biogenesis pathways and biological significance, and further analysis, including the confirmation of their existence, is eagerly anticipated.

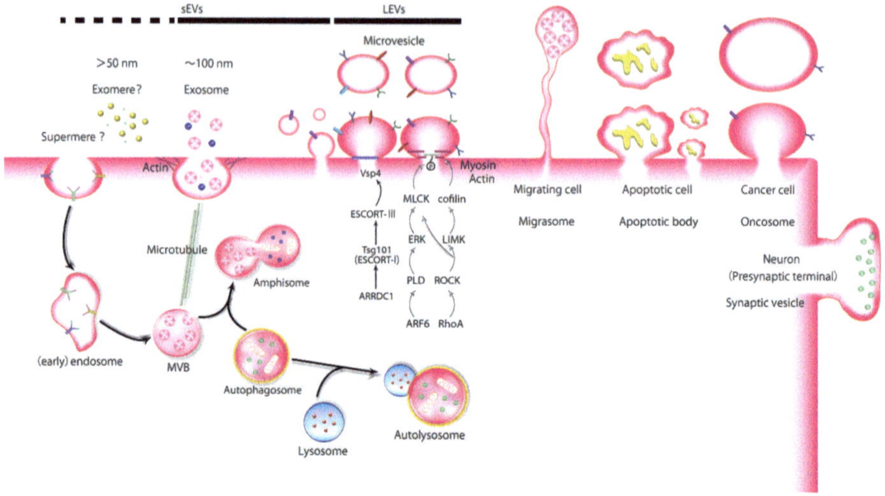

Fig. 1 Biogenesis pathway of EVs
Microvesicle Production Pathway: The ARRDC1 pathway involves a sequence of reactions starting with ARRDC1, followed by Tsg101 (ESCORT-I), ESCORT-III, and Vsp4. The ARF6-RhoA pathway involves reactions of ARF6, PLD, ERK, MLCK, and myosin contraction in that order, and of RhoA, ROCK, LIMK, cofilin, as well as inhibition of actin stretching. ROCK further stimulates MLCK, leading to the production of microvesicles via budding from the cell membrane
Exosome Production Pathway: Exosomes are released through the fusion of multivesicular bodies or amphisomes with the cell membrane via cellular scaffold molecules, such as microtubules. Exomeres and supermeres are vesicles with unknown production pathways. Depending on the cell state and type, various vesicles, such as migrasomes (fibrous structures containing various vesicles in migrating cells), apoptotic bodies (cell fragments in apoptotic cells), oncosomes in cancer cells, and synaptic vesicles in neurons are defined broadly as small EVs (sEVs). Large EVs (LEVs) are also shown

Various types of vesicles are named according to the cell type from which they originate. Apoptotic bodies are relatively large vesicles resulting from cell fragmentation during apoptosis. Migrasomes, observed at the leading edge of migrating cells, are formed by the assembly of proteins such as tetraspanins (TEMs) as mentioned earlier (Huang et al. 2019). In cancer cells, oncosomes serve as particularly large vesicles, whereas in neurons, synaptic vesicles fall under the broader classification of EVs (Fig. 1).

3 Neural EVs and Fine Particulate Matter

Numerous reports have suggested obtaining information on the central nervous system from information on blood using neuronal cell-specific EVs markers. However, caution is advised in their usage due to reports questioning the neuronal specificity of EVs markers, such as L1CAM (Norman et al. 2021), in addition

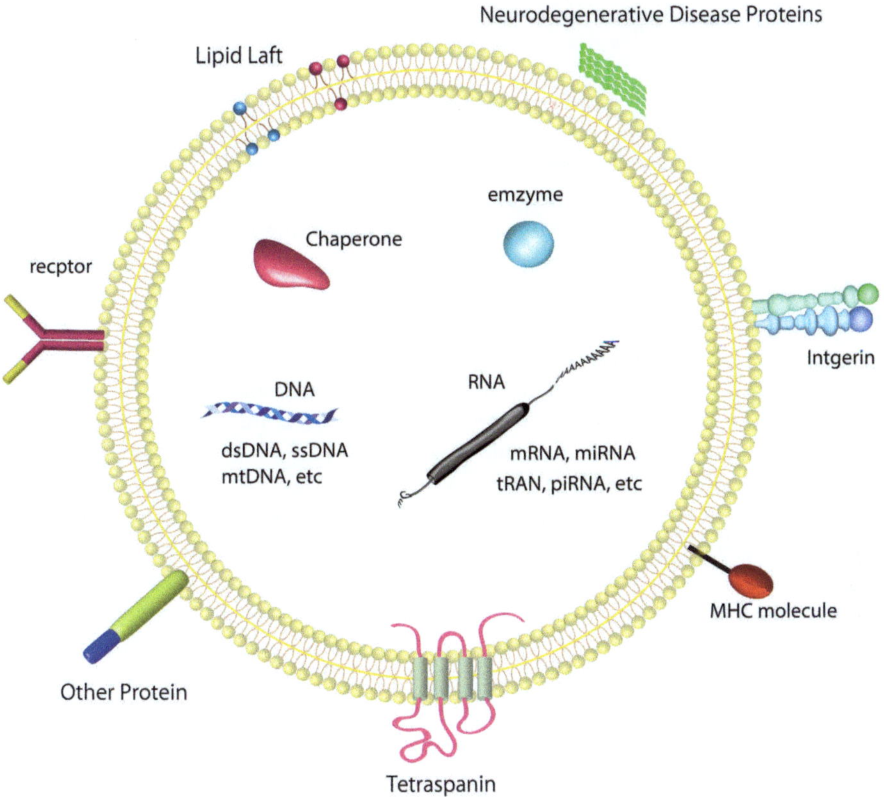

Fig. 2 Schematic structure of an Exosome
Various proteins, nucleic acids, and lipid components are included in EVs

to utilization of crude purification methods. Moreover, recent findings related to ATP1A3 have raised uncertainties regarding their neuronal specificity compared to other organs (You et al. 2023). In this section, we discuss the association between EVs and prominent neurological diseases, taking into account the purity of EV mentioned in literature. Moreover, chronic exposure to ambient fine particulate matter, such as particulate matter 2.5 ($PM_{2.5}$) has been reported to increase the risk of neurodegenerative diseases (Wang et al. 2021a). $PM_{2.5}$ refers to fine particulate matter that consists of tiny particles or droplets with a diameter of 2.5 μm or smaller present in the atmosphere. The mechanisms of action are believed to involve $PM_{2.5}$-induced disruption of the blood–brain barrier, which leads to oxidative stress, inflammation, endoplasmic reticulum stress, mitochondrial dysfunction, impaired autophagy, apoptosis, and other detrimental processes (Nunez et al. 2023).

4 Stroke and EVs

The potential therapeutic application of activated microglia-derived EVs for the treatment of cerebral ischemia has been explored, given the role of microglia, which are immune cells of the nervous system, in neuroinflammation. In a mouse model of ischemic stroke, intravenous injection of EVs derived from primary microglia that had been stimulated with deoxygenation and glucose deprivation into the femoral vein resulted in increased endothelial cell numbers and reduced neuronal cell death in the striatum of the brain (Zhang et al. 2021b). Furthermore, mice treated with these sEVs showed improved motor function. These effects were attributed to microglia-derived sEVs, which contained a high concentration of TGF-β, promoting Smad2/3 pathway activation in the recipient cells (endothelial and neuronal cells). However, the composition of these EVs, including whether they predominantly consist of apoptotic bodies due to deoxygenation and glucose deprivation, and the general notion that sEVs that are delivered via the venous route are less likely to reach the brain warrant further investigation (Sheng et al. 2020). By contrast, in response to demyelinating injuries, oligodendrocyte precursor cells migrate, proliferate, and differentiate into oligodendrocytes to facilitate recovery (Crawford et al. 2016). Microglial activation, including their migration, is enhanced at lesion sites in ischemic animal models. Additionally, microglia-derived sEVs overexpressing IL-4 showed therapeutic effects against neuroinflammation, particularly in an experimental autoimmune encephalomyelitis (EAE) mouse model (Casella et al. 2018). In a mouse model of cerebral cortex ischemia, microglia-derived sEVs overexpressing IL-4 promoted terminal maturation of precursor cells, thereby contributing to remyelination (Raffaele et al. 2021). These findings suggest that microglia-derived sEVs rich in TNF-can influence oligodendrocyte myelination through activation of their receptor, TNFR2. In summary, the potential use of microglia-derived sEVs in the treatment of strokes based on glia-neuron communication has been demonstrated.

The involvement of $PM_{2.5}$, in strokes has been studied extensively. Rats exposed to $PM_{2.5}$ for two months exhibited evidence of microhemorrhages on MRI, and 12 months of exposure led to increased levels of inflammatory cytokines (IL-6, MCP-1, TNF-α) in their blood (Cai et al. 2022). Epidemiologically, $PM_{2.5}$ is recognized as a risk factor for strokes, with a higher risk associated with advanced age and hypertension (Chang et al. 2022)

5 Parkinson's Disease (α-Synuclein) and EVs

α-Synuclein, the first familial Parkinson's disease gene (A53T) discovered, is expressed not only in neurons but also in red blood cells (Barbour et al. 2008). Disorders related to Parkinson's disease (PD), such as Lewy body dementia and multiple system atrophy, are characterized by the accumulation of α-synuclein aggregates in the brain. The presence of α-synuclein in sEVs was confirmed in cultured cells

(Fauré et al. 2006), but its existence in the cerebrospinal fluid remained uncertain (Hasegawa et al. 2011). To address this, attention was drawn to α-synuclein present in red blood cells. Given that patients with Parkinson's disease and α-synuclein-A53T mutant mice exhibit increased blood–brain barrier permeability (Gray and Woulfe 2015), sEVs were concentrated from blood using the red blood cell marker CD235a antibody and injected into the tail veins of A53T mutant mice. The results showed increased formation of α-synuclein aggregates in the brain, which was further exacerbated by sEVs derived from the blood of patients with PD (Sheng et al. 2020). Nevertheless, the mechanism of direct aggregate formation through blood-derived EVs remains unclear, with no data clearly demonstrating the uptake of sEVs from the blood–brain barrier into neuronal cells or indicating their uptake efficiency in tissues other than the brain. In contrast, the propagation of α-synuclein through sEVs may involve inhibition of autolysosome formation via soluble NSF attachment protein receptor complex (SNARE) proteins (Tang et al. 2021). Excessive expression of α-synuclein induces the accumulation of autophagosomes by inhibiting autolysosome fusion, which in turn promotes sEV production through the fusion of autophagosomes with multivesicular bodies (Tang et al. 2021) (Fig. 3). SNARE proteins are involved in the fusion of autolysosomes and lysosomes, as well as in interaction of α-synuclein at synapses. Among these SNARE proteins, the protein level of SNAP29 was found to be reduced in response to excessive α-synuclein expression. Furthermore, as α-synuclein is degraded by autophagy, inhibition of autophagy through SNAP29 is likely to increase α-synuclein levels. Additionally, SNAP29 expression exhibited an inverse correlation with severity of the Lewy body disease as per the Braak Classification system, suggesting that α-synuclein is released more readily through sEVs in a milieu associated with the Lewy body disease.

To investigate the impact of $PM_{2.5}$ on PD, a rotenone-induced PD-like mouse model was subjected to behavioral analysis using a rotarod apparatus, which revealed that exposure to $PM_{2.5}$ exacerbated the symptoms of PD (Wang et al. 2021b). This may be due to the involvement of $PM_{2.5}$, which induces oxidative stress and apoptosis in the substantia nigra of the midbrain. In A53T mice, intracranial and intranasal administration of $PM_{2.5}$ resulted in the expression of phosphorylated SNCA six months later (Yuan et al. 2022). In vitro results indicated that $PM_{2.5}$ may induce α-synuclein aggregation. However, there are conflicting reports, with a study showing no statistically significant association with $PM_{2.5}$; however, NO_2 exposure was significantly associated with the risk of PD (Jo et al. 2021).

6 Amyotrophic Lateral Sclerosis (ALS) and EVs

In ALS, motor neurons derived from iPS cells with the *C9orf72* mutation exhibit reduced release of sEVs (Aoki et al. 2017). However, analysis of sEVs in the cerebrospinal fluid showed no changes in the size or quantity of sEVs in patients with ALS (Gagliardi et al. 2021). Similarly, sEVs derived from the plasma of patients with ALS showed no differences in the size or quantity of secretion compared to those derived

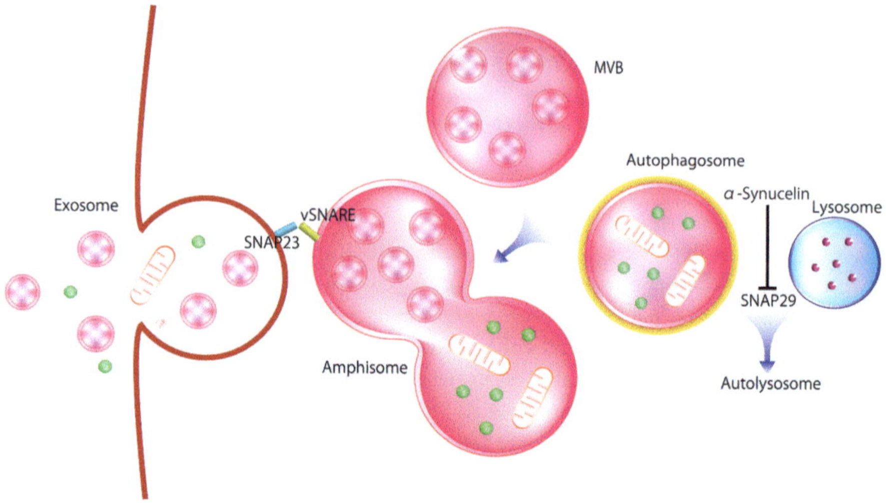

Fig. 3 A model showing propagation of α-synuclein EVs
SNAP29 is involved in the fusion of autophagosomes and lysosomes; however, α-synuclein inhibits this fusion process, promoting the fusion of multivesicular bodies with autophagosomes. Consequently, the cargo is released externally as exosomes (sEVs)

from plasma of healthy individuals (Pasetto et al. 2021). Patients with ALS exhibited decreased levels of the proteasome-like protein BLMH compared to patients with ALS who carried or did not carry the *C9orf72* mutation; however, UBA1, which is involved in ubiquitination, was decreased in patients who carried the mutation (Thompson et al. 2020). In ALS patient-derived muscle cells, there was an increase in cytoplasmic CD63-positive multivesicular bodies, and sEVs obtained from muscle cells of such patients promoted motor neuron death. These muscle cell-derived sEVs contained high amounts of RNA-binding proteins, including the ALS-related RNA-binding protein FUS (Gall et al. 2022). The involvement of sEVs in ALS pathogenesis appears to be associated with protein degradation; however, because *C9orf72* accumulation is thought to be regulated at the RNA level, the RNA profile of sEVs derived from patients with ALS is of interest. Recently, a mechanism for encapsulation of RNA into sEVs that involved an ER membrane protein, vesicle-associated membrane protein-associated protein A (VAP-A), and a complex molecule, CERT (Ceramide Transfer Protein), was proposed (Barman et al. 2022). CERT is involved in the transport of ceramides from the ER to the trans-Golgi apparatus, suggesting its potential involvement in the production and secretion of sEVs through an ESCRT-independent pathway. Interestingly, VAP-B is a major gene associated with ALS.

In epidemiological studies of $PM_{2.5}$, cumulative exposure in the five years before diagnosis has been associated with the onset of ALS (Nunez et al. 2023).

7 AD and EVs

The pathogenesis of AD is associated with the amyloid cascade hypothesis, which suggests that it begins with the formation of senile plaques (Aβ aggregation), which leads to neurofibrillary changes (tau phosphorylation deposition), ultimately causing neuronal death. Aβ interacts with the membranes of neuronal cell-derived sEVs through proteins like prions, inducing their degradation by microglia (Falker et al. 2016). In the context of Aβ deposition, isolation of mesenchymal stem cell-derived sEVs and their injection into a rat AD model resulted in reduced Aβ deposition in the cerebral cortex and hippocampus. This effect was attributed to the activation of the Wnt/β-catenin pathway mediated by *miR-29c-3p* encapsulated in sEVs (Sha et al. 2021). Regarding the propagation of Aβ via microvesicles, when microglia-derived microvesicles treated with Aβ were added to primary cultured mouse cells, they inhibited the formation of mature synaptic spines and induced functional impairment in neuronal cells (Gabrielli et al. 2022). It is believed that this effect is associated with the functionality of Aβ-containing microvesicles. To investigate this hypothesis, microvesicles were injected into the olfactory cortex of the olfactory bulb via the perforant pathway of the hippocampus (dentate gyrus), a region where an early onset of pathology is observed in AD. This resulted in electrophysiological impairment of neuronal cells in the olfactory cortex and the hippocampus (dentate gyrus).

In contrast to Alzheimer's disease (AD), which is characterized by focal amyloidosis that primarily affects the brain, transthyretin amyloidosis (ATTRv) is a type of systemic amyloidosis. In hereditary TTR amyloidosis, aggregates of TTR formed due to mutations (e.g., TTR V30M produced in the liver) leads to their deposition in tissues. Currently, effective siRNA therapies targeting *TTR* mRNA are available; however, they are expensive, and the mechanisms underlying deposition of TTRs in tissues are not well understood. Analysis using blood-derived EVs revealed the promotion of TTR aggregation on the surface of EV membranes and enhancement of TTR deposition into cells through EVs (Yamaguchi et al. 2022) (Fig. 4). Additionally, patients with hereditary TTR amyloidosis showed an increasing trend of the presence of CD63-positive EVs in their serum compared to healthy individuals. Therefore, development of therapies that target EVs and deposition of TTRs in tissues holds promise in the future.

8 Concluding Remark

With the emergence of antibody drugs that have a potential to treat AD, there is increasing hope for the discovery of early diagnostic biomarkers for neurological disorders. Although there are robust ongoing efforts to uncover EV-based biomarkers, the quest for cost-effective and easily applicable molecules for differential diagnosis remains challenging. Advancements in this field are highly awaited. EVs contain various molecules within their membranes and on their surfaces, raising concerns

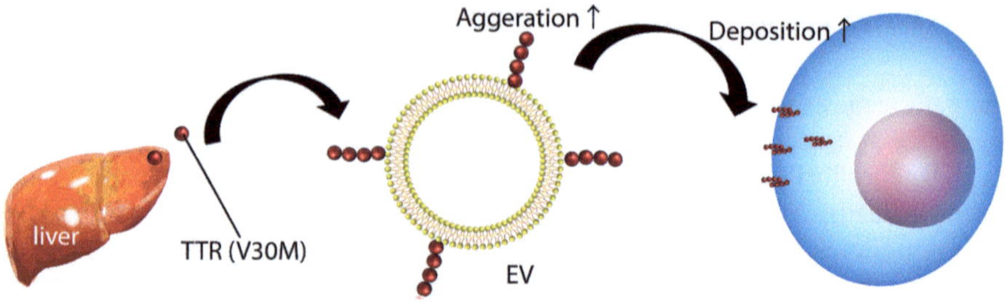

Fig. 4 Transthyretin EV-Mediated Cell Deposition Model
Transthyretin is produced in the liver and released into the bloodstream. Transthyretin amyloidosis with mutations [such as TTR (V30M)] in transthyretin results in a shift from a tetramer to a monomer, promoting formation of aggregates in the membrane of EVs. It is hypothesized that this would make it easier for cells to uptake these aggregates via EVs, thereby facilitating their deposition

about their safety aspects. Particularly, when using EVs derived from cell lines, assessing the risk of oncogenic transformation is a complex issue, and research on 'EV-like particles' based on lipid nanoparticles is also anticipated as a strategy to address these concerns.

References

Aoki Y, Manzano R, Lee Y, Dafinca R, Aoki M, Douglas AGL, Varela MA, Sathyaprakash C, Scaber J, Barbagallo P, Vader P, Mäger I, Ezzat K, Turner MR, Ito N, Gasco S, Ohbayashi N, El Andaloussi S, Takeda S, Fukuda M, Talbot K, Wood MJA (2017) C9orf72 and RAB7L1 regulate vesicle trafficking in amyotrophic lateral sclerosis and frontotemporal dementia. Brain 140:887–897. https://doi.org/10.1093/brain/awx024

Barbour R, Kling K, Anderson JP, Banducci K, Cole T, Diep L, Fox M, Goldstein JM, Soriano F, Seubert P, Chilcote TJ (2008) Red blood cells are the major source of alpha-synuclein in blood. Neurodegener Dis 5:55–59. https://doi.org/10.1159/000112832

Barman B, Sung BH, Krystofiak E, Ping J, Ramirez M, Millis B, Allen R, Prasad N, Chetyrkin S, Calcutt MW, Vickers K, Patton JG, Liu Q, Weaver AM (2022) VAP-A and its binding partner CERT drive biogenesis of RNA-containing extracellular vesicles at ER membrane contact sites. Dev Cell 57:974–994.e8. https://doi.org/10.1016/j.devcel.2022.03.012

Cai L, Yang J, Cosky E, Xin R, Geng X, Ding Y (2022) Enhanced Cerebral Microbleeds by Long-Term Air Pollution Exposure in Spontaneously Hypertensive Rats. Neurol Res 44:196–205. https://doi.org/10.1080/01616412.2021.1968705

Casella G, Colombo F, Finardi A, Descamps H, Ill-Raga G, Spinelli A, Podini P, Bastoni M, Martino G, Muzio L, Furlan R (2018) Extracellular Vesicles Containing IL-4 Modulate Neuroinflammation in a Mouse Model of Multiple Sclerosis. Mol Ther 26:2107–2118. https://doi.org/10.1016/j.ymthe.2018.06.024

Chang C-H, Chen S-H, Liu P-H, Huang K-C, Chiu I-M, Pan H-Y, Cheng F-J (2022) Ambient Air Pollution and Risk for Stroke Hospitalization: Impact on Susceptible Groups. Toxics 10:350. https://doi.org/10.3390/toxics10070350

Colombo M, Moita C, van Niel G, Kowal J, Vigneron J, Benaroch P, Manel N, Moita LF, Théry C, Raposo G (2013) Analysis of ESCRT functions in exosome biogenesis, composition and secretion highlights the heterogeneity of extracellular vesicles. J Cell Sci 126:5553–5565. https://doi.org/10.1242/jcs.128868

Crawford AH, Tripathi RB, Richardson WD, Franklin RJM (2016) Developmental Origin of Oligodendrocyte Lineage Cells Determines Response to Demyelination and Susceptibility to Age-Associated Functional Decline. Cell Rep 15:761–773. https://doi.org/10.1016/j.celrep.2016.03.069

Doyle LM, Wang MZ (2019) Overview of Extracellular Vesicles, Their Origin, Composition, Purpose, and Methods for Exosome Isolation and Analysis. Cells 8:727. https://doi.org/10.3390/cells8070727

Falker C, Hartmann A, Guett I, Dohler F, Altmeppen H, Betzel C, Schubert R, Thurm D, Wegwitz F, Joshi P, Verderio C, Krasemann S, Glatzel M (2016) Exosomal cellular prion protein drives fibrillization of amyloid beta and counteracts amyloid beta-mediated neurotoxicity. J Neurochem 137:88–100. https://doi.org/10.1111/jnc.13514

Fauré J, Lachenal G, Court M, Hirrlinger J, Chatellard-Causse C, Blot B, Grange J, Schoehn G, Goldberg Y, Boyer V, Kirchhoff F, Raposo G, Garin J, Sadoul R (2006) Exosomes are released by cultured cortical neurones. Mol Cell Neurosci 31:642–648. https://doi.org/10.1016/j.mcn.2005.12.003

Gabrielli M, Prada I, Joshi P, Falcicchia C, D'Arrigo G, Rutigliano G, Battocchio E, Zenatelli R, Tozzi F, Radeghieri A, Arancio O, Origlia N, Verderio C (2022) Microglial large extracellular vesicles propagate early synaptic dysfunction in Alzheimer's disease. Brain 145:2849–2868. https://doi.org/10.1093/brain/awac083

Gagliardi D, Bresolin N, Comi GP, Corti S (2021) Extracellular vesicles and amyotrophic lateral sclerosis: from misfolded protein vehicles to promising clinical biomarkers. Cell Mol Life Sci 78:561–572. https://doi.org/10.1007/s00018-020-03619-3

Gray MT, Woulfe JM (2015) Striatal blood-brain barrier permeability in Parkinson's disease. J Cereb Blood Flow Metab 35:747–750. https://doi.org/10.1038/jcbfm.2015.32

Hasegawa T, Konno M, Baba T, Sugeno N, Kikuchi A, Kobayashi M, Miura E, Tanaka N, Tamai K, Furukawa K, Arai H, Mori F, Wakabayashi K, Aoki M, Itoyama Y, Takeda A (2011) The AAA-ATPase VPS4 regulates extracellular secretion and lysosomal targeting of α-synuclein. PLoS ONE 6:e29460. https://doi.org/10.1371/journal.pone.0029460

Hemler ME (2005) Tetraspanin functions and associated microdomains. Nat Rev Mol Cell Biol 6:801–811. https://doi.org/10.1038/nrm1736

Huang Y, Zucker B, Zhang S, Elias S, Zhu Y, Chen H, Ding T, Li Y, Sun Y, Lou J, Kozlov MM, Yu L (2019) Publisher Correction: Migrasome formation is mediated by assembly of micron-scale tetraspanin macrodomains. Nat Cell Biol 21:1301. https://doi.org/10.1038/s41556-019-0389-z

Jo S, Kim Y-J, Park KW, Hwang YS, Lee SH, Kim BJ, Chung SJ (2021) Association of NO2 and Other Air Pollution Exposures With the Risk of Parkinson Disease. JAMA Neurol 78:800–808. https://doi.org/10.1001/jamaneurol.2021.1335

Kowal J, Arras G, Colombo M, Jouve M, Morath JP, Primdal-Bengtson B, Dingli F, Loew D, Tkach M, Théry C (2016) Proteomic comparison defines novel markers to characterize heterogeneous populations of extracellular vesicle subtypes. Proc Natl Acad Sci U S A 113:E968–977. https://doi.org/10.1073/pnas.1521230113

Le Gall L, Duddy WJ, Martinat C, Mariot V, Connolly O, Milla V, Anakor E, Ouandaogo ZG, Millecamps S, Lainé J, Vijayakumar UG, Knoblach S, Raoul C, Lucas O, Loeffler JP, Bede P, Behin A, Blasco H, Bruneteau G, Del Mar Amador M, Devos D, Henriques A, Hesters A, Lacomblez L, Laforet P, Langlet T, Leblanc P, Le Forestier N, Maisonobe T, Meininger V, Robelin L, Salachas F, Stojkovic T, Querin G, Dumonceaux J, Butler Browne G, González De Aguilar J-L, Duguez S, Pradat PF (2022) Muscle cells of sporadic amyotrophic lateral sclerosis patients secrete neurotoxic vesicles. J Cachexia Sarcopenia Muscle 13:1385–1402. https://doi.org/10.1002/jcsm.12945

Li B, Antonyak MA, Zhang J, Cerione RA (2012) RhoA triggers a specific signaling pathway that generates transforming microvesicles in cancer cells. Oncogene 31:4740–4749. https://doi.org/10.1038/onc.2011.636

Minakaki G, Menges S, Kittel A, Emmanouilidou E, Schaeffner I, Barkovits K, Bergmann A, Rockenstein E, Adame A, Marxreiter F, Mollenhauer B, Galasko D, Buzás EI, Schlötzer-Schrehardt U, Marcus K, Xiang W, Lie DC, Vekrellis K, Masliah E, Winkler J, Klucken J (2018) Autophagy inhibition promotes SNCA/alpha-synuclein release and transfer via extracellular vesicles with a hybrid autophagosome-exosome-like phenotype. Autophagy 14:98–119. https://doi.org/10.1080/15548627.2017.1395992

Muralidharan-Chari V, Clancy J, Plou C, Romao M, Chavrier P, Raposo G, D'Souza-Schorey C (2009) ARF6-regulated shedding of tumor cell-derived plasma membrane microvesicles. Curr Biol 19:1875–1885. https://doi.org/10.1016/j.cub.2009.09.059

Nabhan JF, Hu R, Oh RS, Cohen SN, Lu Q (2012) Formation and release of arrestin domain-containing protein 1-mediated microvesicles (ARMMs) at plasma membrane by recruitment of TSG101 protein. Proc Natl Acad Sci U S A 109:4146–4151. https://doi.org/10.1073/pnas.1200448109

Norman M, Ter-Ovanesyan D, Trieu W, Lazarovits R, Kowal EJK, Lee JH, Chen-Plotkin AS, Regev A, Church GM, Walt DR (2021) L1CAM is not associated with extracellular vesicles in human cerebrospinal fluid or plasma. Nat Methods 18:631–634. https://doi.org/10.1038/s41592-021-01174-8

Nunez Y, Balalian A, Parks RM, He MZ, Hansen J, Raaschou-Nielsen O, Ketzel M, Khan J, Brandt J, Vermeulen R, Peters S, Weisskopf MG, Re DB, Goldsmith J, Kioumourtzoglou M-A (2023) Exploring Relevant Time Windows in the Association Between PM2.5 Exposure and Amyotrophic Lateral Sclerosis: A Case-Control Study in Denmark. Am J Epidemiol 192:1499–1508. https://doi.org/10.1093/aje/kwad099

Pasetto L, Callegaro S, Corbelli A, Fiordaliso F, Ferrara D, Brunelli L, Sestito G, Pastorelli R, Bianchi E, Cretich M, Chiari M, Potrich C, Moglia C, Corbo M, Sorarù G, Lunetta C, Calvo A, Chiò A, Mora G, Pennuto M, Quattrone A, Rinaldi F, D'Agostino VG, Basso M, Bonetto V (2021) Decoding distinctive features of plasma extracellular vesicles in amyotrophic lateral sclerosis. Mol Neurodegener 16:52. https://doi.org/10.1186/s13024-021-00470-3

Raffaele S, Gelosa P, Bonfanti E, Lombardi M, Castiglioni L, Cimino M, Sironi L, Abbracchio MP, Verderio C, Fumagalli M (2021) Microglial vesicles improve post-stroke recovery by preventing immune cell senescence and favoring oligodendrogenesis. Mol Ther 29:1439–1458. https://doi.org/10.1016/j.ymthe.2020.12.009

Schlienger S, Campbell S, Claing A (2014) ARF1 regulates the Rho/MLC pathway to control EGF-dependent breast cancer cell invasion. Mol Biol Cell 25:17–29. https://doi.org/10.1091/mbc.E13-06-0335

Sha S, Shen X, Cao Y, Qu L (2021) Mesenchymal stem cells-derived extracellular vesicles ameliorate Alzheimer's disease in rat models via the microRNA-29c-3p/BACE1 axis and the Wnt/β-catenin pathway. Aging (Albany NY) 13:15285–15306. https://doi.org/10.18632/aging.203088

Sheng L, Stewart T, Yang D, Thorland E, Soltys D, Aro P, Khrisat T, Xie Z, Li N, Liu Z, Tian C, Bercow M, Matsumoto J, Zabetian CP, Peskind E, Quinn JF, Shi M, Zhang J (2020) Erythrocytic α-synuclein contained in microvesicles regulates astrocytic glutamate homeostasis: a new perspective on Parkinson's disease pathogenesis. Acta Neuropathol Commun 8:102. https://doi.org/10.1186/s40478-020-00983-w

Tang Q, Gao P, Arzberger T, Höllerhage M, Herms J, Höglinger G, Koeglsperger T (2021) Alpha-Synuclein defects autophagy by impairing SNAP29-mediated autophagosome-lysosome fusion. Cell Death Dis 12:854. https://doi.org/10.1038/s41419-021-04138-0

Thompson AG, Gray E, Mäger I, Thézénas M-L, Charles PD, Talbot K, Fischer R, Kessler BM, Wood M, Turner MR (2020) CSF extracellular vesicle proteomics demonstrates altered protein homeostasis in amyotrophic lateral sclerosis. Clin Proteomics 17:31. https://doi.org/10.1186/s12014-020-09294-7

Trajkovic K, Hsu C, Chiantia S, Rajendran L, Wenzel D, Wieland F, Schwille P, Brügger B, Simons M (2008) Ceramide triggers budding of exosome vesicles into multivesicular endosomes. Science 319:1244–1247. https://doi.org/10.1126/science.1153124

Wang J, Ma T, Ma D, Li H, Hua L, He Q, Deng X (2021a) The Impact of Air Pollution on Neurodegenerative Diseases. Ther Drug Monit 43:69–78. https://doi.org/10.1097/FTD.0000000000000818
Wang Y, Li C, Zhang X, Kang X, Li Y, Zhang W, Chen Y, Liu Y, Wang W, Ge M, Du L (2021b) Exposure to PM2.5 aggravates Parkinson's disease via inhibition of autophagy and mitophagy pathway. Toxicology 456:152770. https://doi.org/10.1016/j.tox.2021.152770
Wei D, Zhan W, Gao Y, Huang L, Gong R, Wang W, Zhang R, Wu Y, Gao S, Kang T (2021) RAB31 marks and controls an ESCRT-independent exosome pathway. Cell Res 31:157–177. https://doi.org/10.1038/s41422-020-00409-1
Yamaguchi H, Kawahara H, Kodera N, Kumaki A, Tada Y, Tang Z, Sakai K, Ono K, Yamada M, Hanayama R (2022) Extracellular Vesicles Contribute to the Metabolism of Transthyretin Amyloid in Hereditary Transthyretin Amyloidosis. Front Mol Biosci 9:839917. https://doi.org/10.3389/fmolb.2022.839917
You Y, Zhang Z, Sultana N, Ericsson M, Martens YA, Sun M, Kanekiyo T, Ikezu S, Shaffer SA, Ikezu T (2023) ATP1A3 as a target for isolating neuron-specific extracellular vesicles from human brain and biofluids. Sci Adv 9:eadi3647. https://doi.org/10.1126/sciadv.adi3647
Yuan X, Yang Y, Liu C, Tian Y, Xia D, Liu Z, Pan L, Xiong M, Xiong J, Meng L, Zhang Z, Ye K, Jiang H, Zhang Z (2022) Fine Particulate Matter Triggers α-Synuclein Fibrillization and Parkinson-like Neurodegeneration. Mov Disord 37:1817–1830. https://doi.org/10.1002/mds.29181
Zhang H, Lyden D (2019) Asymmetric-flow field-flow fractionation technology for exomere and small extracellular vesicle separation and characterization. Nat Protoc 14:1027–1053. https://doi.org/10.1038/s41596-019-0126-x
Zhang Q, Jeppesen DK, Higginbotham JN, Graves-Deal R, Trinh VQ, Ramirez MA, Sohn Y, Neininger AC, Taneja N, McKinley ET, Niitsu H, Cao Z, Evans R, Glass SE, Ray KC, Fissell WH, Hill S, Rose KL, Huh WJ, Washington MK, Ayers GD, Burnette DT, Sharma S, Rome LH, Franklin JL, Lee YA, Liu Q, Coffey RJ (2021a) Supermeres are functional extracellular nanoparticles replete with disease biomarkers and therapeutic targets. Nat Cell Biol 23:1240–1254. https://doi.org/10.1038/s41556-021-00805-8
Zhang L, Wei W, Ai X, Kilic E, Hermann DM, Venkataramani V, Bähr M, Doeppner TR (2021b) Extracellular vesicles from hypoxia-preconditioned microglia promote angiogenesis and repress apoptosis in stroke mice via the TGF-β/Smad2/3 pathway. Cell Death Dis 12:1068. https://doi.org/10.1038/s41419-021-04363-7

Extracellular Vesicle Isolation and Analysis Using Nanowires

Kunanon Chattrairat and Takao Yasui

Abstract Extracellular vesicles (EVs) have been considered as biomarkers for diagnosis and prognosis in therapeutic treatments. Although applications of EVs in health care are being started and more are expected, EV studies have not yet provided a deep understanding of EV biogenesis and functions. So, comprehensive study of EVs continues. To gain a better understanding of EVs, the reliable isolation and analysis of EVs are necessary. In addition, the quality and quantity of isolated EVs play a crucial role in EV analysis. Large numbers of studies have been conducted to isolate or analyze EVs with high yield, purity, sensitivity, and selectivity. During the past decade, EV studies have been largely enabled by existing analytical techniques. However, these techniques are still deficient for the isolation of high-yield and high-purity homogeneous EVs and their specific subtypes from complex matrices such as physiological fluids. Nanowires are two-dimensional nanomaterials that have unique properties, such as a high volume-to-surface ratio, a high aspect ratio, and a capacity for integration with microfluidics. Demonstrations of nanowires capturing EVs from various biological samples, such as plasma, serum, and urine, have been made. Additionally, EVs captured on nanowires can be analyzed with their biomolecules, such as miRNAs and membrane proteins, for identification of various diseases. This chapter explores nanowire technology in EV studies, and looks at its challenges and future prospects.

Keywords Nanotechnology · Nanowires · Extracellular vesicles · Microfluidics

---

K. Chattrairat · T. Yasui (✉)
Department of Life Science and Technology, Tokyo Institute of Technology, Nagatsuta 4259, Midori-ku, Yokohama 226-8501, Japan
e-mail: yasuit@bio.titech.ac.jp

K. Chattrairat
e-mail: chattrairat.k.aa@m.titech.ac.jp

T. Yasui
Institutes of Innovation for Future Society, Institute of Nano-Life-Systems, Nagoya University, Furo-Cho, Chikusa-ku, Nagoya 464-8603, Japan

Institute of Quantum Life Science, National Institutes for Quantum Science and Technology (QST), Anagawa 4-9-1, Inage-ku, Chiba 263-8555, Japan

Y. Baba et al. (eds.), *Extracellular Fine Particles*,
https://doi.org/10.1007/978-981-97-7067-0_15

1 Introduction

1.1 Background and Motivation

Extracellular vesicles (EVs), including exosomes, microvesicles, and apoptotic bodies, have been recognized as having important functions in cellular communication. EVs serve as a carriage for bioactive cargos, including proteins, nucleic acids, and lipids, thereby facilitating intercellular signaling (Muthu et al. 2021). Investigating EV compositions and functions has great potential for revealing physiological processes and diagnosing diseases with diverse symptoms (Vlaeminck-Guillem 2018; Kogure et al. 2020). Conventional isolation methods, however, have faced various limitations for studies of EVs because EVs are nanosized and highly heterogeneous, which leads to them being obtained at low yield, purity, sensitivity, and selectivity (Zhou et al. 2023). Therefore, a novel isolation technology is required to overcome the limitations.

Nanotechnology materials have transformed biomedical research due to their unique characteristics and properties, including their high surface-to-volume ratio, controllability of their structure, and their non-bulk structure property. Nanotechnology materials in the form of nanowires have provided numerous promising biomedical platforms (Rahong et al. 2016). In addition to their high surface-to-volume ratio, nanowires have a high aspect ratio, and a potential capability for microfluidic integration. Because of these characteristics, nanowires are perspective candidates for capturing and detecting EVs (Suthar et al. 2023). The integration of the nanowire technology and EV study not only enhances researchers' ability to explore the fundamental aspects of biological roles and intercellular communication but also opens new avenues for developing robust diagnostic tools (Freitas et al. 2021; Lyu et al. 2023; Chattrairat et al. 2023). This chapter navigates the convergence of these two domains by exploring the application of nanowires in the analysis of extracellular vesicles and their potential impact on advancement of diagnostics and therapeutic interventions.

1.2 Brief Overview of Extracellular Vesicles

When microparticles were first discovered in 1967, Peter Wolf used the term platelet dust to describe the subcellular coagulant material found in studying blood coagulation (Couch et al. 2021; Hargett and Bauer 2013). For about four decades after that discovery, microparticles and, synonymously, microvesicles were defined as 0.1–1 μm cell-derived vesicles that lack a nucleus or synthetic capacity, may contain cytoskeletal proteins, and expose some quantity of phosphatidylserine on their surfaces. From the late 2000s EVs came to be defined by lipid bilayer vesicles released by cells from prokaryotes to eukaryotes and plants (Shehzad et al. 2021; Nikoloff et al. 2021; Nemati et al. 2022). However, researchers have focused on the human EVs secreted by most cells into body fluids such as blood, urine, saliva, milk, and tears because the EVs provide the potential for disease diagnoses and therapeutics (Vismara et al. 2022; Wu et al. 2022). EVs contain various biomolecules, such as DNAs, lipids, proteins, RNAs, and membrane proteins (Fig. 1). EVs are divided into three groups: exosomes, microvesicles (MVs), and apoptotic bodies. Exosomes and MVs are widely studied because of their specific functions in biological processes of both homeostatic and pathological conditions, whereas apoptotic bodies are still in their infancy in biological applications. Additionally, exosomes and MVs both encapsulate proteins, lipids, and various nucleic acids, to serve as potent mediators of cell-to-cell communication.

The origin of exosomes and MVs are different: exosomes are secretions of intraluminal vesicles (ILVs) produced by the fusion of multivesicular bodies (MVBs) containing ILVs and plasma membrane (PM), and MVs are released directly into the extracellular matrix through external budding of the PM. Some of the physical properties of exosomes and MVs are overlapped such as size, density, and membrane protein. Exosomes range in size from 30 to 200 nm, while MVs range from 100 nm to 1 μm (Thery et al. 2018). EVs play crucial roles in physiological conditions, such as tissue remodeling, intercellular material transport, and metabolic regulation, and in pathological conditions, such as cancers, infections, and neurodegenerative disorders. Studying the composition and functions of EVs provides a window into the intricacies of cellular signaling and opens avenues for understanding disease mechanisms at a molecular level.

Studies of EVs are aimed at identifying and understanding various EV functions, and EVs are recognized as biomarkers in both homeostasis and pathological conditions. However, EVs are small in size and highly heterogeneous; therefore, effective isolation methods are necessary to separate EVs from complex biological samples,

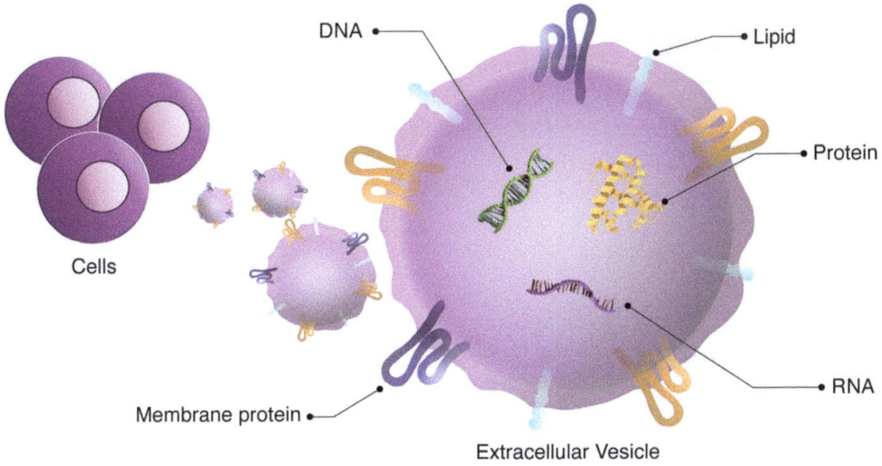

Fig. 1 The schematic image shows the secreted EV compositions from the cells including DNA, RNA, lipid, protein, and membrane protein

such as blood, urine, saliva, and tears. A reliable isolation method can ensure high purity and yield, which is essential for accurate analysis and characterization.

1.3 Role of Nanowires in Biomedical Analysis

In his well-known 1959 lecture "There's plenty of room at the bottom," Richard Feynman (1960) unveiled the start of a new era in the scientific and nanotechnological world. During this era, nanoscience and nanotechnology saw the microelectronic industry manufacturing devices of submicron dimensions with incredible control. Today, nanotechnology is of great interest in medical applications, such as biomedical engineering, and nanomedicine. Nanotechnology originates from nanomaterials that have unique characteristics and properties. The term "nanomaterials" refers to bulk materials that are remodeled in either one or more of their physical dimensions to a nanoscale, and they are nanoparticles, nanosheets, and nanowires.

Nanoparticles (or nanodots) are zero-dimensional nanostructures, for example, metal and metal oxide nanoparticles which have unusual optical properties. Nanowires are one-dimensional nanostructures that have a high surface-to-volume ratio and a high aspect ratio. Nanosheets are two-dimensional nanostructures such as graphene that have excellent electrical properties. Among these three nanostructure types, nanowires have broad potential applications in biomedical analysis and clinical diagnosis because of their nano size, intrinsic properties, flexible fabrication process, and device-integrated ability. Some of their properties are not dependent solely on the material type, but also on material geometry, including the high surface-to-volume ratio, biomolecular interactions, and enhanced electrical conductivity. When compared with other nanostructures, nanowires are more attractive because of their high aspect ratio which provides a high interaction surface area to biomolecules and the potential to communicate electrical and optical signals across cellular membranes. Additionally, the chemical composition, morphology, size, structure, and doping of nanowires can be controlled, leading to adjustable properties in biomedical analysis use. All these features contribute towards enhanced sensitivity, selectivity, and precision in biomedical analysis.

1.3.1 Sensing Platforms

One of the key applications of nanowires in biomedical analysis is as a sensing platform. The high surface-to-volume ratio of nanowires allows for efficient biomolecule detection and functionalization with biological receptors, utilizing the specific binding to detect target biomolecules. These capabilities lead to the development of nanowire-based biosensors with applications ranging from detecting disease biomarkers to monitoring cellular responses.

1.3.2 Diagnostic Tools

Nanowires also play a crucial role in the development of advanced diagnostic tools. Nanowires are capable of converting biological molecules or signals into measurable electrical or optical outputs, leading to rapid and accurate detection. Specifically, nanowire-based microfluidic devices have promise in the context of point-of-care diagnostics.

1.3.3 Imaging and Monitoring

Finally, nanowires are applied in imaging and monitoring cellular processes. Some nanowire structures are suitable for an advanced magnetic resonance imaging application as a non-invasive cell monitoring technique (Martinez-Banderas et al. 2020). With their good electrical conductivity and high aspect ratio, nanowires are also capable of being inserted into a cell where they can monitor real-time cellular events on a nanoscale (Wang et al. 2022). These capabilities are instrumental in understanding dynamic cellular behaviors and responses to external stimuli.

1.4 Objectives of the Chapter

This chapter focuses on semiconductor nanowires in biomedical analysis, specifically, in the field of EVs including isolation and detection. First, nanowire technology in biomedical analysis is briefly outlined; this includes nanowire synthesis and characterization, and nanowire applications. Next, both conventional and nanowire-based methods for EV isolation are discussed in terms of techniques, analysis results, and challenges. In the final section, applications of nanowire-based platforms, challenges, and future perspectives that come with this nanowire-based technology are summarized. The contents of this chapter are expected to shed light on the remaining barriers and future research directions for nanowire-based technology.

2 Nanowire Technology

2.1 Synthesis of Nanowires

Researchers are exploring methods for developing, manipulating, visualizing, and analyzing nanoscale materials and structures. The methods that synthesize nanomaterials are mainly divided into two approaches: top-down and bottom-up fabrications (Fig. 2). This section focuses on methods that can synthesize the nanowire structure.

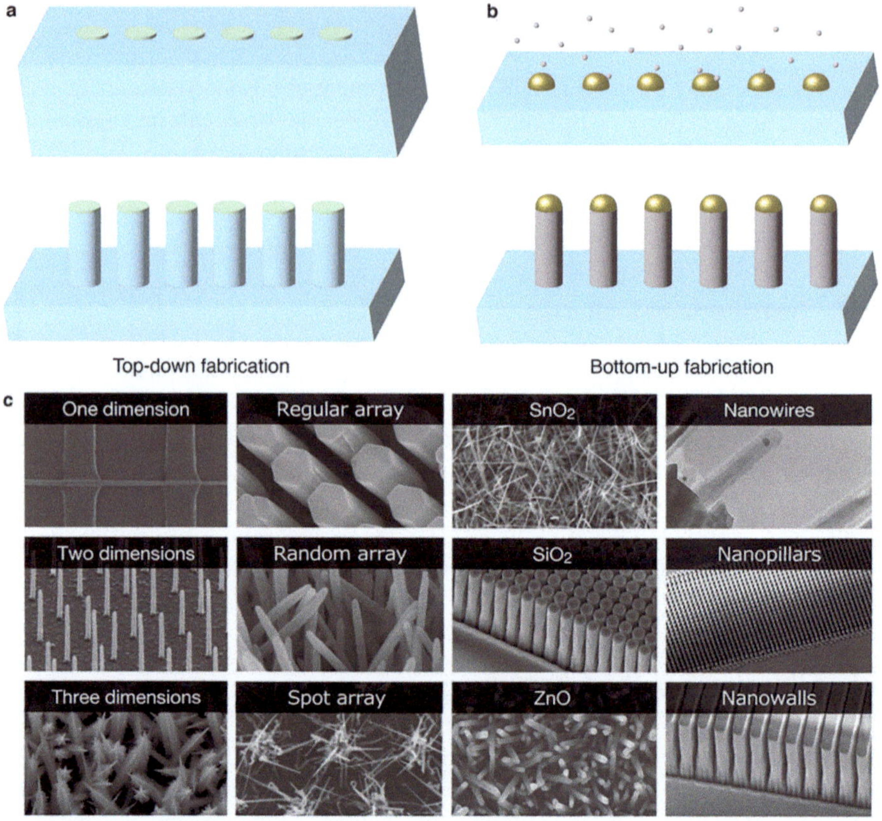

Fig. 2 Nanowire fabrication and structure. **a** Top-down and **b** bottom-up fabrications. **c** (Columns left to right) Nanowire structures in different: dimensions; arrangements; materials and surfaces; and spatial locations and structures of microfluidic channels (Takahashi et al. 2021). Reprinted, with permission, from referenced sources

The fundamentals of nanowire fabrication by top-down and bottom-up paradigms have been previously reviewed by Hobbs et al. (2012).

2.1.1 Top-Down Fabrication for Nanowires

Many methods can synthesize nanowires that offer a distinct advantage. Conventional top-down nanowire fabrication represents the early stage of nanowire fabrication. Top-down fabrication is a subtractive technique, like carving a statue. This fabrication makes wide use of advanced lithography, electron-beam writing, chemical etching, and nanoimprinting techniques to directly turn bulk structures into nanowire structures.

Wet etching

The wet etching method was developed to get silicon nanowires by using the principle of a micro-electrochemical redox reaction. Specifically, metal atoms from an etching solution, such as HF solution, deposited on a silicon surface can form nuclei that behave as a cathode, and the area surrounding these nuclei behaves as an anode and will be etched away and dissolved into the solution (Peng et al. 2002; Fülöp et al. 2016). This method was developed to tailor the grown nanowires to decrease the diameter of the grown nanowires, resulting in an increase in aspect ratio (Baraissov et al. 2019; Salhi et al. 2018).

Dry etching

Inductively coupled plasma reactive ion etching (ICP-RIE) is optimized to fabricate silicon nanowires (Refino et al. 2021; Janssen and Gheeraert 2011; Mn et al. 2016). Briefly, within the generated ICP, reactive gases are dissociated into radicals and electrons. Then, the vertically accelerated ions bombard the surface with high kinetic energy and physically etch the passivation layers. Subsequently, etching continues with radicals chemically bonding to exposed atoms of the surface and detaching as stable compounds.

Photolithography and electron-beam lithography

The substrate of interest is coated with a photoresist, a photoactive polymer, by using a spin coater. Positive photoresists typically produce photoacids which attack the polymer backbone. Then, the exposed regions are etched by a developer. Conversely, negative photoresists use different photochemical reactions so that light exposure produces cross-links or causes polymerization, making the exposed regions more resistant to a developer than unexposed regions. The resolution limit for projection photolithography is often set by the optical system, the wavelength of UV light, and the properties of the photoresist. The standard far-field projection lithography using UV wavelength light can provide ~200 nm resolution, while typical electron-beam lithography using accelerated electrons can provide a resolution of ~20 nm. Lithography alone cannot synthesize nanowires; however, it can help to manipulate and control nanowire structures and patterns (Shi et al. 2016; Liu et al. 2020).

2.1.2 Bottom-Up Fabrication for Nanowires

The bottom-up technique exploits atoms and molecules that spontaneously assemble into organized structures, without human intervention. This technique is particularly useful for constructing structures with nanoscale precision, and it relies on thermodynamics, self-organization, and self-assembly of materials. Some bottom-up fabrications for nanowires are briefly presented here; Christesen et al. (2016) have comprehensively reviewed nanowire synthesis by bottom-up fabrication.

Physical vapor deposition (PVD) and chemical vapor deposition (CVD)

Both PVD and CVD deposit the elements on a substrate. However, the way they accomplish the deposition differs: PVD uses plasma to vaporize the solid material and condense it onto the target substrate, whereas CVD injects gas or vapor to react with the target substrate to form a solid thin film. The nanowires can be synthesized by using metal-assisted seeding.

Hydrothermal method

This method includes various techniques of crystallizing substances from high-temperature aqueous solutions at high vapor pressures. As the hydrothermal method utilizes a self-assembly property, the nanowires can be synthesized under controlled conditions without metal-assisted seeding. However, the material used for the synthesis is the key to obtaining the desired nanowire structure because the crystalline structure for each element is different. It has been demonstrated that nanowires of ZnO which is a hexagonal crystal can be synthesized as nanowire structures using a lower temperature and pressure compared to other materials.

2.2 *Nanowire Structures*

The variety of synthesis methods for nanowires allows control for realizing desired properties, especially when integrating the nanowires in microfluidic channels. Nanowire structures can be developed and controlled for different: dimensions; arrangements; materials and surfaces; and spatial locations and structures of microfluidic channels (Fig. 2c). With the desired controls, nanowires are being developed in EV studies for isolation, extraction, and detection that have been difficult to achieve with conventional techniques.

2.2.1 Characterization of Nanowires

The nanowires have been mainly characterized using imaging techniques to observe morphology of the nanowire structures. Two frequently used techniques, electron microscopy (EM) and atomic force microscopy (AFM) are presented here; they provide morphological details and surface profiles on a nanoscale.

Electron microscopy

The most fundamental method for observing nanomaterials is EM. Unlike visible light, which is limited by wavelength, electrons can be controlled to generate electron beams with small (sub-nm) spot sizes, providing scientists with the most widely used nanoscale microscopy. There are two main types of EM, scanning electron microscopy (SEM) and transmission electron microscopy (TEM). Both generate electrons from an electron source and accelerate the excited electrons to a desired beam energy. The beam is focused by a magnetic objective lens to impinge on a sample surface. The ability to get images from electrons using magnetic condenser lenses is useful for observing nano-sized materials and makes both SEM and TEM powerful tools to observe nanowire structures and morphology. Moreover, the EM equipped with an X-ray technique can provide elemental and structural information about the observed samples. One limitation of EM is for quantitative measurement. Due to the small observation area, getting only a few EM images cannot represent whole samples. However, this limitation can be overcome by making a large number of measurements for the multiple observation areas.

Atomic force microscopy

A sharp stylus is dragged over a surface of interest being observed by AFM, and deflection of the stylus is used to map the surface topography. The AFM was originally developed by using a microfabricated tip instead of a macroscopic stylus (Binnig et al. 1986). Due to the tip radii of curvature ranging from a few to tens of nanometers, the AFM can provide good resolution in nano size measurements. The stylus is located at the end of a cantilever. Deflection of the cantilever occurs because of the force between the stylus and the surface of interest. A short-range repulsive force occurs when the tip is sufficiently close to the surface. The signal can be measured either optically (e.g. detection of a laser beam) or electrically (e.g. measuring electrical resistance of the cantilever).

3 Extracellular Vesicle Isolation

The EVs are secreted from normal cells and abnormal cells such as epithelial and cancer cells. Researchers believe that these EVs systematically picked up active biomolecules to transport them to targets as a form of intercellular communication. Therefore, isolation and enrichment of EVs represent a critical step in downstream analysis to achieve accurate analytic data. Conventional methods have long been employed for this purpose, for example, ultracentrifugation (UC), size exclusive chromatography (SEC), ultrafiltration, and immunoaffinity. Although these conventional methods can obtain EVs from biological samples, they come with inherent limitations such as being time-consuming procedures, having low yields, and having a potential to alter EV integrity. Nanowire-based technology has offered a path to overcoming the limitations; however, there is still plenty of room for material improvement and clinical development.

3.1 *Methods for Extracellular Vesicle Isolation and Their Challenges*

The number of EV studies has rapidly increased since the early 2000s due to attractive EV functions and properties, such as cell-to-cell communication, the ability to reflect parent cells, and involvement in both homeostasis and pathology. Although EVs are highly heterogeneous, researchers have been trying to isolate the EVs with high purity and yield. The International Society for Extracellular Vesicles (ISEV) was formed in 2011 to bring together scientists around the world working in the field, to give them the opportunity to learn and share their work and to provide guidance on the best methodological practices in the field. ISEV has published a list of minimal information for studies of extracellular vesicles (MISEV), covering EV separation and isolation, characterization, and functional studies (Thery et al. 2018). For reproducibility of EV studies, such information is critical.

The EV research field began with studies of coagulation by using UC (Couch et al. 2021; Hargett and Bauer 2013). Most biological fluids contain EVs such as blood, plasma, serum, urine, saliva, and tears (Bano et al. 2021). Due to high heterogeneity in the physicochemical properties and functions of EVs, studies of EVs remain challenging. In addition to this challenge, various particles with properties overlapping those of the EVs also exist in biological fluids. Therefore, the isolation methods used influence the purity and yield of isolated EV samples, and the methods impact the downstream analyses of EV physicochemical properties (Sharma et al. 2020; Deun et al. 2014): the most suitable EV isolation method should be selected according to the application. Although EV isolation methods are rapidly updated, some of the current methods usually do not guarantee purity and yield and they destroy the vesicle integrity. Some methods are multi-step and time-consuming, and have low repeatability. Figure 3 shows conventional methods for EV isolation, including UC, SEC, ultrafiltration, and immunoaffinity, as well as the novel nanowire-based isolation method. In this section, the advantages and disadvantages of each method are discussed before focusing on the nanowire-based isolation method.

3.1.1 Ultracentrifugation

The most common method to isolate a precipitate in solution is centrifugation (Fig. 3a). The centrifugation method is also commonly used to isolate precipitates in biofluids such as precipitates of red blood cells from blood. UC relies on the principles of differential centrifugation and utilizes high-speed centrifugal forces (G force) to separate particles based on their size and density. UC (including differential centrifugation) is the primary EV isolation method; it is used in 81% of all applications (Gardiner et al. 2016). Centrifugation speed depends on the experiment, and the temperature should be maintained at 4 °C during the process to ensure that protease, DNase, and RNase are inactive (Wang et al. 2021). To isolate the EVs from biological samples, UC is required with multiple rounds at increasing speeds

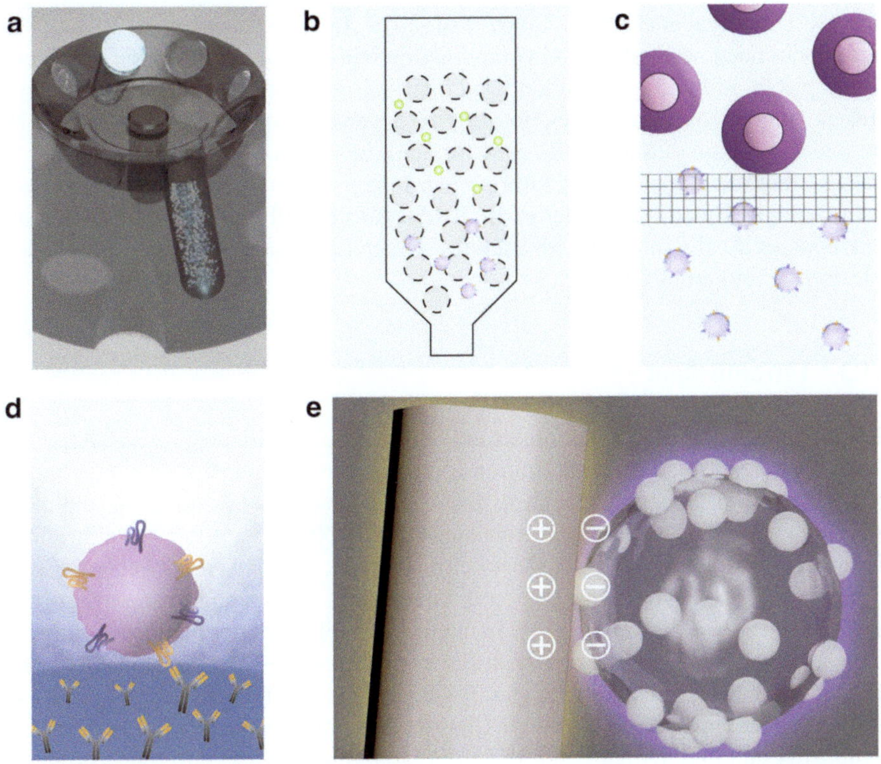

Fig. 3 EV isolation methods. **a** UC. **b** SEC. **c** Ultrafiltration. **d** Immunoaffinity. **e** Nanowire-based isolation

because of the nano sizes of the EVs. After removing cells and their debris by low-speed centrifugation, EVs are precipitated and purified by additional centrifugation (100,000 g) for at least 70 min or longer (Cvjetkovic et al. 2014). However, due to their high heterogeneity, different EV sources (e.g., blood, serum, and urine) and UC conditions affect EV purity, integrity, and downstream analysis. And, the physical high centrifuge force isolating EVs also may cause damage to them. An additional purification may be needed to isolate EV subpopulations after using the UC method (Nocera et al. 2019).

3.1.2 Size Exclusion Chromatography

The widely used SEC employs polymers to form a porous stationary phase in a chromatographic column (Fig. 3b). EVs are isolated according to differences in the path length of the different-sized molecules or particles. SEC is effective for isolating EVs from soluble contaminants and is suitable for various biological fluids. In addition, SEC is more efficient for downstream proteomic analysis compared with UC (Baranyai et al. 2015; Takov et al. 2019; Lozano-Ramos et al. 2015). This method has been normally used in combination with UC or ultrafiltration to achieve higher purity and shorter isolation time than only one selected method. Although SEC is simple, robust and does not require expensive equipment, it is limited by the low resolution, dilution of EV isolates, and presence of contaminants in vesicle isolates, including lipoproteins, viral particles, free proteins, and protein complexes from biological matrices (Sodar et al. 2016). Additionally, SEC stationary phases may nonspecifically interact with analytes, which can result in changes in isolation selectivity (Tengattini 2018).

3.1.3 Ultrafiltration

Filtration is a size-based isolation, usually with cellulose filters (Fig. 3c). After the first filtration, subsequent ultrafiltration is included to remove contaminants smaller than a specific size and to concentrate the vesicle sample, providing a fast and inexpensive method (Lobb et al. 2015; Konoshenko et al. 2018). The term "ultrafiltration" should refer exclusively to pore sizes below 0.1 μm, rather than the general use as filtration of 0.1, 0.22, or 0.45 μm pore diameter sizes (Taylor and Shah 2015; Sousa et al. 2023). The EVs prepared by this method are likely intact particles compared to the UC method which requires high forces and pressures. However, this isolation technique greatly depends on the quality of filter membranes and the uniformity of the membrane pore size distribution. Therefore, method use may be limited by high complexity and high dynamic range, leading to a low recovery rate and insufficient efficiency of isolation from high-abundance contaminants (Taylor and Shah 2015; Wei et al. 2017).

3.1.4 Immunoaffinity

Although a rigorous definition of specific EV markers remains unknown, the immunoaffinity-based EV isolation technique has been developed to isolate EVs based on EVs membrane surface protein markers (e.g. CD9, CD63, and CD81) (Fig. 3d). This technique is suitable for isolating EV subtypes based on markers rather than isolating all EVs, leading to a relatively low yield but high purity (Stam et al. 2021). This technique can be easily combined with other isolation methods to increase the efficiency, specificity and integrity of isolated EVs. Ideally, the markers should be present only on the surface of the EVs (without soluble variants). A combination of markers may be required to increase the opportunities for isolating specific EVs. To use this technique, magnetic nanoparticles are usually coated with specific antibodies against the EV-associated surface molecules (called immunomagnetic nanoparticles). Next, these immunomagnetic nanoparticles are incubated with the samples, forming EV-magnetic nanoparticle complexes. Finally, a magnetic field induces the movement of complexes and the complexes are isolated from the samples.

The immunoaffinity-based EV isolation is also combined with other isolation techniques to increase the high-yield homogenous population of EVs (Zarovni et al. 2015; Mathivanan et al. 2010).

3.2 Nanowire-Based Isolation Techniques

Nanowire-based isolation is an emerging nanotechnology that has been applied to EV studies due to the remarkable properties of the nanowires and their capability to integrate with microfluidics as noted earlier in this chapter. Furthermore, EVs that have been isolated by nanowires can be extracted to obtain biomarker molecules (e.g. miRNAs) for further downstream analysis, or to detect specific surface membrane biomarkers (e.g. surface membrane proteins) (Fig. 3e).

Nanowires can capture EVs from biological samples based on the surface charge that is possible through electrostatic interaction between negatively charged EVs and positively charged nanowires. In 2017, Yasui et al. proposed a device composed of nanowires anchored into a microfluidic substrate (Fig. 4a) (Yasui et al. 2017). This device enables urinary EV collections with high efficiency and miRNA extraction from the collected EVs with a significant number of sequences (around 1000 types). This number of sequences was much larger than that by UC despite the fact that the nanowire device used a smaller sample volume and shorter treatment time than the latter method. Additionally, they identified urinary miRNAs that could potentially serve as biomarkers for various cancers (bladder, prostate, lung, pancreas, and liver). Suwatthanarak et al. (2021) further developed the nanowire microfluidics device to improve the EV capture on nanowires by using peptide surface functionalization (Fig. 4b). Compared with unmodified nanowires, the surface-modified nanowires provided a nearly twofold increase in EV capture. Subsequently, the captured EVs could be released from the nanowires under a neutral salt condition for downstream analysis.

Since nanowires can be simply integrated into the microfluidic, Kitano et al. (2021) developed a sterilizable and mass-producible nanowire-based device that can extract a significantly greater variety and quantity of miRNAs from urine (Fig. 4c). Kitano et al. analyzed urinary miRNAs from patients with central nervous system (CNS) tumors ($n = 119$) and noncancer subjects ($n = 100$). Their finding indicated that the glioblastoma organoid-derived differentially expressed microRNAs (DEMs) were highly similar to DEMs in the urine of patients, suggesting that many CNS tumor-derived microRNAs could be identified in urine directly. Furthermore, they constructed a diagnostic model based on selected miRNAs and found that the model could differentiate patients and noncancer subjects in an independent dataset at a sensitivity and specificity of 100 and 97%, respectively.

Yasui et al. (2021) developed a methodology with nanowires to capture EVs through surface charge-based capture with diverse isoelectric points of the oxide nanowires (Fig. 5). The membranes of EVs consist of various proteins and proteoglycans, which determine EV surface charge. Although EV subsets obtained through

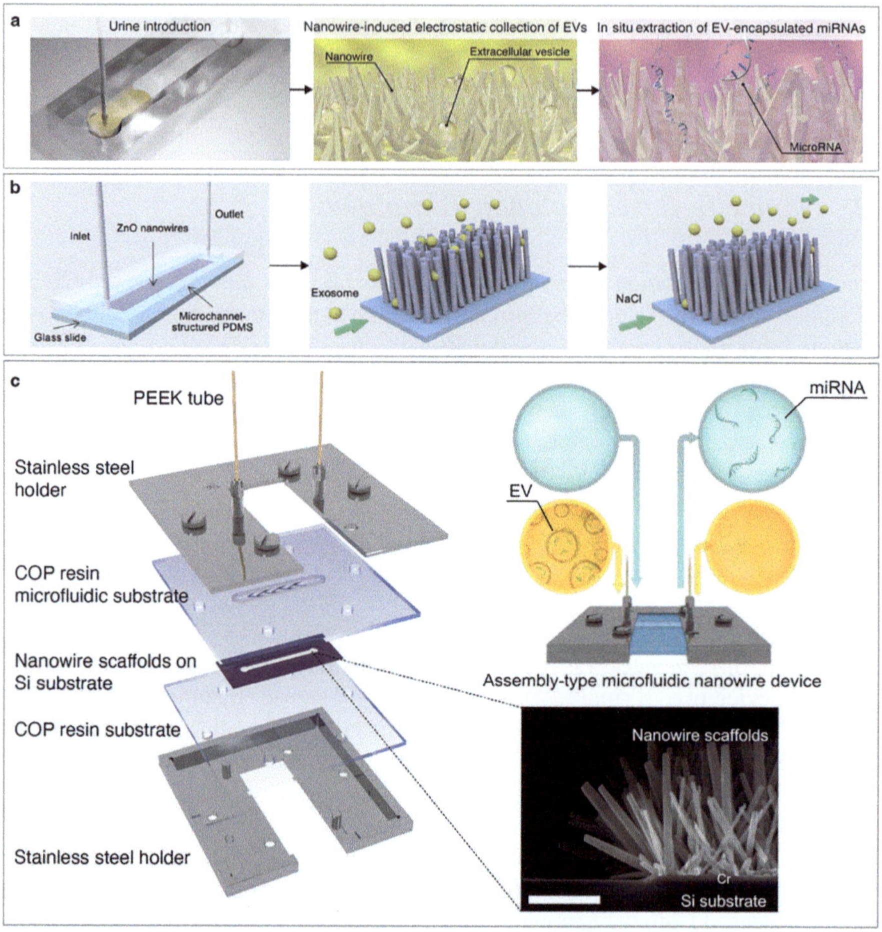

Fig. 4 Nanowire microfluidic isolation technique. **a** ZnO nanowires captured EVs and extracted the extracellular vesicle miRNAs from urine (Yasui et al. 2017). **b** Nanowire captured intact EVs and released them by using NaCl solution (Suwatthanarak et al. 2021). **c** The assembly-type of nanowire microfluidics captured and extracted urinary EVs to obtain urinary miRNAs and the extracted miRNAs could be identified for central nervous system tumors (Kitano et al. 2021). Reprinted, with permission, from referenced sources

density-, size-, and immunoaffinity-based captures and expressed membrane proteins have been investigated, the correlation between EV subsets obtained through surface charge-based capture and expressed membrane proteins was lacking. Therefore, this study confirmed the correlation between negatively charged EV subsets and expressed membrane proteins derived from cells. It also determined that a colon cancer-related membrane protein was overexpressed on negatively charged EVs derived from colon cancer cells.

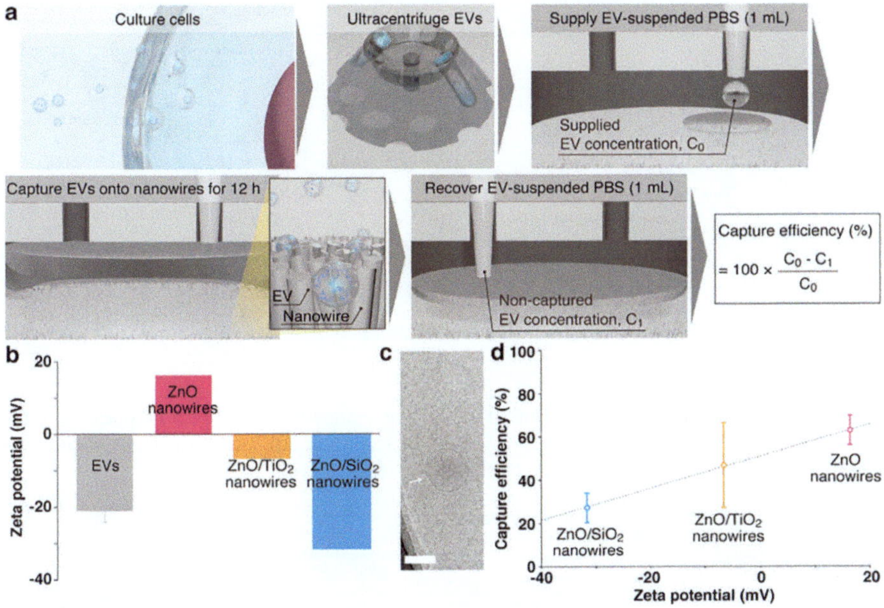

Fig. 5 **a** Schematic showing steps for surface charge-based separation of EVs (Yasui et al. 2021). **b** Zeta potentials of the EVs in PBS, and ZnO (bare), ZnO/TiO_2 (core/shell) and ZnO/SiO_2 (core/shell) nanowires in water (Yasui et al. 2021). **c** A cryo-TEM image of the EVs; scale bar, 100 nm. The white arrow indicates the EVs (Yasui et al. 2021). **d** Capture efficiency of EVs using different material nanowires (pink, ZnO (bare); orange, ZnO/TiO_2 (core/shell); cyan, ZnO/SiO_2 (core/shell)) (Yasui et al. 2021). Reprinted, with permission, from referenced sources

In addition to chemically synthesized nanowire microfluidics, bacteria-derived nanowires could serve as binding sites for antibody modification. Pham et al. (2023) developed magnetic iron oxide nanowires derived from bacterial biofilms conjugated with placental alkaline phosphatase (PLAP) antibodies to capture specific EV subtypes (Fig. 6). Subsequently, they eluted or directly lysed the captured EVs for downstream analyses (e.g., NTA and mass spectrometry). A high number of EV proteins was obtained by mass spectrometry with comparable performance to that when using immune-magnetic beads. The iron oxide magnetic nanowires provide an efficient method for EV isolation and EV proteomic studies.

4 Extracellular Vesicle Analysis

The EVs are recognized as novel biomarker diagnostic tools because EVs systematically carry active biomolecules both inside and outside the EV membranes such as RNAs, DNAs, and proteins. After isolating the EVs, the biomolecules contained in and on EVs are detected or analyzed by various methods (Fig. 7).

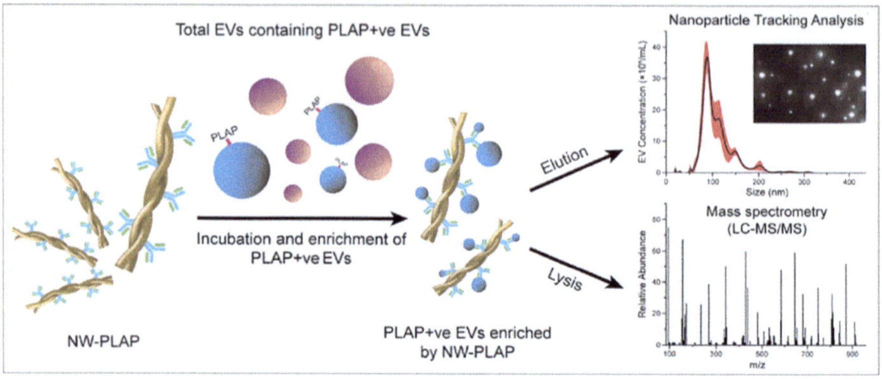

Fig. 6 Immunoaffinity nanowire isolation technique. Nanowires modified with placental alkaline phosphatase (PLAP) could capture EVs and conduct a further downstream analysis (Pham et al. 2023). Reprinted, with permission, from referenced sources

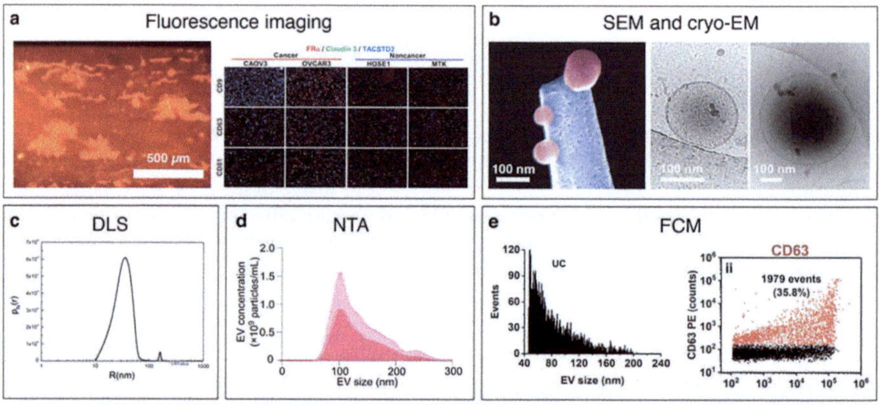

Fig. 7 Overview of EV analysis techniques. **a** Fluorescence imaging by using PKH26 labeled EVs (Yasui et al. 2017) and the ExoView platform (Yokoi et al. 2023b). **b** SEM image of EVs captured on a nanowire (Paisrisarn et al. 2022) and cryogenic-electron microscopy (cryo-EM) images of small and large EVs (Yokoi et al. 2023b). Results obtained by size distribution analysis techniques including **c** dynamic light scattering (DLS) (Palmieri et al. 2014), **d** nanoparticle tracking analysis (NTA) (Yokoi et al. 2023a), and **e** flow cytometry (FCM) (Tian et al. 2019). Reprinted, with permission, from referenced sources

4.1 *Methods for Extracellular Vesicle Analysis*

4.1.1 Conventional EV Analysis Methods and Challenges

Microscopy and imaging

EV imaging helps researchers to understand the physical properties of EVs, such as their morphology and biological mechanisms. Using confocal microscopy can monitor EV activities by fluorescently labeled membranes with about 200 nm resolution (Fig. 7a) (Zomer et al. 2015; Lai et al. 2015). To detect EVs by fluorescence microscopy, they need to be stained by one of two approaches: protein-based staining or lipid-based staining. These labels can be conjugated to nanobodies or antibodies that can modify the final size of the EVs upon binding (Colombo et al. 2021). The advantage of using fluorescence microscopy depends on where the lipids and proteins are in the host cells during EV formation and take up. However, a common dye used to label EVs (PKH67) can aggregate, forming micelles that result in false positives. This causes overestimation of the EV population. There are challenges to get highly specific and long-lasting stable reporters.

EM also plays an important role in EV imaging. EM is a powerful tool to investigate EVs due to its high resolution, ranging from 1 to 3 nm for TEM and approximately 5 nm for SEM (Fig. 7b). The MVBs inside cells were first discovered by EM (Harding et al. 1983). In addition to TEM, cryo-TEM (TEM done at cryogenic conditions) can directly observe EV samples, reduce damage (Milne et al. 2013), and minimize changes in morphology due to dehydration (Milne et al. 2013), and it does not require a staining process. Cryo-TEM observed spherical-shaped EVs (middle and right images in Fig. 7b) (Yokoi et al. 2023b), while TEM observed cup-shaped EV due to dehydration (Lobb et al. 2015; Thery et al. 2009; Chernyshev et al. 2015). However, there are some limitations for cryo-TEM including complicated sample preparation, low signal-to-noise ratio, high cost, and low throughput. Also, TEM lacks capability for quantitative measurement. Thus, if a concentration is required, other measurements should be made together with TEM.

Dynamic light scattering

Dynamic light scattering (DLS), also known as photon correlation spectroscopy, is a simple and time-saving technique to determine the size distribution of particles dispersed in a fluid medium using scattering of a laser beam (Fig. 7c). By recording the intensity of the scattered light as a function of time and analyzing the autocorrelation functions of the intensity spectra, the distribution of the particle size can be obtained (Szatanek et al. 2017). DLS provides the size and particle size distribution, or polydispersity index (PDI), for particle sizes ranging from 1 nm to 6 μm (Maguire et al. 2018). But DLS is ideal for the size measurement of monodispersed sample populations because the scattering intensity is proportional to d^6 where d is the diameter of the particle. In the case of biological fluid samples, the samples are polydisperse because they contain different sizes of vesicles, particles, and biomolecules, which range from several nanometers to about a thousand nanometers (Labrie et al. 2013; Lawrie et al. 2009). Therefore, a reliable isolation method for EVs is required to remove large particles (>1 μm), e.g. dead cells and cell debris, and possibly also remove protein complexes and aggregates. The drawback of DLS is that its resolution is poor when particle sizes differ by less than three times (e.g., 100 nm and 300).

Nanoparticle tracking analysis

Nanoparticle tracking analysis (NTA) is one of the most used methods in EV research because it can provide both particle size and concentration which are the main classifiers for EVs (Fig. 7d). By using a combination of a light scattering microscope and a camera, NTA tracks the Brownian motion of individual particles in a solution using a recorded video and it calculates their size and total concentration of the nanoparticles (Dragovic et al. 2011). Researchers can classify the EVs as small and large EVs using the obtained sizes. By tracking a single particle, software can determine a theoretical hydrodynamic diameter by the Stokes–Einstein equation. Compared with DLS, NTA provides better resolution for heterogeneous mixtures of particles that vary in size. The smallest detectable EV size is around 50 nm. However, the upper limit is about 1 μm due to the limitation of the Brownian motion principle. The main disadvantage of NTA is that it cannot determine chemical compositions or distinguish particle types in a sample, leading to possible interference from lipoproteins, debris, and protein aggregates of similar size (Buzás et al. 2017; Akers et al. 2016). Therefore, proper isolation is required.

Flow cytometry

Flow cytometry (FCM) is an optical analysis technique that detects the photons emitted or scattered from a single particle (Fig. 7e). Fluorescence signals (forward-scattered light, FSC) and scattered light signals (side-scattered light, SSC) in two directions can be measured for a single particle. Conventional FCM is a helpful tool to measure the size, concentration, and phenotype of particles at a micron size level with high reproducibility and accuracy. Conversely, when the samples are at a nano-size level, i.e. EVs, the conventional FCM cannot provide accurate results. Therefore, nano-flow cytometry (nano-FCM or nFCM) has been developed. Nano-FCM uses the same principle as FCM, but the collected angles of the light scatter change, enabling detection down to 40 nm (Tian et al. 2019; Bertolini et al. 2016; Morales-Kastresana et al. 2019; Choi et al. 2019; Arraud et al. 2015). A suitable concentration of samples of about 10^7–10^9 particles/mL is required to ensure the accuracy of quantification analysis because excessively high concentration causes a swarm effect that results in coincidental detection.

4.1.2 Nanowire-Based Extracellular Vesicle Analysis Technique

In 2023, Chattrairat et al. (2023) developed an all-in-one nanowire assay system for the capture and analysis of EVs (Fig. 8). EVs contain essential biological information not only from inside the vesicles (e.g., nucleic acids and proteins) but also from

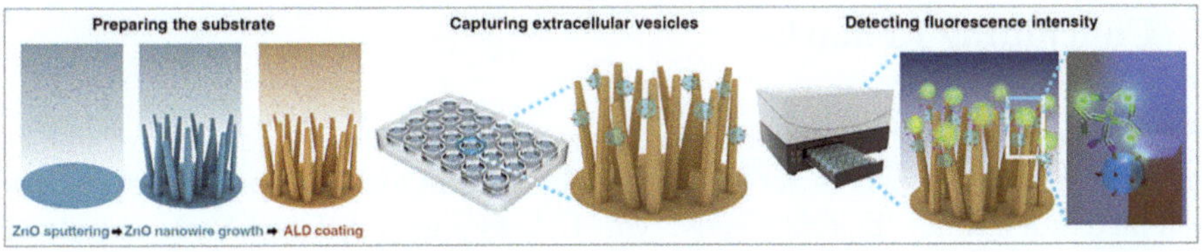

Fig. 8 All-in-one nanowire assay system for capture and analysis of EVs. The core–shell nanowires enable simple and time-effective EV capture and detection

outside the vesicles (e.g., membrane proteins and glycolipids). And, the nanowire-based approach has shown to be more effective for capturing EVs via surface charge interactions compared to conventional methods. Although isolated EVs can be obtained by UC, SEC, and immunoaffinity, additional measurement systems are needed to analyze EVs after that isolation. Therefore, Chattrairat et al. upgraded the conventional well plate assay to an all-in-one nanowire-integrated well plate assay system (i.e., a nanowire assay system) that enables charge-based EV capture and EV analysis of membrane proteins. The all-in-one nanowire assay system can capture EVs based on electrostatic force and detect EVs based on fluorescence detection. With this assay system, they determined that the membrane protein expression ratio of CD31/CD63 was higher in urine samples from glioblastoma patients than that from noncancer subjects.

Although conventional methods can isolate and separately analyze EVs, this all-in-one system is advantageous with respect to being simple and requiring only a small sample volume (0.3 mL). Moreover, this system provides a nonspecific capture with the potential to capture more information about EVs compared to the partial capture by immunoaffinity (Mateescu et al. 2017). Since users capture EVs and profile EV membrane proteins in one assay system, they have an opportunity to develop a powerful tool for cancer diagnosis with high accuracy.

The all-in-one nanowire assay system can provide isolation and detection in just one assay, whereas conventional methods require at least two steps, one for isolation and another for detection. It is significant that a capture approach for a specific EV subpopulation by this assay is possible by using surface modification. Suwatthanarak et al. (2021) developed the nanowire microfluidics device by using peptide surface functionalization, and it showed a higher capture efficiency than bare nanowires. Therefore, surface modification of nanowires by using peptides and antibodies is one option to improve the selectivity which means that nanowires can capture homogenous EVs.

5 Concluding Remarks

EVs are like cars inside human bodies to transport things from one place to another place. Specifically, the EVs act as intercellular communication and tumor progression mediators. Since the first discovery of EVs as dust platelets, EVs have been of particular note for their potential in clinical applications. EVs not only have various components, depending on the cell line but also have heterogeneity in the same cell line. Moreover, their heterogeneity in both physical and chemical properties is also challenging. Since EVs were recognized to have a potential as biomarkers for diseases, researchers have been using existing analytical techniques and developing novel technologies for EV isolation and characterization. While conventional techniques increase the depth of knowledge about structures, functions, and biogenesis of EVs, the understanding of the homeostatic and pathophysiological biology of EVs remains limited. This is due to deficiencies in current techniques for the isolation of high-purity EVs and their specific subtypes from complex matrices (e.g., biological fluids). Also, the analysis of EVs is challenging due to their nano-size particle and heterogeneity.

Reliable and reproducible isolation must be prioritized for both total EVs and specific EV subpopulations. The isolation of EVs is based: on morphological (e.g., size and shape), biophysical (e.g., surface charge), and biological (e.g., cells of origin) properties; biochemical materials (e.g., membranes, proteins, and internal markers); and functional properties (e.g., biological activity). Additionally, reproducible isolation needs prioritization because isolation techniques with different settings may provide different analysis results for the EVs. Besides reliability and reproducibility, the miniaturization of EV isolation would enable effective and robust isolation (e.g., finger-prick blood specimens). This would give the opportunity for fast and sensitive diagnostic applications. The nanowire-based capture approach has the advantages of: being rapid and requiring a small sample volume while getting a high EV yield; having the ability to select between nonspecific and specific captures; and being suited to mass production.

For analysis, conventional EV isolation methods (e.g., UC and SEC) require measurement techniques (e.g., fluorescence imaging and FCM) to obtain biological compositions that reflect EV information. The nanowire-based capture approach has the advantages of rapid and simple operation with high EV yield, the ability to discover massive numbers of miRNAs, and the ability to obtain the correlation between surface charge and membrane proteins.

As EVs have the potential to be efficient for diagnostic purposes and therapeutic strategies, EV studies are going to grow in various aspects such as isolation, analytical characterization, and standardization of methods. With these methods, understanding of EV biology and cargo components can enhance and help in further translating the application of EVs into clinical applications.

Acknowledgements This research was supported by the Japan Agency for Medical Research and Development (AMED) JP21he2302007, the JSPS Grant-in-Aid for Scientific Research (B) 21H01960, the Japan Science and Technology Agency (JST) PRESTO (JPMJPR19H9), the New

Energy and Industrial Technology Development Organization (NEDO) JPNP20004, and the Moonshot Research and Development Program (Grant nos. 22zf0127004s0902 and JP22zf0127009) from the AMED.

References

Akers JC, Ramakrishnan V, Nolan JP, Duggan E, Fu CC, Hochberg FH, *et al.* Comparative Analysis of Technologies for Quantifying Extracellular Vesicles (EVs) in Clinical Cerebrospinal Fluids (CSF). PLoS One. 2016;11(2):e0149866. https://doi.org/10.1371/journal.pone.0149866.

Arraud N, Gounou C, Turpin D, Brisson AR. Fluorescence triggering: A general strategy for enumerating and phenotyping extracellular vesicles by flow cytometry. Cytometry Part A. 2015;89(2):184–95. https://doi.org/10.1002/cyto.a.22669.

Bano R, Ahmad F, Mohsin M. A perspective on the isolation and characterization of extracellular vesicles from different biofluids. RSC Adv. 2021;11(32):19598–615. https://doi.org/10.1039/d1ra01576a.

Baraissov Z, Pacco A, Koneti S, Bisht G, Panciera F, Holsteyns F, *et al.* Selective Wet Etching of Silicon Germanium in Composite Vertical Nanowires. ACS Appl Mater Interfaces. 2019;11(40):36839–46. https://doi.org/10.1021/acsami.9b11934.

Baranyai T, Herczeg K, Onodi Z, Voszka I, Modos K, Marton N, *et al.* Isolation of Exosomes from Blood Plasma: Qualitative and Quantitative Comparison of Ultracentrifugation and Size Exclusion Chromatography Methods. PLoS One. 2015;10(12):e0145686. https://doi.org/10.1371/journal.pone.0145686.

Bertolini F, Danielson KM, Estanislau J, Tigges J, Toxavidis V, Camacho V, *et al.* Diurnal Variations of Circulating Extracellular Vesicles Measured by Nano Flow Cytometry. PLoS One. 2016;11(1). https://doi.org/10.1371/journal.pone.0144678.

Binnig G, Quate CF, Gerber C. Atomic force microscope. Phys Rev Lett. 1986;56(9):930–3. https://doi.org/10.1103/PhysRevLett.56.930.

Buzás EI, Gardiner C, Lee C, Smith ZJ. Single particle analysis: Methods for detection of platelet extracellular vesicles in suspension (excluding flow cytometry). Platelets. 2017;28(3):249–55. https://doi.org/10.1080/09537104.2016.1260704.

Chattrairat K, Yasui T, Suzuki S, Natsume A, Nagashima K, Iida M, *et al.* All-in-One Nanowire Assay System for Capture and Analysis of Extracellular Vesicles from an ex Vivo Brain Tumor Model. ACS Nano. 2023;17(3):2235–44. https://doi.org/10.1021/acsnano.2c08526.

Chernyshev VS, Rachamadugu R, Tseng YH, Belnap DM, Jia Y, Branch KJ, *et al.* Size and shape characterization of hydrated and desiccated exosomes. Anal Bioanal Chem. 2015;407(12):3285–301. https://doi.org/10.1007/s00216-015-8535-3.

Choi D, Montermini L, Jeong H, Sharma S, Meehan B, Rak J. Mapping Subpopulations of Cancer Cell-Derived Extracellular Vesicles and Particles by Nano-Flow Cytometry. ACS Nano. 2019;13(9):10499–511. https://doi.org/10.1021/acsnano.9b04480.

Christesen JD, Pinion CW, Hill DJ, Kim S, Cahoon JF. Chemically Engraving Semiconductor Nanowires: Using Three-Dimensional Nanoscale Morphology to Encode Functionality from the Bottom Up. J Phys Chem Lett. 2016;7(4):685–92. https://doi.org/10.1021/acs.jpclett.5b02444.

Colombo F, Norton EG, Cocucci E. Microscopy approaches to study extracellular vesicles. Biochim Biophys Acta Gen Subj. 2021;1865(4):129752. https://doi.org/10.1016/j.bbagen.2020.129752.

Couch Y, Buzas EI, Di Vizio D, Gho YS, Harrison P, Hill AF, *et al.* A brief history of nearly EV-erything - The rise and rise of extracellular vesicles. J Extracell Vesicles. 2021;10(14):e12144. https://doi.org/10.1002/jev2.12144.

Cvjetkovic A, Lotvall J, Lasser C. The influence of rotor type and centrifugation time on the yield and purity of extracellular vesicles. J Extracell Vesicles. 2014;3. https://doi.org/10.3402/jev.v3.23111.

De Sousa KP, Rossi I, Abdullahi M, Ramirez MI, Stratton D, Inal JM. Isolation and characterization of extracellular vesicles and future directions in diagnosis and therapy. Wiley Interdiscip Rev Nanomed Nanobiotechnol. 2023;15(1):e1835. https://doi.org/10.1002/wnan.1835.

Van Deun J, Mestdagh P, Sormunen R, Cocquyt V, Vermaelen K, Vandesompele J, *et al.* The impact of disparate isolation methods for extracellular vesicles on downstream RNA profiling. J Extracell Vesicles. 2014;3. https://doi.org/10.3402/jev.v3.24858.

Dragovic RA, Gardiner C, Brooks AS, Tannetta DS, Ferguson DJ, Hole P, *et al.* Sizing and phenotyping of cellular vesicles using Nanoparticle Tracking Analysis. Nanomedicine. 2011;7(6):780–8. https://doi.org/10.1016/j.nano.2011.04.003.

Feynman RP. There's Plenty of Room at the Bottom. Caltech Engineering and Science. 1960;23(5):22–36.

de Freitas RCC, Hirata RDC, Hirata MH, Aikawa E. Circulating Extracellular Vesicles As Biomarkers and Drug Delivery Vehicles in Cardiovascular Diseases. Biomolecules. 2021;11(3). https://doi.org/10.3390/biom11030388.

Fülöp G, d'Hollosy S, Hofstetter L, Baumgartner A, Nygård J, Schönenberger C, *et al.* Wet etch methods for InAs nanowire patterning and self-aligned electrical contacts. Nanotechnology. 2016;27(19):195303. https://doi.org/10.1088/0957-4484/27/19/195303.

Gardiner C, Di Vizio D, Sahoo S, Thery C, Witwer KW, Wauben M, *et al.* Techniques used for the isolation and characterization of extracellular vesicles: results of a worldwide survey. J Extracell Vesicles. 2016;5:32945. https://doi.org/10.3402/jev.v5.32945.

Harding C, Heuser J, Stahl P. Receptor-mediated endocytosis of transferrin and recycling of the transferrin receptor in rat reticulocytes. J Cell Biol. 1983;97(2):329–39. https://doi.org/10.1083/jcb.97.2.329.

Hargett LA, Bauer NN. On the origin of microparticles: From "platelet dust" to mediators of intercellular communication. Pulm Circ. 2013;3(2):329–40. https://doi.org/10.4103/2045-8932.114760.

Hobbs RG, Petkov N, Holmes JD. Semiconductor Nanowire Fabrication by Bottom-Up and Top-Down Paradigms. Chem Mater. 2012;24(11):1975–91. https://doi.org/10.1021/cm300570n.

Janssen W, Gheeraert E. Dry etching of diamond nanowires using self-organized metal droplet masks. Diamond and Related Materials. 2011;20(3):389–94. https://doi.org/10.1016/j.diamond.2011.01.037.

Kitano Y, Aoki K, Ohka F, Yamazaki S, Motomura K, Tanahashi K, *et al.* Urinary microRNA-based diagnostic model for central nervous system tumors using nanowire scaffolds. ACS Appl Mater Interfaces. 2021;13(15):17316–29. https://doi.org/10.1021/acsami.1c01754.

Kogure A, Yoshioka Y, Ochiya T. Extracellular Vesicles in Cancer Metastasis: Potential as Therapeutic Targets and Materials. Int J Mol Sci. 2020;21(12). https://doi.org/10.3390/ijms21124463.

Konoshenko MY, Lekchnov EA, Vlassov AV, Laktionov PP. Isolation of Extracellular Vesicles: General Methodologies and Latest Trends. Biomed Res Int. 2018;2018:8545347. https://doi.org/10.1155/2018/8545347.

Labrie A, Marshall A, Bedi H, Maurer-Spurej E. Characterization of platelet concentrates using dynamic light scattering. Transfus Med Hemother. 2013;40(2):93–100. https://doi.org/10.1159/000350362.

Lai CP, Kim EY, Badr CE, Weissleder R, Mempel TR, Tannous BA, *et al.* Visualization and tracking of tumour extracellular vesicle delivery and RNA translation using multiplexed reporters. Nat Commun. 2015;6:7029. https://doi.org/10.1038/ncomms8029.

Lawrie AS, Albanyan A, Cardigan RA, Mackie IJ, Harrison P. Microparticle sizing by dynamic light scattering in fresh-frozen plasma. Vox Sang. 2009;96(3):206–12. https://doi.org/10.1111/j.1423-0410.2008.01151.x.

Liu Q, Yasui T, Nagashima K, Yanagida T, Horiuchi M, Zhu Z, *et al.* Photolithographically Constructed Single ZnO Nanowire Device and Its Ultraviolet Photoresponse. Anal Sci. 2020;36(9):1125–9. https://doi.org/10.2116/analsci.20N002.

Lobb RJ, Becker M, Wen SW, Wong CS, Wiegmans AP, Leimgruber A, *et al.* Optimized exosome isolation protocol for cell culture supernatant and human plasma. J Extracell Vesicles. 2015;4:27031. https://doi.org/10.3402/jev.v4.27031.

Lozano-Ramos I, Bancu I, Oliveira-Tercero A, Armengol MP, Menezes-Neto A, Del Portillo HA, *et al.* Size-exclusion chromatography-based enrichment of extracellular vesicles from urine samples. J Extracell Vesicles. 2015;4:27369. https://doi.org/10.3402/jev.v4.27369.

Lyu F, Wu K, Wu SY, Deshpande RP, Tyagi A, Ruiz I, *et al.* Functional evaluation of dendritic cells and extracellular vesicles as immunotherapy for breast cancer. Oncogene. 2023. https://doi.org/10.1038/s41388-023-02893-2.

Maguire CM, Rosslein M, Wick P, Prina-Mello A. Characterisation of particles in solution - a perspective on light scattering and comparative technologies. Sci Technol Adv Mater. 2018;19(1):732–45. https://doi.org/10.1080/14686996.2018.1517587.

Martinez-Banderas AI, Aires A, Plaza-Garcia S, Colas L, Moreno JA, Ravasi T, *et al.* Magnetic core-shell nanowires as MRI contrast agents for cell tracking. J Nanobiotechnology. 2020;18(1):42. https://doi.org/10.1186/s12951-020-00597-3.

Mateescu B, Kowal EJ, van Balkom BW, Bartel S, Bhattacharyya SN, Buzas EI, *et al.* Obstacles and opportunities in the functional analysis of extracellular vesicle RNA - an ISEV position paper. J Extracell Vesicles. 2017;6(1):1286095. https://doi.org/10.1080/20013078.2017.1286095.

Mathivanan S, Lim JW, Tauro BJ, Ji H, Moritz RL, Simpson RJ. Proteomics analysis of A33 immunoaffinity-purified exosomes released from the human colon tumor cell line LIM1215 reveals a tissue-specific protein signature. Mol Cell Proteomics. 2010;9(2):197–208. https://doi.org/10.1074/mcp.M900152-MCP200.

Milne JL, Borgnia MJ, Bartesaghi A, Tran EE, Earl LA, Schauder DM, *et al.* Cryo-electron microscopy--a primer for the non-microscopist. FEBS J. 2013;280(1):28–45. https://doi.org/10.1111/febs.12078.

MN MN, Hashim U, Md Arshad MK, Rahim Ruslinda A, Rahman SF, Fathil MF, *et al.* Top-Down Nanofabrication and Characterization of 20 nm Silicon Nanowires for Biosensing Applications. PLoS One. 2016;11(3):e0152318. https://doi.org/10.1371/journal.pone.0152318.

Morales-Kastresana A, Musich TA, Welsh JA, Telford W, Demberg T, Wood JCS, *et al.* High-fidelity detection and sorting of nanoscale vesicles in viral disease and cancer. J Extracell Vesicles. 2019;8(1). https://doi.org/10.1080/20013078.2019.1597603.

Muthu S, Bapat A, Jain R, Jeyaraman N, Jeyaraman M. Exosomal therapy-a new frontier in regenerative medicine. Stem Cell Investig. 2021;8:7. https://doi.org/10.21037/sci-2020-037.

Nemati M, Singh B, Mir RA, Nemati M, Babaei A, Ahmadi M, *et al.* Plant-derived extracellular vesicles: a novel nanomedicine approach with advantages and challenges. Cell Commun Signal. 2022;20(1):69. https://doi.org/10.1186/s12964-022-00889-1.

Nikoloff JM, Saucedo-Espinosa MA, Kling A, Dittrich PS. Identifying extracellular vesicle populations from single cells. Proc Natl Acad Sci U S A. 2021;118(38). https://doi.org/10.1073/pnas.2106630118.

Nocera AL, Mueller SK, Stephan JR, Hing L, Seifert P, Han X, *et al.* Exosome swarms eliminate airway pathogens and provide passive epithelial immunoprotection through nitric oxide. J Allergy Clin Immunol. 2019;143(4):1525–35 e1. https://doi.org/10.1016/j.jaci.2018.08.046.

Paisrisarn P, Yasui T, Zhu Z, Klamchuen A, Kasamechonchung P, Wutikhun T, *et al.* Tailoring ZnO nanowire crystallinity and morphology for label-free capturing of extracellular vesicles. Nanoscale. 2022;14:4484–94. https://doi.org/10.1039/d1nr07237d.

Palmieri V, Lucchetti D, Gatto I, Maiorana A, Marcantoni M, Maulucci G, *et al.* Dynamic light scattering for the characterization and counting of extracellular vesicles: a powerful noninvasive tool. Journal of Nanoparticle Research. 2014;16(9). https://doi.org/10.1007/s11051-014-2583-z.

Peng K-Q, Yan Y-J, Gao S-P, Zhu J. Synthesis of Large-Area Silicon Nanowire Arrays via Self-Assembling Nanoelectrochemistry. Advanced Materials. 2002;14(16):1164–7. https://doi.org/10.1002/1521-4095(20020816)14:16<1164::AID-ADMA1164>3.0.CO;2-E.

Pham QN, Winter M, Milanova V, Young C, Condina MR, Hoffmann P, *et al.* Magnetic enrichment of immuno-specific extracellular vesicles for mass spectrometry using biofilm-derived iron oxide nanowires. Nanoscale. 2023;15(3):1236–47. https://doi.org/10.1039/d2nr05619d.

Rahong S, Yasui T, Kaji N, Baba Y. Recent developments in nanowires for bio-applications from molecular to cellular levels. Lab Chip. 2016;16(7):1126–38. https://doi.org/10.1039/c5lc01306b.

Refino AD, Yulianto N, Syamsu I, Nugroho AP, Hawari NH, Syring A, *et al.* Versatilely tuned vertical silicon nanowire arrays by cryogenic reactive ion etching as a lithium-ion battery anode. Sci Rep. 2021;11(1):19779. https://doi.org/10.1038/s41598-021-99173-4.

Salhi B, Hossain MK, Al-Sulaiman F. Wet-chemically etched silicon nanowire: Effect of etching parameters on the morphology and optical characterizations. Solar Energy. 2018;161:180–6. https://doi.org/10.1016/j.solener.2017.12.038.

Sharma S, LeClaire M, Wohlschlegel J, Gimzewski J. Impact of isolation methods on the biophysical heterogeneity of single extracellular vesicles. Sci Rep. 2020;10(1):13327. https://doi.org/10.1038/s41598-020-70245-1.

Shehzad A, Islam SU, Shahzad R, Khan S, Lee YS. Extracellular vesicles in cancer diagnostics and therapeutics. Pharmacol Ther. 2021;223:107806. https://doi.org/10.1016/j.pharmthera.2021.107806.

Shi R, Huang C, Zhang L, Amini A, Liu K, Shi Y, *et al.* Three Dimensional Sculpturing of Vertical Nanowire Arrays by Conventional Photolithography. Sci Rep. 2016;6:18886. https://doi.org/10.1038/srep18886.

Sodar BW, Kittel A, Paloczi K, Vukman KV, Osteikoetxea X, Szabo-Taylor K, *et al.* Low-density lipoprotein mimics blood plasma-derived exosomes and microvesicles during isolation and detection. Sci Rep. 2016;6:24316. https://doi.org/10.1038/srep24316.

Stam J, Bartel S, Bischoff R, Wolters JC. Isolation of extracellular vesicles with combined enrichment methods. J Chromatogr B Analyt Technol Biomed Life Sci. 2021;1169:122604. https://doi.org/10.1016/j.jchromb.2021.122604.

Suthar J, Taub M, Carney RP, Williams GR, Guldin S. Recent developments in biosensing methods for extracellular vesicle protein characterization. Wiley Interdiscip Rev Nanomed Nanobiotechnol. 2023;15(1):e1839. https://doi.org/10.1002/wnan.1839.

Suwatthanarak T, Thiodorus IA, Tanaka M, Shimada T, Takeshita D, Yasui T, *et al.* Microfluidic-based capture and release of cancer-derived exosomes via peptide-nanowire hybrid interface. Lab Chip. 2021;21(3):597–607. https://doi.org/10.1039/d0lc00899k.

Szatanek R, Baj-Krzyworzeka M, Zimoch J, Lekka M, Siedlar M, Baran J. The Methods of Choice for Extracellular Vesicles (EVs) Characterization. Int J Mol Sci. 2017;18(6). https://doi.org/10.3390/ijms18061153.

Takahashi H, Baba Y, Yasui T. Oxide nanowire microfluidics addressing previously-unattainable analytical methods for biomolecules towards liquid biopsy. Chem Commun. 2021;57(98):13234–45. https://doi.org/10.1039/d1cc05096f.

Takov K, Yellon DM, Davidson SM. Comparison of small extracellular vesicles isolated from plasma by ultracentrifugation or size-exclusion chromatography: yield, purity and functional potential. J Extracell Vesicles. 2019;8(1):1560809. https://doi.org/10.1080/20013078.2018.1560809.

Taylor DD, Shah S. Methods of isolating extracellular vesicles impact down-stream analyses of their cargoes. Methods. 2015;87:3–10. https://doi.org/10.1016/j.ymeth.2015.02.019.

Tengattini S. Chromatographic Approaches for Purification and Analytical Characterization of Extracellular Vesicles: Recent Advancements. Chromatographia. 2018;82(1):415–24. https://doi.org/10.1007/s10337-018-3637-7.

Thery C, Ostrowski M, Segura E. Membrane vesicles as conveyors of immune responses. Nat Rev Immunol. 2009;9(8):581–93. https://doi.org/10.1038/nri2567.

Thery C, Witwer KW, Aikawa E, Alcaraz MJ, Anderson JD, Andriantsitohaina R, *et al.* Minimal information for studies of extracellular vesicles 2018 (MISEV2018): a position statement of the International Society for Extracellular Vesicles and update of the MISEV2014 guidelines. J Extracell Vesicles. 2018;7(1):1535750. https://doi.org/10.1080/20013078.2018.1535750.

Tian Y, Gong M, Hu Y, Liu H, Zhang W, Zhang M, *et al.* Quality and efficiency assessment of six extracellular vesicle isolation methods by nano-flow cytometry. J Extracell Vesicles. 2019;9(1). https://doi.org/10.1080/20013078.2019.1697028.

Vismara M, Manfredi M, Zara M, Trivigno SMG, Galgano L, Barbieri SS, *et al.* Proteomic and functional profiling of platelet-derived extracellular vesicles released under physiological or tumor-associated conditions. Cell Death Discov. 2022;8(1):467. https://doi.org/10.1038/s41420-022-01263-3.

Vlaeminck-Guillem V. Extracellular Vesicles in Prostate Cancer Carcinogenesis, Diagnosis, and Management. Front Oncol. 2018;8:222. https://doi.org/10.3389/fonc.2018.00222.

Wang F, Ren F, Ma Z, Qu L, Gourgues R, Xu C, *et al.* In vivo non-invasive confocal fluorescence imaging beyond 1,700 nm using superconducting nanowire single-photon detectors. Nat Nanotechnol. 2022;17(6):653–60. https://doi.org/10.1038/s41565-022-01130-3.

Wang J, Ma P, Kim DH, Liu BF, Demirci U. Towards Microfluidic-Based Exosome Isolation and Detection for Tumor Therapy. Nano Today. 2021;37. https://doi.org/10.1016/j.nantod.2020.101066.

Wei Z, Batagov AO, Schinelli S, Wang J, Wang Y, El Fatimy R, *et al.* Coding and noncoding landscape of extracellular RNA released by human glioma stem cells. Nat Commun. 2017;8(1):1145. https://doi.org/10.1038/s41467-017-01196-x.

Wu A, Wolley MJ, Fenton RA, Stowasser M. Using human urinary extracellular vesicles to study physiological and pathophysiological states and regulation of the sodium chloride cotransporter. Front Endocrinol (Lausanne). 2022;13:981317. https://doi.org/10.3389/fendo.2022.981317.

Yasui T, Yanagida T, Ito S, Konakade Y, Takeshita D, Naganawa T, *et al.* Unveiling massive numbers of cancer-related urinary-microRNA candidates via nanowires. Sci Adv. 2017;3(12):e1701133. https://doi.org/10.1126/sciadv.1701133.

Yasui T, Paisrisarn P, Yanagida T, Konakade Y, Nakamura Y, Nagashima K, *et al.* Molecular profiling of extracellular vesicles via charge-based capture using oxide nanowire microfluidics. Biosens Bioelectron. 2021;194:113589. https://doi.org/10.1016/j.bios.2021.113589.

Yokoi A, Yoshida K, Koga H, Kitagawa M, Nagao Y, Iida M, *et al.* Spatial exosome analysis using cellulose nanofiber sheets reveals the location heterogeneity of extracellular vesicles. Nat Commun. 2023a;14(1):6915. https://doi.org/10.1038/s41467-023-42593-9.

Yokoi A, Ukai M, Yasui T, Inokuma Y, Hyeon-Deuk K, Matsuzaki J, *et al.* Identifying high-grade serous ovarian carcinoma specific extracellular vesicles by polyketone-coated nanowires. Sci Adv. 2023b;9(27):eade6958. https://doi.org/10.1126/sciadv.ade6958.

Zarovni N, Corrado A, Guazzi P, Zocco D, Lari E, Radano G, *et al.* Integrated isolation and quantitative analysis of exosome shuttled proteins and nucleic acids using immunocapture approaches. Methods. 2015;87:46–58. https://doi.org/10.1016/j.ymeth.2015.05.028.

Zhou X, Huang Q, Jiang Y, Tang H, Zhang L, Li D, *et al.* Emerging technologies for engineering of extracellular vesicles. Front Bioeng Biotechnol. 2023;11:1298746. https://doi.org/10.3389/fbioe.2023.1298746.

Zomer A, Maynard C, Verweij FJ, Kamermans A, Schafer R, Beerling E, *et al.* In Vivo imaging reveals extracellular vesicle-mediated phenocopying of metastatic behavior. Cell. 2015;161(5):1046–57. https://doi.org/10.1016/j.cell.2015.04.042.

Particulars of Oral Cavity

Kiyotaka Shiba

Abstract Oral fluids (OFs) contain a diverse array of extracellular vesicles (EVs) that hold promise as a source of diagnostic information. Developing EV-based diagnostics using OFs requires an understanding of the physicochemical properties and heterogeneity of these EVs. This review explores strategies for differentiating EVs in OFs, including differential centrifugation, density gradient centrifugation, and a novel method based on sedimentation patterns. These techniques have revealed distinct subpopulations of EVs in OFs, each associated with specific biological functions and potential diagnostic utility. However, the complexity of EVs in OFs presents challenges, and a comprehensive understanding of their biogenesis and composition is still emerging. Future research should focus on refining EV isolation methods and exploring the diagnostic potential of both EV and non-EV particles in OFs.

Keywords EVs · EVs · sEVs · L-EVs · OFs · Saliva · Diagnosis · Differentiation

1 Introduction

Extracellular vesicles (EVs) are commonly defined as vesicles enclosed by a lipid bilayer, naturally secreted by cells (Lotvall et al. 2014; Thery et al. 2018; Welsh et al. 2024). EVs describe not a single molecular species but rather collectively represent the entities that broadly encompasses diverse populations released through various secretory pathways, with different sizes and physicochemical properties. The distinction of EVs based on their physicochemical characteristics, as well as their biological functions derived from different secretion pathways, is currently a focus of vigorous research in laboratories around the world. Establishing more versatile differentiation methods for EVs, as well as associating their biogenesis pathways with their physicochemical characteristics, will take further time and effort. Along with this basic research, development research aimed at clinical applications is also actively underway. EVs are thought to be released from almost all cells and the

K. Shiba (✉)
Cancer Institute, Japanese Foundation for Cancer Research, Tokyo, Japan
e-mail: kshiba@jfcr.or.jp

Y. Baba et al. (eds.), *Extracellular Fine Particles*,
https://doi.org/10.1007/978-981-97-7067-0_16

released EVs carry molecules from their parent cells. EVs are stably present in various bodily fluids such as prostatic fluid, blood, urine, ascites and oral fluids (OFs) (Skog et al. 2008; Brody et al. 1983; Pisitkun et al. 2004; Runz et al. 2007; Marzesco et al. 2005). Thus, analyzing the molecules carried by EVs in body fluids potentially provides information for diagnostic purposes. While blood (plasma, serum) is often the first type of body fluid considered for samples, urine and OFs, among others, also present attractive sources. From this perspective, the isolation of EVs from various body fluids and their potential applications in the diagnostic field are being reported (Yanez-Mo et al. 2015). Each body fluid has its strengths and weaknesses in terms of extracting diagnostic information, and there are specific considerations due to the physicochemical characteristics of each fluid during isolation (Zonneveld et al. 2014; Correll et al. 2022). This review introduces the uniqueness of isolating EVs from OFs, with a focus on our research findings.

2 What is OFs?

OFs is a body fluid collected from the oral cavity. To envision the implemented form of EV-diagnostics using OFs, we can consider the fluid collected in a collection tube through expectoration as the OFs referred to here. OFs is often expressed as "whole saliva," but if saliva is defined as the fluid secreted by the salivary glands (considered as narrow-sense saliva), then OFs encompasses not only this narrow-sense saliva but also includes gingival fluid, oral mucosal transudate, mucous from the pharynx, and nasal cavity (Llena-Puy 2006). Moreover, OFs contains particulates such as oral bacteria, desquamated epithelial cells, and blood cells. In this report, we will uniformly refer to the fluid from the oral cavity as OFs, but the main component of OFs must be (narrow-sense) saliva, which is secreted by the three major salivary glands and countless minor salivary glands, amounting to 0.5–1.5 L daily (Llena-Puy 2006). This abundantly secreted saliva component originates from blood plasma components. Although it is unlikely that blood plasma components appear directly in the oral cavity, the fact that many diagnostic markers found in plasma are also detected in OFs is intriguing (Wong 2006; Yan et al. 2009; Shiiki et al. 2011). Furthermore, whether the extracellular vesicle components in the blood move to the OFs and if so, through what mechanism, are points of interest and are currently under analysis (Gallo et al. 2012; Zlotogorski-Hurvitz et al. 2015; Nonaka and Wong 2017).

3 The Advantages of Focusing on EVs in OFs

(i) ***The advantage of focusing on OFs rather than other body fluids***

As mentioned above, since diagnostic markers circulating in the blood can also be detected in OFs (Wong 2006; Yan et al. 2009; Shiiki et al. 2011), comparable diagnostic information can be expected to be obtained from OFs, which do not need invasive blood collection. OFs diagnostics techniques are often touted for their "non-invasiveness." Considering that diagnostics such as genetic testing, hormone testing, and virus testing using OFs collected at home and sent to testing laboratories via mail have been put into practical use, it is expected that similarly non-invasive OF-derived EV-diagnostic systems will be established in the near future. However, we have noted that the collection of OFs using expectoration, such as gathering 5 mL of OFs, can surprisingly be a time-consuming task (unpublished experience). This is particularly pertinent for older adults who produce less OFs, and thus the collection process might not be minimally invasive for them. When considering collection of OFs in hospitals, it is important to find diagnostic advantages of the EV-based testing over conventional blood tests.

(ii) ***The advantage of focusing on the EVs in OFs, not just the OFs as a whole***

It goes without saying, specificity and sensitivity are crucial for diagnostics, but cost and simplicity of diagnostic systems are also important points to consider. Isolating specific subclasses of EVs from OFs is a time-consuming and costly process, so it must yield a clinical impact that justifies these expenses. The advantages of diagnostics focused on EVs can be summarized as (a) concentration of diagnostic information, (b) analysis of information molecules as a package, and (c) setting the correct window for analysis as shown in Fig. 1.

(a) **Concentration of diagnostic information**

Due to the wide dynamic range of molecule presence in body fluids, when the diagnostic information is carried by only trace amounts of a molecule, it is important to achieve a high signal-to-noise ratio amidst the background of more abundant molecules. If it is known that the diagnostic molecule is carried by a specific subclass of EVs, analyzing the targeted diagnostic molecule after concentrating that subclass of EVs can be expected to yield a high signal-to-noise ratio (Fig. 1a).

(b) **Analysis of information molecules as a package**

A significant characteristic of cellular vesicles is that they carry molecules from the parent cell as a package. Therefore, analyzing the group of molecules uniquely carried by a subclass of EVs can provide more sophisticated diagnostic information (Fig. 1b). The simplest form of such analysis is the sandwich ELISA method, and the ultimate form is analysis at the single EV level (Hilton and White 2021). However, it should be noted that recent studies have shown that the molecules carried by extracellular vesicles, especially in the protein corona covering their surface, are molecules that associate after secretion from the parent cell rather than being derived from the parent cell, and in some cases, this protein corona that associates after secretion may play an important role in the physiological activity of the extracellular vesicles (Toth et al. 2021; Buzas 2022), which may make it difficult to define "a single EV". Advancements in the basic understanding of EV genesis are expected to resolve this issue.

(c) **Setting the correct window**

With the clarification of the presence of EVs, it has been discovered that molecules previously thought to be small soluble molecules are in fact associated with larger particles in body fluids (Hagey et al. 2021; Zhang and Wrana 2014). This finding plays a crucial role in guiding the fractionation of target diagnostic molecules from body fluids. For example, cancer cell-derived diagnostic DNA fragments circulate in the bloodstream as complexes with large EVs (L-EVs), small EVs (sEVs) or soluble protein depending on disease stage (Hagey et al. 2021), which emphasizes appropriate setting of size-window for sample preparation to include informative molecules (Fig. 1c). Furthermore, molecules inside EVs cannot be accurately quantified without the proper disruption of the lipid bilayer (f for example, by treating with detergent) of the vesicles.

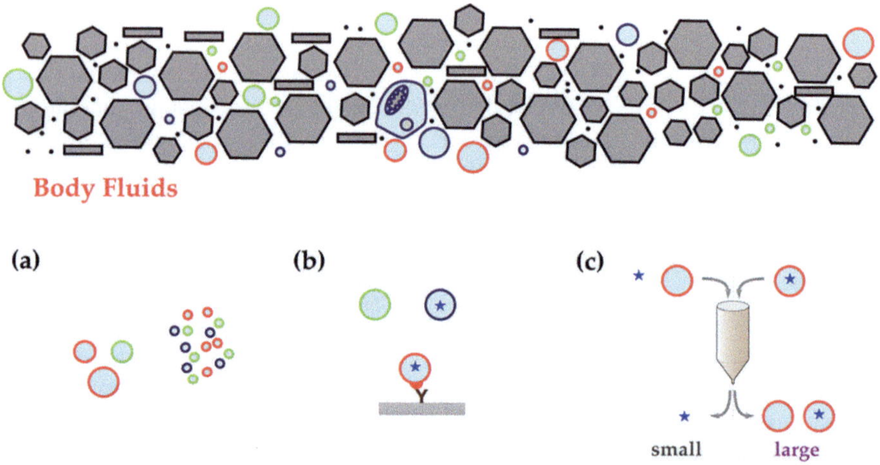

Fig. 1 Schematic illustration of body fluids and the advantage of focusing on the EVs in body fluids (or OFs in this review). The upper figure schematically illustrates the diversity of particles contained in OFs. The circles represent EVs, and the differences in color represent differences in the physicochemical properties of EVs. **a** In addition to EVs, OFs contain a large amount of non-EV substances, which overwhelmingly outnumber EVs in terms of quantity. If the desired diagnostic information is associated with EVs, concentrating the EV fraction from this mixture by some method would separate the large amount of background molecules, and can obtain diagnostic information with a higher signal to noise (S/N) ratio. **b** The most significant feature of EVs is that they carry multiple information molecules (proteins, lipids, nucleic acids, metabolites, etc.) as packages. For example, if a certain EV surface molecule (schematically shown as a red semicircle) is related to disease information, a combination of other molecules carried simultaneously by the EV carrying this molecule may provide more accurate diagnostic information. **c** We may have to reviewing the existing diagnostic protocols under the conditions that the presence of EVs are recognized. For example, even diagnostic molecules (blue ★ on the left in the figure) that were believed to exist as soluble molecules before may require pre-treatment as large particles (right in the figure) if they exist in association with EVs (right in the figure), otherwise attempting to obtain diagnostic information in the wrong window. Even newly discovered diagnostic molecules are required to carefully examine their mode of existence in body fluids

4 The Heterogeneity of EVs in OFs

According to the experimental guidelines for extracellular vesicles published (minimum information for studies of extracellular vesicles, MISEV) by the International Society for Extracellular Vesicles, EVs are defined as "small vesicles" delimited by a "lipid bilayer" that cannot replicate on their own (Lotvall et al. 2014; Thery et al. 2018; Welsh et al. 2024). We generally collect "particles resembling EVs" from a body fluid, but it is not easy to conclude these are EVs (as MISEV defined). In particular, the criteria of "delimited by a lipid bilayer" is a difficult point to prove. Electron microscopy provides a relatively direct method for confirming the bilayer, but it is a time-consuming and laborious task for conventional researchers. Body fluids contain various particulate matter within them. For instance, OFs includes, apart from EVs, blood cells, cell debris from desquamated epithelium, oral bacteria, outer membrane vesicles (OMVs) derived from oral bacteria, and remnants of food. The potential cellular sources of EVs include oral epithelial cells, salivary gland cells, and blood cells. There is also the possibility that EVs contained in circulating blood or interstitial fluid may be directly released into the oral cavity through some mechanism. In addition to the diversity of cellular sources, a single cell can release EVs through multiple pathways (Colombo et al. 2014; van Niel et al. 2018; Dixson et al. 2023). While the overall picture of these mechanisms is not yet fully understood, they can be broadly classified into three categories (Fig. 2). (i) pathways involving intracellular organelles such as (Johnstone 1987; Raposo et al. 1996), lysosomes, and autophagosomes (Jeppesen et al. 2019); (ii) pathways related to cellular protrusions and cell motility, which, in a sense, are created by the tearing of cells (Rilla 2021; Sung et al. 2021); and (iii) pathways originating from controlled cell death or inflammatory responses (Poon et al. 2019). The detailed elucidation of each pathway and the characteristics of the EVs produced by each are significant challenges in basic research on EVs, and it will take more efforts for the complete picture to become clear.

Ideal development of diagnostic methods would involve identifying the cells and pathways associated with the release of disease-related EVs and what characteristics should be used to differentiate the subclasses of EVs that carry the diagnostic information. However, the current situation is such that these basic insights are still insufficient. Therefore, a practical strategy is to differentially prepare various oral EVs by various methods and investigate where the target diagnostic information is concentrated which is a realistic strategy for establishing a diagnostic system.

5 Strategies for Differentiating EVs

The differentiation of EVs is primarily divided into two strategies: (i) methods that use differences in the physicochemical properties of EVs, and (ii) methods that separate EVs focusing specific molecular species displayed on them using molecular

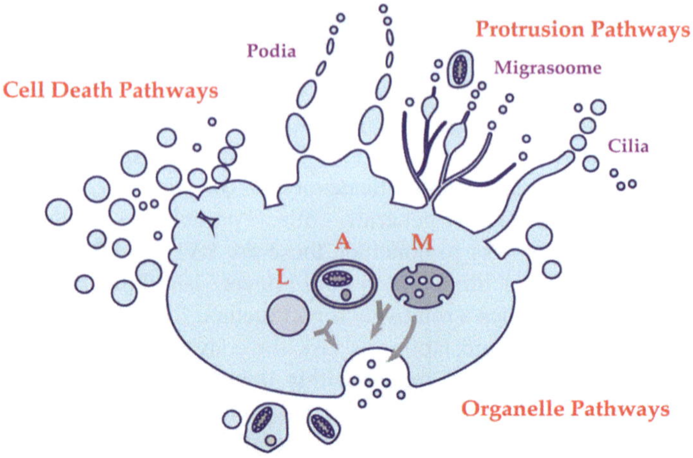

Fig. 2 Schematic illustration showing that various EVs are generated through various pathways. The generation pathways are broadly categorized into (i) pathways involving intracellular organelles ("Organelle Pathways"), (ii) pathways related to cellular protrusions and cell motility ("Protrusion Pathways"), and (iii) pathways originating from controlled cell death or inflammatory responses ("Cell Death Pathways"). In addition to the involvement of the classical Johnstone's multivesicular bodies (MVBs) (denoted as "M" in the figure) (Johnstone et al. 1987), recent studies have suggested the possible involvement of autophagosomes (denoted as "A" in the figure), lysosomes (denoted as "L" in the figure), and amphisomes formed by the fusion of these organelles (not shown in the figure) (Jeppesen et al. 2019), but further research is needed to elucidate the details. "Protrusion Pathways" are closely related to cell motility and are thought to be generated from filopodia, microvilli, cilia, etc., and are considered to be closely related to cell motility (Rilla 2021; Sung et al. 2021). Nomenclatures more focusing on the motility, such as migrasomes and retraction fibers, have also been proposed (Taylor and Robbins 1963; Ma et al. 2015), and there is a possibility that EVs may also arise from tunneling nanotubes (or cytonemes) that connect cells (Ramirez-Weber and Kornberg 1999; Rustom et al. 2004). In addition to classical apoptosis (Poon et al. 2019), several different pathways named "regulated cell death," such as necroptosis and pyroptosis, are involved in "Cell Death Pathways" (Wallach et al. 2016). In a sense, they are pathways deeply related to "Organelle Pathways," and there may be some overlap, but here we classify them as different categories. There is also a possibility that various inflammatory reactions may be involved in this category (Wallach et al. 2016; Linkermann et al. 2014). Future research will reveal a more detailed picture of the various inflammatory reactions that may be involved in this category. Ectosomes and microvesicles, which are thought to be generated by budding of the cell plasma membrane, are considered to involve generation mechanisms in all three categories and are not explicitly shown in this figure

biological techniques. The latter is characterized by affinity separation methods that use molecules, such as antibodies, which bind specifically to particular molecules; various affinity EV separation kits are already commercially available. In our case, we have been developing various methods primarily using centrifugation, based on the differentiation techniques that utilize the physicochemical property differences (i). Diagnostic methods using centrifuges, especially ultracentrifuges, are considered difficult for clinical implementation, but we consider it as development stage for clinical implementation. Ultimately, we need to complete a system that isolate specific subclasses of EVs using more convenient columns, filters, and microfluidic systems. Below, we introduce the methods we have reported so far. The techniques described here for differentiating 'EVs' are, in reality, used to distinguish various forms of 'particulate matter.' It should be noted that this includes not only EVs but also other particulate matter.

6 Pentapartite Fractionation of OFs by Differential Centrifugation

Differential centrifugation is based on a straightforward concept: large particles are pelleted at lower speeds, while smaller particles require higher speeds to sediment. Several protocols designed to isolate sEVs, typically around 100 nm in diameter, employ this method of differential centrifugation, which removes larger particles unnecessary for sEV isolation (including L-EVs) at low-speed centrifugation before collecting small particles in the pellet at high speed. Using this simple method, Hiraga et al. reported a study in which human OFs were precipitated at 300 g for 10 min, 2,000 g for 10 min, 10,000 g for 30 min, and 160,000 *g* for 60 min (Hiraga et al. 2021). Following centrifugation, the resulting fractions are referred to as 0.3, 2, 10, 160 K and Sup, respectively, which constitute pentapartite fractionation of OFs (Fig. 3). The 160 K fraction is where the so-called sEVs are concentrated. In the early stages of the EV research of this century, sEVs monopolized attention, but now it is recognized that, besides sEVs, L-EVs also exist and are likely to play an important biological role (Jeppesen et al. 2019; Kowal et al. 2016; Melentijevic et al. 2017; Kilinc et al. 2021; Mathieu et al. 2021; Rai e al. 2021; Skovronova et al. 2021; Weems et al. 2023; Karbanova et al. 2024; Ortmann et al. 2024; Rossi et al. 2024; Schone et al. 2024; Kawano et al. 2023). In this sense, L-EVs are an indispensable source for extracting diagnostic information. The presence of vesicles enclosed by a lipid bilayer in each fraction divided by this differential centrifugation method has been confirmed by embedding each fraction in resin, creating ultrathin sections, and observing them under an electron microscope with positive staining (Hiraga et al. 2021). Furthermore, the distribution of molecules frequently used as EV markers, such as CD63 and CD81 has been confirmed to be widespread across each fraction. These molecules have been often referred to as exosome markers. Although 'exosome' is not strictly defined and can lead to confusion (Couch et al. 2021; MacDonald and Salem 2023), if we were to assume exosomes as EVs originating through multivesicular body (MVB) as used by Johnstone (1987), then they should indeed be concentrated in the 160 K fraction based on size. However, our research, along with many other reports (Kowal et al. 2016; Lischnig et al. 2022; Xu et al. 2016), has detected these molecules in the fractions containing L-EVs, suggesting that calling CD63 and CD81 exosome markers is erroneous. Moreover, miRNAs, which are often and uncritically thought to be primarily carried by sEVs, are found not only in the 160 K fraction but also in other fractions. There are numerous reports that miRNAs and DNAs, in addition to being inside EVs, are also abundantly associated with their exterior (Lazaro-Ibanez et al. 2019; Zand Karimi et al. 2022). When using nucleic acids in OFs for diagnostic information, it would be narrow the window to nucleic acids associated with sEVs (Fig. 1c). It is conceivable that the desired diagnostic information may be concentrated in the fraction containing L-EVs.

We have not specifically analyzed EVs in the oral cavity with a particular oral disease in mind, but we have investigated the distribution of epidermal growth factor receptor (EGFR), whose expression changes have been reported in association with oral squamous cell carcinoma (OSCC) (Zlotogorski-Hurvitz et al. 2015; Sanderson et al. 2008; Jyothi Meka et al. 2015; Higginbotham et al. 2016). The

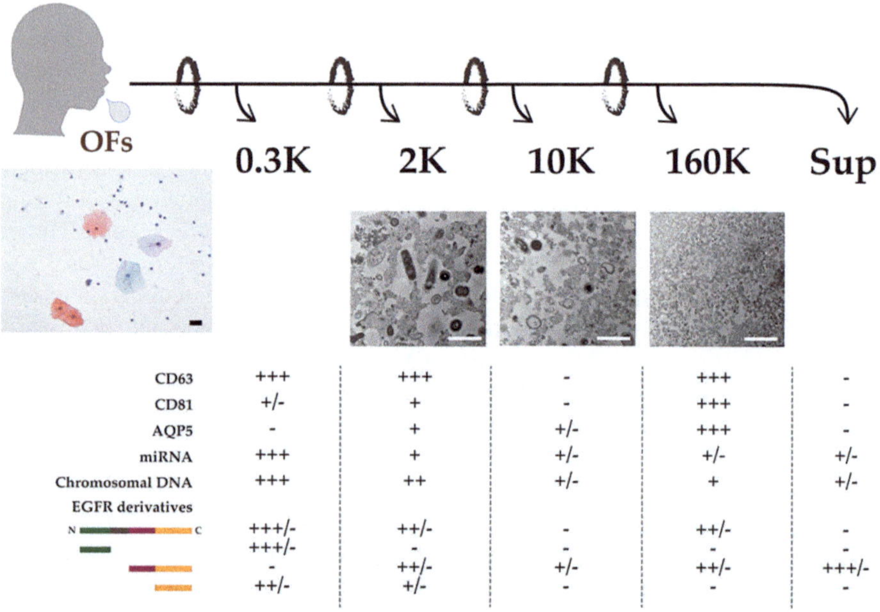

Fig. 3 Summary of the pentapartite fractionation experiment of OFs by differential centrifugation (Hiraga et al. 2021). Human OFs were divided into five fractions by differential centrifugation, and the characteristics of each fraction were summarized. The leftmost image is a light microscopic view of whole OFs stained with Papanicolaou stain (Bibbo and Wilbur 2014). The scale bar is 20 μm. The blue and red stained objects are squamous epithelial cells. The small particulate objects are hematopoietic cells, and EV-equivalent objects are not observed at this magnification. The three images on the right show TEM observations after positive staining of ultrathin sections of the 2, 10, and 160 K fractions after resin embedding. The scale bars are all 2 μm. Particles of various sizes are observed, and some have a lipid bilayer structure. In addition, large EVs with complex multilayered structures are also observed. The distribution of major molecules is shown below. Those separated by slashes showed differences in the results obtained from healthy individuals on the left and OSCC patients on the right (see text for details; refer to the original literature for more details (Hiraga et al. 2021). In this experiment, the Sup fraction is diluted compared to other fractions; therefore, caution is required when comparing molecular concentrations

samples were collected from healthy individuals and OSCC patients, and interestingly, some processed fragments of EGFR were detected in the fraction containing EVs, which agreed with the previous studies reporting some cancer cell lines release EGFR-containing EVs (Jeppesen et al. 2019).

It is of interest to determine which sells the EVs contained in the oral cavity originate from. Aquaporin 5 (AQP5) is a channel protein for water molecules and a membrane protein that characterizes salivary glands in the oral space (Ishikawa and Ishida 2000; Delporte 2014). There are reports that it is not expressed in the oral epithelium. When we examined the distribution of AQP5, it was found to be enriched in the 160 K fraction with less expression in the 2 and 10 K fractions. This indicates that the 160 K fraction con-tains at least EVs released from salivary glands. Thus, the oral cavity contains EVs of various sizes, and when developing diagnostic methods for specific diseases, it is necessary to first determine which fraction to focus on for research. It is essential to avoid uncritically narrowing the window to fractions containing only sEVs when devising a development strategy.

7 High-Throughput Analysis of sEVs in OFs

Prior to the above-mentioned research exploring the overall picture of EVs contained in OFs, our initial research reported a study focusing on sEVs in OFs (Iwai et al. 2016). As the gold standard for sEV preparation, there is a method using equilibrium density gradient centrifugation, which involves long-time centrifugation through a medium with a density gradient to separate and prepare particles based on density. Since EVs are vesicular structures containing a large amount of water molecules, they have relatively low density, and this property is utilized to separate them from other non-vesicular components (Thery et al. 2006).

In order to use this method, which has been established as a method for preparing sEVs from the supernatant of cultured cells, for the isolation of sEVs from OFs, we first focused on (i) the problem of OF-specific viscosity and (ii) high-throughput processing. OFs is characterized by its viscosity, primarily derived from mucin (Zalewska et al. 2000), but when applying equilibrium density gradient centrifugation to OFs, we encountered problems with particle deployment within the density gradient, presumably due to its viscosity (Momen-Heravi et al. 2012). Using the linear particle fractionation in the density gradient as an indicator, we found that sonication treatment immediately after collecting the OFs resolved the issue. There are concerns that physical treatments like sonication might artificially affect subsequent vesicle separation (Liu et al. 2020), but at least for small EVs, no results were obtained indicating that sonication had an effect (Yamamoto et al. 2021; Hiraga et al. 2021). Other methods to solve the viscosity issue of OFs include dilution and the use of mucinase (Ogawa e al. 2011; Shon et al. 2020). Dilution is simple and seems to have less physical impact (Ogawa e al. 2011), but it increases the volume of the sample. The method is suitable for ion-exchange column-based separation but not for centrifugation-based methods. Enzymatic treatment with mucinase is also an attractive method (Shon et al. 2020), which may become more widespread as enzymes become commercially available.

In Iwai's report (2016), sucrose, iodixanol (OptiPrep), and iohexol (Nycodenz) were compared as media for creating the density gradient, and iodixanol showed the best separation capabilities. Similar reports have been made by other studies (Neves et al. 2009), suggesting that iodixanol, rather than sucrose, should be used for density gradient separation of EVs.

For equilibrium density gradient centrifugation, it was essential to use a swinging-bucket rotor to avoid disturbance of the density gradient. However, swinging-bucket rotors allow only a limited number of samples to be centrifuged at once and require long centrifugation times due to its high k-factor value (Cvjetkovic et al. 2014), which presents problems for diagnostic analysis that requires multiple clinical samples. To determine whether the predicted disturbances in the density gradient post-centrifugation indeed occur with an angle rotor, we tested using iodixanol for density gradient centrifugation. Contrary to expectations, it was found that angle rotors could also perform density gradient separation without issues. This finding, increasing the number of samples that can be centrifuged at once and reducing centrifugation time, is reported as a protocol (Iwai et al. 2017). Using this protocol, sEV fractions from six healthy human OF samples were deployed in the density gradient, and the expression of 96 miRNAs was investigated. Many were found to be enriched in the fraction with a density of 1.10–1.13 g/ml, which is known to be the fraction of sEVs (Raposo et al. 1996) and expression of CD63 and AQP5 were confirmed in the fraction. However, it should be noted that although this study has shown the association of several miRNAs with the sEV fraction having densities of 1.10–1.13 g/ml, other fractions, including those containing larger EVs, have not been investigated.

8 Utilizing Specimen-Specific Drift of Densities for Differentiation of EVs in OFs

The principle of equilibrium density gradient centrifugation is that as particles move within the density gradient medium, they reach a point where their intrinsic density matches the surrounding medium, balancing the buoyant force with the centrifugal force, causing them to stop moving. There are two methods: floating the particle up from the bottom of the tube and spinning it down from the top to the bottom. Either way, it takes time for the particles to move through the density gradient medium. While it might be possible to calculate the time to reach equilibrium for an ideal particle, the bulk and fine structures of actual biological samples are quite complex, making it difficult to hastily determine if equilibrium has been reached.

Yamamoto et al. took a different perspective (Yamamoto et al. 2021) and performed fractionation in both directions, floating the same sample from the bottom of a tube and spinning it down from the top of a tube, and after sufficient centrifugation time, the particles collected from the same density fraction were analyzed (Fig. 4a). “Sufficient time” is a tricky point; most proteins do not reach the same density from both directions even after the 17 h that many protocols adopt. Similar reports have already been made by other groups (Kowal et al. 2016; Iwai et al. 2016; Palma et al. 2012; Aalberts et al. 2012), indicating that caution is needed when interpreting results from extracellular vesicle concentration experiments using equilibrium density gradient centrifugation.

As it seemed unlikely that EVs reached equilibrium after only 17 h, Yamamoto et al. set a time limit of 96 h. Under these conditions, OFs from three healthy individuals were centrifuged in both directions, and mass spectrometry (MS) analysis was conducted on the proteins in each fraction. As a result, out of a total of 11,749 detected proteins, 1,429 non-overlapping proteins were identified. Out of them, 476 were common among all three specimens, among which 111 proteins had identical density over all three specimens. Interestingly, some proteins out of them showed specimen-specific densities. For an example, densities of CD63 containing fraction were 1.12, 1.10 and 1.09 g/ml for specimen-1, -2 and -3, respectively. We classified 16 proteins that have the same density-drift pattern Group IA-1 (Fig. 4b), which contained CD81, CD9 and other membrane proteins (Fig. 4c). Group IA-2

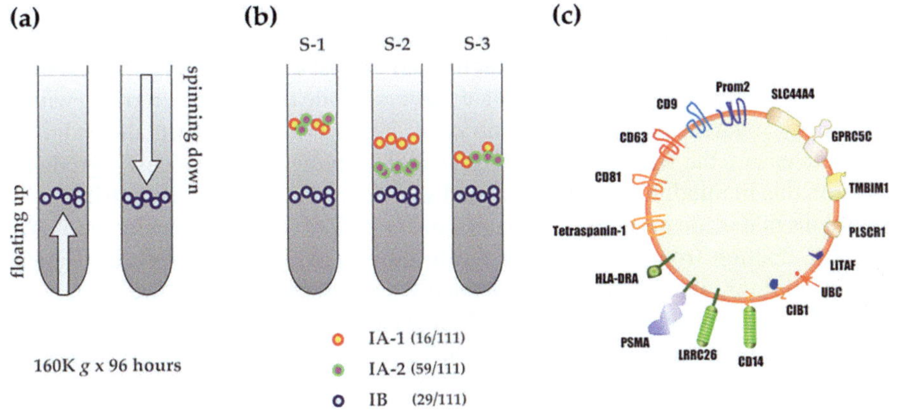

Fig. 4 **a** In density gradient ultracentrifugation experiments, there are two protocols: floating the specimen up from the bottom of high density and spinning the specimen down from the top of low density. Yamamoto et al.‘s method was to ultracentrifuge the same sample in both directions for a long time (96 h), and then perform MS analysis for each density fraction (Yamamoto et al. 2021). Proteins that reach the same density fraction in both directions are considered to be in an equilibrium state within the EVs. **b** Analysis of the L-EVs + sEVs fraction from three healthy individuals (S-1, S-2, S-3) revealed that in addition to 29 proteins (Group IB) that reached equilibrium, 16 proteins in Group IA-1 and 59 proteins in Group IA-2 were observed to reach equilibrium at specimen-specific densities (the remaining 7 had other minor specimen-specific densities). **c** The expected mode of existence of the 16 proteins of Group IA-1. Except for polyubiquitin-C (UBC), all are membrane proteins or proteins anchored to the membrane. Refer to the literature (Yamamoto et al. 2021) for protein abbreviations, etc.

comprised other 59 proteins that had distinct density-drift pattern than Group IA-1 (Fig. 4b), many of which are involved in membrane dynamics, membrane trafficking and innate immune system. Group IB, which comprised 29 proteins, had identical densities among specimens, which contained immunoglobulin, HSPB1, S100A14 etc. (Yamamoto et al. 2021). Differences in oral pH or salt concentration may account for the observed sample-specific density drift of EVs. In any case, it was an interesting finding that the groups of proteins that are recovered from the same density fraction across the 3 samples and those that are recovered at sample-specific densities are associated with extracellular vesicles with different physicochemical properties.

9 EV Differentiation by Sedimentation Patterns (ESP) Revealed Three Groups of L-EVs

In Yamamoto's experiments (2021), the investigated OFs were the supernatant fraction obtained from sonication followed by 2,600 g × 30 min centrifugation, which corresponds to combined fraction of the 10 and 160 K fractions of Hiraga's pentapartite analyses. Under these centrifugation conditions, L-EVs, which should be sedimented during 2,000 g × 1 h centrifugation, are thus excluded. Kawano et al. have focused on L-EVs contained in the 2 K fraction of OFs and have developed the method named "EV differentiation by sedimentation patterns (ESP)" (Fig. 5). In this ESP method, instead of waiting for equilibrium, snapshot-pictures of sedimentation patterns are captured for EVs (particulate matters) while they are still moving within the density gradient medium (Kawano et al. 2023). The experimental conditions sought to ensure the widest possible distribution of proteins in the density gradient, and with this in mind, Kawano et al. found short-duration, low-speed centrifugation conditions that made proteins fractionated over as wide a range of density gradients. The 2 K fraction from the OFs of three healthy individuals was separated under these conditions, and the proteins contained in each density fraction were analyzed using MS. Then, the common proteins detected among the three healthy individuals were divided into three groups based on their movement patterns. The differences in migration patterns are presumed to be due to the generation mechanisms and physicochemical properties of the EVs (or large particles not defined as EVs) with which each protein is associated, but the details will have to await future research. Surprisingly, when the biological functions of the proteins belonging to these three groups were categorized, there was a striking alignment between the groupings and related functions. Namely, Group 1 proteins were mostly shared with small EV cargos and enriched in immuno-related proteins, Group 2 proteins were associated with energy metabolism and protein synthesis, and Group 3 proteins were associated with vesicle trafficking. This alignment of functions suggests that the three groups classified by this ESP method are indeed related to some generative functions. Notably, there was little overlap between the proteins in the L-EV-containing fractions identified by Kawano et al. and those in the non-L-EV fractions analyzed by Yamamoto et al.

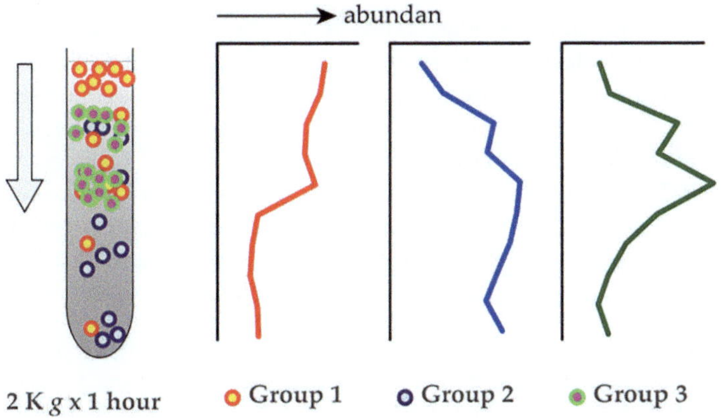

Fig. 5 An experiment to differentiate L-EVs by utilizing differences in sedimentation patterns in low-speed, short-time density gradient centrifugation (Kawano et al. 2023). The 2 K fraction was spun down from the top to the bottom of the tube at 2,000 g for 1 h in a density gradient centrifugation, and the proteins contained in each density fraction were analysed by MS. Based on the distribution pattern of proteins in each fraction, they could be divided into three groups, and this grouping was related to Group 1 (denoted by red circles and line): shared markers with sEVs and immunological reaction, Group 2 (blue circles and line): energy metabolism and protein synthesis, and Group 3 (green circles and line): vesicle trafficking, respectively

(2021). If we assume all these proteins are associated with EVs, it becomes apparent that EVs of different sizes are associated with different proteins.

10 Conclusion Remarks

Even in EV research using supernatants from cultured cell lines, it is surprising how different the results can be when varying the type of cell line or culture conditions (Nordin 2022; Scheiber et al. 2022; Rayamajhi et al. 2023). Given that various EVs derived from different cells are contained within the body's fluids, the complexity only increases. While there is no doubt that new diagnostic methods and therapies based on EVs will eventually come to fruition, there are two key points to bear in mind in such development research: (i) do not uncritically narrow the focus to sEVs. In the early stages of the rise of EV research in this century, so much attention was focused on sEVs that it was as if only EVs generated via MBVs were EVs. While there is no doubt that sEVs are interesting entities, and ultimately diagnostic and therapeutic methods might converge around sEVs, at the initial stages of development research, it is crucial to survey without prejudice which subclass of the diverse EVs possesses the desired function and then narrow down the window of focus. And (ii) do not uncritically narrow the focus to just EVs. Particulate matter in body fluids is not solely composed of EVs. Preparing a sample consisting entirely of EVs is extremely difficult, and there is no need to uncritically believe that the diagnostic information or therapeutic potential resides in a particular EV subclass. There is no issue if the ultimately desired diagnostic or therapeutic method is realized by non-EV particles. It is also essential to carefully consider non-EV particles as potential carriers of the desired function. Recent reports of protein coronas covering EV surfaces (many of which likely associate after secretion from parent cells independently) (Toth et al. 2021; Buzas 2022), association of EV with lipoprotein (Lozano-Andres et al. 2023) and extracellular matrix (Debnath et al. 2023), and reports of super-large non-membrane-bound assemblies composed of multiple organelles (Zaffagnini et al. 2024) suggest that our knowledge of biogenic particles is still quite limited.

Acknowledgements The research results introduced were obtained through collaborative research with Drs. Kazuya Iwai, Satoshi Yamamoto, Chiho Hiraga, Takamasa Kawano, Koji Ueda, Sachiko Matsumura as well as Ms. Tamiko Minamisawa at Japanese Foundation for Cancer Research and Dr. Kohji Okamura at National Center for Child Health and Development. I am also grateful for Dr. Yasutomo Yajima, Takeshi Nomura and Sadamitsu Hashimoto at Tokyo Dental College for their helpful supports and discussions.

References

Aalberts, M. *et al.* Identification of distinct populations of prostasomes that differentially express prostate stem cell antigen, annexin A1, and GLIPR2 in humans. *Biol Reprod* **86**, 82 (2012). https://doi.org/10.1095/biolreprod.111.095760

Bibbo, M. & Wilbur, D. *Comprehensive cytopathology, 4th edition.* (Saunders, 2014).

Brody, I., Ronquist, G. & Gottfries, A. Ultrastructural localization of the prostasome - an organelle in human seminal plasma. *Ups J Med Sci* **88**, 63–80 (1983). https://doi.org/10.3109/03009738309178440

Buzas, E. I. Opportunities and challenges in studying the extracellular vesicle corona. *Nat Cell Biol* **24**, 1322–1325 (2022). https://doi.org/10.1038/s41556-022-00983-z

Colombo, M., Raposo, G. & Thery, C. Biogenesis, secretion, and intercellular interactions of exosomes and other extracellular vesicles. *Annu Rev Cell Dev Biol* **30**, 255–289 (2014). https://doi.org/10.1146/annurev-cellbio-101512-122326

Correll, V. L. *et al.* Optimization of small extracellular vesicle isolation from expressed prostatic secretions in urine for in-depth proteomic analysis. *J Extracell Vesicles* **11**, e12184 (2022). https://doi.org/10.1002/jev2.12184

Couch, Y. *et al.* A brief history of nearly EV-erything - The rise and rise of extracellular vesicles. *J Extracell Vesicles* **10**, e12144 (2021). https://doi.org/10.1002/jev2.12144

Cvjetkovic, A., Lotvall, J. & Lasser, C. The influence of rotor type and centrifugation time on the yield and purity of extracellular vesicles. *J Extracell Vesicles* **3** (2014). https://doi.org/10.3402/jev.v3.23111

Debnath, K., Heras, K. L., Rivera, A., Lenzini, S. & Shin, J. W. Extracellular vesicle-matrix interactions. *Nat Rev Mater* **8**, 390–402 (2023). https://doi.org/10.1038/s41578-023-00551-3

Delporte, C. Aquaporins in salivary glands and pancreas. *Biochim Biophys Acta* **1840**, 1524–1532 (2014). https://doi.org/10.1016/j.bbagen.2013.08.007

Dixson, A. C., Dawson, T. R., Di Vizio, D. & Weaver, A. M. Context-specific regulation of extracellular vesicle biogenesis and cargo selection. *Nat Rev Mol Cell Biol* **24**, 454–476 (2023). https://doi.org/10.1038/s41580-023-00576-0

Gallo, A., Tandon, M., Alevizos, I. & Illei, G. G. The majority of microRNAs detectable in serum and saliva is concentrated in exosomes. *PLoS One* **7**, e30679 (2012). https://doi.org/10.1371/journal.pone.0030679

Hagey, D. W. *et al.* Extracellular vesicles are the primary source of blood-borne tumour-derived mutant KRAS DNA early in pancreatic cancer. *J Extracell Vesicles* **10**, e12142 (2021). https://doi.org/10.1002/jev2.12142

Higginbotham, J. N. *et al.* Identification and characterization of EGF receptor in individual exosomes by fluorescence-activated vesicle sorting. *J Extracell Vesicles* **5**, 29254 (2016). https://doi.org/10.3402/jev.v5.29254

Hilton, S. H. & White, I. M. Advances in the analysis of single extracellular vesicles: A critical review. *Sens Actuators Rep* **3** (2021). https://doi.org/10.1016/j.snr.2021.100052

Hiraga, C. *et al.* Pentapartite fractionation of particles in oral fluids by differential centrifugation. *Sci Rep* **11**, 3326 (2021). https://doi.org/10.1038/s41598-021-82451-6

Ishikawa, Y. & Ishida, H. Aquaporin water channel in salivary glands. *Jpn J Pharmacol* **83**, 95–101 (2000). https://doi.org/10.1254/jjp.83.95

Iwai, K., Minamisawa, T., Suga, K., Yajima, Y. & Shiba, K. Isolation of human salivary extracellular vesicles by iodixanol density gradient ultracentrifugation and their characterizations. *J Extracell Vesicles* **5**, 30829 (2016). https://doi.org/10.3402/jev.v5.30829

Iwai, K., Yamamoto, S., Yoshida, M. & Shiba, K. Isolation of Extracellular Vesicles in Saliva Using Density Gradient Ultracentrifugation. *Methods Mol Biol* **1660**, 343–350 (2017). https://doi.org/10.1007/978-1-4939-7253-1_27

Jeppesen, D. K. *et al.* Reassessment of Exosome Composition. *Cell* **177**, 428–445 e418 (2019). https://doi.org/10.1016/j.cell.2019.02.029

Johnstone, R. M., Adam, M., Hammond, J. R., Orr, L. & Turbide, C. Vesicle formation during reticulocyte maturation. Association of plasma membrane activities with released vesicles (exosomes). *J Biol Chem* **262**, 9412–9420 (1987).

Jyothi Meka, N. *et al.* Quantitative Immunoexpression of EGFR in Oral Potentially Malignant Disorders: Oral Leukoplakia and Oral Submucous Fibrosis. *J Dent Res Dent Clin Dent Prospects* **9**, 166–174 (2015). https://doi.org/10.15171/joddd.2015.031

Karbanova, J. *et al.* Extracellular lipidosomes containing lipid droplets and mitochondria are released during melanoma cell division. *Cell Commun Signal* **22**, 57 (2024). https://doi.org/10.1186/s12964-024-01471-7

Kawano, T. *et al.* Differentiation of Large Extracellular Vesicles in Oral Fluid: Combined Protocol of Small Force Centrifugation and Pattern Analysis. *bioRxiv*, 2023.2004.2029.537961 (2023). https://doi.org/10.1101/2023.04.29.537961

Kilinc, S. *et al.* Oncogene-regulated release of extracellular vesicles. *Dev Cell* **56**, 1989–2006 e1986 (2021). https://doi.org/10.1016/j.devcel.2021.05.014

Kowal, J. *et al.* Proteomic comparison defines novel markers to characterize heterogeneous populations of extracellular vesicle subtypes. *Proc Natl Acad Sci U S A* **113**, E968–977 (2016). https://doi.org/10.1073/pnas.1521230113

Lazaro-Ibanez, E. *et al.* DNA analysis of low- and high-density fractions defines heterogeneous subpopulations of small extracellular vesicles based on their DNA cargo and topology. *J Extracell Vesicles* **8**, 1656993 (2019). https://doi.org/10.1080/20013078.2019.1656993

Linkermann, A., Stockwell, B. R., Krautwald, S. & Anders, H. J. Regulated cell death and inflammation: an auto-amplification loop causes organ failure. *Nat Rev Immunol* **14**, 759–767 (2014). https://doi.org/10.1038/nri3743

Lischnig, A., Bergqvist, M., Ochiya, T. & Lasser, C. Quantitative Proteomics Identifies Proteins Enriched in Large and Small Extracellular Vesicles. *Mol Cell Proteomics* **21**, 100273 (2022). https://doi.org/10.1016/j.mcpro.2022.100273

Liu, L. *et al.* Cancer associated fibroblasts-derived exosomes contribute to radioresistance through promoting colorectal cancer stem cells phenotype. *Exp Cell Res* **391**, 111956 (2020). https://doi.org/10.1016/j.yexcr.2020.111956

Llena-Puy, C. The role of saliva in maintaining oral health and as an aid to diagnosis. *Med Oral Patol Oral Cir Bucal* **11**, E449–455 (2006).

Lotvall, J. *et al.* Minimal experimental requirements for definition of extracellular vesicles and their functions: a position statement from the International Society for Extracellular Vesicles. *J Extracell Vesicles* **3**, 26913 (2014). https://doi.org/10.3402/jev.v3.26913

Lozano-Andres, E. *et al.* Physical association of low density lipoprotein particles and extracellular vesicles unveiled by single particle analysis. *J Extracell Vesicles* **12**, e12376 (2023). https://doi.org/10.1002/jev2.12376

Ma, L. *et al.* Discovery of the migrasome, an organelle mediating release of cytoplasmic contents during cell migration. *Cell Res* **25**, 24–38 (2015). https://doi.org/10.1038/cr.2014.135

MacDonald, E. & Salem, N., Jr. Dr. Eberhard Trams-The man who coined the name "exosomes" A prescient but largely forgotten pioneer. *J Extracell Vesicles* **12**, e12370 (2023). https://doi.org/10.1002/jev2.12370

Marzesco, A. M. *et al.* Release of extracellular membrane particles carrying the stem cell marker prominin-1 (CD133) from neural progenitors and other epithelial cells. *J Cell Sci* **118**, 2849–2858 (2005). https://doi.org/10.1242/jcs.02439

Mathieu, M. *et al.* Specificities of exosome versus small ectosome secretion revealed by live intracellular tracking of CD63 and CD9. *Nat Commun* **12**, 4389 (2021). https://doi.org/10.1038/s41467-021-24384-2

Melentijevic, I. *et al.* C. elegans neurons jettison protein aggregates and mitochondria under neurotoxic stress. *Nature* **542**, 367–371 (2017). https://doi.org/10.1038/nature21362

Momen-Heravi, F. *et al.* Impact of biofluid viscosity on size and sedimentation efficiency of the isolated microvesicles. *Front Physiol* **3**, 162 (2012). https://doi.org/10.3389/fphys.2012.00162

Neves, J. S., Perez, S. A., Spencer, L. A., Melo, R. C. & Weller, P. F. Subcellular fractionation of human eosinophils: isolation of functional specific granules on isoosmotic density gradients. *J Immunol Methods* **344**, 64–72 (2009). https://doi.org/10.1016/j.jim.2009.03.006

Nonaka, T. & Wong, D. T. W. Saliva-Exosomics in Cancer: Molecular Characterization of Cancer-Derived Exosomes in Saliva. *Enzymes* **42**, 125–151 (2017). https://doi.org/10.1016/bs.enz.2017.08.002

Nordin, J. Z. Transfection reagents affect Extracellular Vesicle cargo transfer to recipient cells: The importance of appropriate controls in EV research. *J Extracell Vesicles* **11**, e12227 (2022). https://doi.org/10.1002/jev2.12227

Ogawa, Y. *et al.* Proteomic analysis of two types of exosomes in human whole saliva. *Biol Pharm Bull* **34**, 13–23 (2011). https://doi.org/10.1248/bpb.34.13

Ortmann, W. *et al.* Large extracellular vesicle (EV) and neutrophil extracellular trap (NET) interaction captured in vivo during systemic inflammation. *Sci Rep* **14**, 4680 (2024). https://doi.org/10.1038/s41598-024-55081-x

Palma, J. *et al.* MicroRNAs are exported from malignant cells in customized particles. *Nucleic Acids Res* **40**, 9125–9138 (2012). https://doi.org/10.1093/nar/gks656

Pisitkun, T., Shen, R. F. & Knepper, M. A. Identification and proteomic profiling of exosomes in human urine. *Proc Natl Acad Sci U S A* **101**, 13368–13373 (2004). https://doi.org/10.1073/pnas.0403453101

Poon, I. K. H. *et al.* Moving beyond size and phosphatidylserine exposure: evidence for a diversity of apoptotic cell-derived extracellular vesicles in vitro. *J Extracell Vesicles* **8**, 1608786 (2019). https://doi.org/10.1080/20013078.2019.1608786

Rai, A., Fang, H., Claridge, B., Simpson, R. J. & Greening, D. W. Proteomic dissection of large extracellular vesicle surfaceome unravels interactive surface platform. *J Extracell Vesicles* **10**, e12164 (2021). https://doi.org/10.1002/jev2.12164

Ramirez-Weber, F. A. & Kornberg, T. B. Cytonemes: cellular processes that project to the principal signaling center in Drosophila imaginal discs. *Cell* **97**, 599–607 (1999). https://doi.org/10.1016/s0092-8674(00)80771-0

Raposo, G. *et al.* B lymphocytes secrete antigen-presenting vesicles. *J Exp Med* **183**, 1161–1172 (1996). https://doi.org/10.1084/jem.183.3.1161

Rayamajhi, S., Sulthana, S., Ferrel, C., Shrestha, T. B. & Aryal, S. Extracellular vesicles production and proteomic cargo varies with incubation time and temperature. *Exp Cell Res* **422**, 113454 (2023). https://doi.org/10.1016/j.yexcr.2022.113454

Rilla, K. Diverse plasma membrane protrusions act as platforms for extracellular vesicle shedding. *J Extracell Vesicles* **10**, e12148 (2021). https://doi.org/10.1002/jev2.12148

Rossi, I. V., de Almeida, R. F., Sabatke, B., de Godoy, L. M. F. & Ramirez, M. I. Trypanosoma cruzi interaction with host tissues modulate the composition of large extracellular vesicles. *Sci Rep* **14**, 5000 (2024). https://doi.org/10.1038/s41598-024-55302-3

Runz, S. *et al.* Malignant ascites-derived exosomes of ovarian carcinoma patients contain CD24 and EpCAM. *Gynecol Oncol* **107**, 563–571 (2007). https://doi.org/10.1016/j.ygyno.2007.08.064

Rustom, A., Saffrich, R., Markovic, I., Walther, P. & Gerdes, H. H. Nanotubular highways for intercellular organelle transport. *Science* **303**, 1007–1010 (2004). https://doi.org/10.1126/science.1093133

Sanderson, M. P. *et al.* Generation of novel, secreted epidermal growth factor receptor (EGFR/ErbB1) isoforms via metalloprotease-dependent ectodomain shedding and exosome secretion. *J Cell Biochem* **103**, 1783–1797 (2008). https://doi.org/10.1002/jcb.21569

Scheiber, A. L. *et al.* Culture Condition of Bone Marrow Stromal Cells Affects Quantity and Quality of the Extracellular Vesicles. *Int J Mol Sci* **23** (2022). https://doi.org/10.3390/ijms23031017

Schone, N. *et al.* PD-L1 on large extracellular vesicles is a predictive biomarker for therapy response in tissue PD-L1-low and -negative patients with non-small cell lung cancer. *J Extracell Vesicles* **13**, e12418 (2024). https://doi.org/10.1002/jev2.12418

Shiiki, N. *et al.* Association between saliva PSA and serum PSA in conditions with prostate adenocarcinoma. *Biomarkers* **16**, 498–503 (2011). https://doi.org/10.3109/1354750X.2011.598566

Shon, D. J. *et al.* An enzymatic toolkit for selective proteolysis, detection, and visualization of mucin-domain glycoproteins. *Proc Natl Acad Sci U S A* **117**, 21299–21307 (2020). https://doi.org/10.1073/pnas.2012196117

Skog, J. *et al.* Glioblastoma microvesicles transport RNA and proteins that promote tumour growth and provide diagnostic biomarkers. *Nat Cell Biol* **10**, 1470–1476 (2008). https://doi.org/10.1038/ncb1800

Skovronova, R. *et al.* Surface Marker Expression in Small and Medium/Large Mesenchymal Stromal Cell-Derived Extracellular Vesicles in Naive or Apoptotic Condition Using Orthogonal Techniques. *Cells* **10** (2021). https://doi.org/10.3390/cells10112948

Sung, B. H., Parent, C. A. & Weaver, A. M. Extracellular vesicles: Critical players during cell migration. *Dev Cell* **56**, 1861–1874 (2021). https://doi.org/10.1016/j.devcel.2021.03.020

Taylor, A. C. & Robbins, E. Observations on microextensions from the surface of isolated vertebrate cells. *Dev Biol* **6**, 660–673 (1963). https://doi.org/10.1016/0012-1606(63)90150-7

Thery, C. *et al.* Minimal information for studies of extracellular vesicles 2018 (MISEV2018): a position statement of the International Society for Extracellular Vesicles and update of the MISEV2014 guidelines. *J Extracell Vesicles* **7**, 1535750 (2018). https://doi.org/10.1080/20013078.2018.1535750

Thery, C., Amigorena, S., Raposo, G. & Clayton, A. Isolation and characterization of exosomes from cell culture supernatants and biological fluids. *Curr Protoc Cell Biol* **Chapter 3**, Unit 3 22 (2006). https://doi.org/10.1002/0471143030.cb0322s30

Toth, E. A. *et al.* Formation of a protein corona on the surface of extracellular vesicles in blood plasma. *J Extracell Vesicles* **10**, e12140 (2021). https://doi.org/10.1002/jev2.12140

van Niel, G., D'Angelo, G. & Raposo, G. Shedding light on the cell biology of extracellular vesicles. *Nat Rev Mol Cell Biol* **19**, 213–228 (2018). https://doi.org/10.1038/nrm.2017.125

Wallach, D., Kang, T. B., Dillon, C. P. & Green, D. R. Programmed necrosis in inflammation: Toward identification of the effector molecules. *Science* **352**, aaf2154 (2016). https://doi.org/10.1126/science.aaf2154

Weems, A. D. *et al.* Blebs promote cell survival by assembling oncogenic signalling hubs. *Nature* **615**, 517–525 (2023). https://doi.org/10.1038/s41586-023-05758-6

Welsh, J. A. *et al.* Minimal information for studies of extracellular vesicles (MISEV2023): From basic to advanced approaches. *J Extracell Vesicles* **13**, e12404 (2024). https://doi.org/10.1002/jev2.12404

Wong, D. T. Salivary diagnostics powered by nanotechnologies, proteomics and genomics. *J Am Dent Assoc* **137**, 313–321 (2006). https://doi.org/10.14219/jada.archive.2006.0180

Xu, R., Greening, D. W., Zhu, H. J., Takahashi, N. & Simpson, R. J. Extracellular vesicle isolation and characterization: toward clinical application. *J Clin Invest* **126**, 1152–1162 (2016). https://doi.org/10.1172/JCI81129

Yamamoto, S. *et al.* Specimen-specific drift of densities defines distinct subclasses of extracellular vesicles from human whole saliva. *PLoS One* **16**, e0249526 (2021). https://doi.org/10.1371/journal.pone.0249526

Yan, W. *et al.* Systematic comparison of the human saliva and plasma proteomes. *Proteomics Clin Appl* **3**, 116–134 (2009). https://doi.org/10.1002/prca.200800140

Yanez-Mo, M. *et al.* Biological properties of extracellular vesicles and their physiological functions. *J Extracell Vesicles* **4**, 27066 (2015). https://doi.org/10.3402/jev.v4.27066

Zaffagnini, G. *et al.* Mouse oocytes sequester aggregated proteins in degradative super-organelles. *Cell* **187**, 1109–1126 e1121 (2024). https://doi.org/10.1016/j.cell.2024.01.031

Zalewska, A., Zwierz, K., Zolkowski, K. & Gindzienski, A. Structure and biosynthesis of human salivary mucins. *Acta Biochim Pol* **47**, 1067–1079 (2000).

Zand Karimi, H. *et al.* Arabidopsis apoplastic fluid contains sRNA- and circular RNA-protein complexes that are located outside extracellular vesicles. *Plant Cell* **34**, 1863–1881 (2022). https://doi.org/10.1093/plcell/koac043

Zhang, L. & Wrana, J. L. The emerging role of exosomes in Wnt secretion and transport. *Curr Opin Genet Dev* **27**, 14–19 (2014). https://doi.org/10.1016/j.gde.2014.03.006

Zlotogorski-Hurvitz, A. *et al.* Human saliva-derived exosomes: comparing methods of isolation. *J Histochem Cytochem* **63**, 181–189 (2015). https://doi.org/10.1369/0022155414564219

Zonneveld, M. I. *et al.* Recovery of extracellular vesicles from human breast milk is influenced by sample collection and vesicle isolation procedures. *J Extracell Vesicles* **3**, 24215 (2014). https://doi.org/10.3402/jev.v3.24215

High Dimensional Cytometry for Studying Heterogeneous Small Particles

Kazuki Hattori, Yuichiro Iwamoto, Ryosuke Kojima, Yusuke Yoshioka, and Sadao Ota

Abstract Extracellular vesicles (EVs) play a crucial role in intercellular communication by transferring functional molecules such as nucleic acids and proteins. However, their small size and heterogeneity present significant challenges for analysis, requiring new, standardized methods with high sensitivity and scalability. This chapter introduces cutting-edge technologies, focusing on optical measurements as a promising, non-destructive approach for enhancing EV analysis.

1 Introduction

Extracellular vesicles (EVs) are lipid vesicles secreted by cells. In the past, extracellular vesicles were thought to be carriers for the disposal of unwanted cellular components. However, recent studies have shifted the focus to their important role in intercellular communication; EVs are now recognized for their ability to transfer functional nucleic acids, lipids, and proteins between cells, and their function as signal transducers in both normal physiological and pathological conditions has attracted attentions (Zaborowski et al. 2015). This has stimulated research on the role of EVs in cell-to-cell communication and cancer metastasis, as well as their potential as biomarkers in diagnosis (Chang et al. 2021).

Despite this growing interest, the analysis of EVs, including exosomes, membrane vesicles, and microvesicles, presents unique challenges. Conventional cellular analysis methods are inadequate because of the small size and high heterogeneity of EVs.

K. Hattori · Y. Iwamoto · S. Ota (✉)
Research Center for Advanced Science and Technology, The University of Tokyo, Komaba 4-6-1, Meguro 153-8904, Tokyo, Japan
e-mail: sadaota@g.ecc.u-tokyo.ac.jp

R. Kojima
Graduate School of Medicine, The University of Tokyo, Hongo 7-3-1, Bunkyo 113-0033, Tokyo, Japan

Y. Yoshioka
Department of Molecular and Cellular Medicine, Institute of Medical Science, Tokyo Medical University, Nishishinjuku 6-7-1, Shinjuku 160-0023, Tokyo, Japan

Y. Baba et al. (eds.), *Extracellular Fine Particles*,
https://doi.org/10.1007/978-981-97-7067-0_17

Standardized methods for EV isolation and analysis have not yet been established, resulting in inconsistencies and practical difficulties. For example, the proteomic profiles of EVs from the same source have been found to vary widely depending on the isolation method employed (Doyle and Wang 2019). Furthermore, the clinical application of EVs for diagnostic purposes requires the development of a highly reproducible and cost-effective analytical pipeline.

Gaps in the analysis results obtained from existing methods indicate the need for new technological advances that can become standard. The ideal techniques are expected to simultaneously have the sensitivity, throughput, scalability, and ability to detect a variety of particles, including physical and chemical properties, surface proteins, and nucleic acid content. In addition, the spatial and temporal distribution of the EVs, which reflects the complexity of intracellular mechanisms and cell–cell interactions, requires high-content, high-throughput cytometry techniques that go beyond conventional cell reporter systems. The integration of optical analysis and sequencing technologies is also expected to play an important role in understanding the multimodal nature of EVs and the mechanisms underlying their function.

In this chapter, we provide an overview of the current landscape in nanoparticle analysis methods and introduce our recently developed technologies. These technologies synergistically combine diverse methods, molecular tools, and information analysis to enhance the resolution and dimensionality of analysis. We particularly emphasize optical measurement as a promising, non-destructive, and high-throughput approach in the analysis of EVs and cells.

2 Nanoparticle Analyzers

This section reviews advances in optical methods for analyzing extracellular vesicles (EVs), from bulk to single-particle analytical methods, and discusses necessary techniques that have not yet been achieved.

2.1 *Bulk Measurement Methods*

In conventional bulk measurements, optical illumination is applied to a solution in which EVs are dispersed. The resulting scattering and fluorescence signals from these EVs are collected and statistically analyzed. The general advantages of bulk measurements include simplicity, high signal-to-noise ratio (SNR) due to large sample throughput, and the ability to measure at high concentrations. It can only provide average sample measurements, making it difficult to analyze mixtures or perform detailed characterization of minor or heterogeneous EVs.

Dynamic light scattering (DLS) is a technique used to determine the size distribution profile of small particles in suspension (Yamashita et al. 2023). It measures the fluctuations in the scattered signal caused by the Brownian motion of the particles, with larger particles showing smaller fluctuations in the signal and smaller particles showing larger fluctuations. From these fluctuations, an autocorrelation function is calculated to deduce the diffusion coefficient and, subsequently, estimate particle size. Recent advances in DLS have focused on improving particle size resolution, often in combination with separation methods such as centrifugation or asymmetric flow field flow fractionation before DLS measurements (Measuring particle concentration of multimodal synthetic reference materials and extracellular vesicles with orthogonal techniques: Who is up to the challenge—Vogel 2021).

Fluorescence Correlation Spectroscopy (FCS) (Yu et al. 2021) is a technique that operates on principles similar to DLS but measures fluctuations caused by molecules diffusing through focused light. Due to the time-correlation of the fluorescence, FCS can be used to quantitatively evaluate the concentration, diffusion coefficient (size), and interaction of molecules in vitro. It enables the analysis of specific targets of interest by using fluorescent labeling. Recent developments in FCS-based EV analysis include Fluorescence Cross-Correlation Spectroscopy (FCCS) for observing particle interactions by measuring two colors simultaneously. In addition, improvements in FCS temporal signal analysis methods can now provide even single-particle information on the content and biophysical properties of thousands of nanoparticles (Sych et al. 2023).

The ExoScreen assay provides a new liquid biopsy procedure for sensitive and rapid detection of disease-specific EVs (Yoshioka et al. 2014). This assay system was developed on the AlphaLISA technique based on an amplified luminescent proximity homogeneous assay using photosensitizer beads. In this assay, EVs are captured by two antibodies modified in different ways. One is a biotinylated antibody, and the other is an antibody conjugated to AlphaLISA acceptor beads. ExoScreen technology detected CD9 and CD63 present on EVs in 5 μl of serum without purification steps.

2.2 *Single Particle Measurement Methods*

Single particle measurements analyze the scattering and fluorescence signals from each EV. These methods have the advantage of allowing multiparametric measurements from each particle and can identify rare subpopulations in heterogeneous particle mixtures. On the other hand, signals from small single particles are inherently weak. Therefore, it is difficult to satisfy both high sensitivity and high throughput at the same time, and it is difficult to perform scalable measurements and statistically robust subpopulation analysis, which limits their practical application in the medical and pharmaceutical fields.

Dark-field microscopy is a technique that excludes non-scattered light from the image and typically detects high-frequency components, effectively reducing background noise and weak scattering from nanoparticles detected within the dark optical image field. The measurement capability of this technique is limited by the field of view (FOV). Recent developments include single-particle scattering spectroscopy, which combines scattering and absorption measurements using white light excitation (Weigel et al. 2014; Al-Zubeidi et al. 2021).

Nanoparticle Tracking Analysis (NTA) adopts dark-field microscopy and captures videos of the Brownian motion of particles in suspension, and calculates particle size using the Stokes–Einstein equation. Additionally, by creating calibration curves in advance, NTA can simultaneously measure particle size and refractive index (Kashkanova et al. 2022). It also allows for time-lapse observation. The number of particles that can be measured is limited by the field of view of the video capture; NTA relies on the random motion of Brownian movement, which can lead to significant variability in measurements. To address this, methods using statistical processing, such as likelihood estimation, have been developed to correct particle size distribution. Moreover, advancements in NTA include interferometric NTA (iNTA), which utilizes interference to enhance the detection sensitivity of particle scattering.

Single-photon-counting-based flow cytometry (spFCM) (Zhu et al. 2014). overcomes the throughput limitations of microscope-based techniques like NTA. With a specially designed hydrodynamic focusing device, it slows down the rate of flows to detect more scattering and fluorescence of particles through single-photon counting. Advances in acoustic focusing of nanoparticles at low flow velocities could further help improve the throughput and scalability of spFCM technology (Ugawa et al. 2022; Tsuyama et al. 2023).

Microfluidic Resistive Pulse Sensing (MRPS) (Fraikin et al. 2011) is another technique that detects particles suspended in an electrolytic solvent as they pass through a nanopore, measuring changes in electric current. While offering high throughput, the detection sensitivity has been limited. Recent advancements are ongoing to integrate fluorescence and other measurements for multiparametric analysis.

2.3 *Conclusions and Perspectives*

To enable EV population analysis and uncover their heterogeneous nature at higher resolutions, single-particle analysis techniques have significantly advanced, providing detailed insights into their heterogeneity. Techniques such as darkfield microscopy, NTA, Single-photon-counting-based flow cytometry, and MRPS each offer unique capabilities in the detailed and accurate analysis of EVs, enhancing our understanding and potential applications of these nanoparticles in various fields. Yet, the current methods still struggle with the tradeoff between measurement scalability and sensitivity, which significantly limits the practical application especially when identifying rare nanoparticles in heterogeneous mixtures.

3 Nanoparticle Sorters

Enrichment is an important approach to examine specific EVs because EVs are dispersed in complex mixtures that contain more non-vesicular particles such as lipoproteins. In this section, we review advances in methods for isolating or sorting extracellular vesicles (EVs) and discuss necessary techniques that have not yet been achieved.

3.1 Conventional Methods

In this overview, separation methods that operate without the use of microfluidic devices are classified as conventional methods.

The ultracentrifugation (UC) method is considered the gold standard for exosome separation and uses differential centrifugation (DC) to separate exosomes based on their size and density (Dragovic et al. 2011). The process involves continuous and step-wise centrifugation at increasing speed to remove cells, cell debris, larger extracellular vesicles (EVs), and proteins, eventually yielding exosomes. Despite its widespread use, this method has limitations, including high equipment costs, long processing times, low yields, and the potential for exosome damage due to high centrifugal forces.

The ultrafiltration (UF) and size-exclusion chromatography (SEC) methods are size-based separation techniques for exosomes (Shu et al. 2021). UF, using a nanomembrane, offers quick isolation and low equipment cost but risks exosome damage and is susceptible to contaminants and membrane blocking. SEC utilizes a porous stationary phase to separate particles based on size. SEC maintains the integrity of exosomes and has high yield and reproducibility, but it may co-isolate albumin and lipoproteins.

The precipitation-based isolation method involves adding a reagent such as polyethylene glycol to the sample to precipitate exosomes (Coughlan et al. 2020). Its advantages are its ease of use, high yield, and the fact that it does not require special equipment. However, isolated exosomes often coexist with lipoproteins or proteins, making it difficult to remove the precipitating reagent.

Immunoaffinity-based isolation is also a powerful approach to capturing EVs. This method faces challenges such as the high heterogeneity of exosomes, irreversible damage during elution from magnetic particles, and the use of expensive reagents.

Utilizing microfluidic devices is another emerging approach to the EV isolation based on their physical and biochemical properties (Ding et al. 2021). A series of the aforementioned EV isolation approaches have been implemented in micro-device. Size-based separation methods often designed filters or sieves with specific pore sizes that allowed exosomes to pass through while trapping larger particles. Immunoaffinity-based separation methods utilize specific interactions between antibodies and exosome surface proteins to capture target exosomes with high specificity in the device, and this method is particularly useful for isolating exosomes with specific biomarkers. Other methods developed include dielectrophoresis, which uses an electric field to manipulate EVs based on the dielectric properties of the particles, and acoustophoresis, which uses an acoustic sound field to control the position of particles within microchannels that form resonant cavities of sound. Enhanced purity and yield of isolated exosomes and lower sample volume requirements leading to reduced processing time are the advantages of these microdevice-based approaches. On the other hand, scaling up the technology for high-throughput uses of a large number of EVs remains challenging. Integration of multiple steps (isolation, detection, and analysis) in a single device is complex but necessary for comprehensive analysis.

3.2 Fluorescence-Activated Vesicle Sorting

Flow cytometry is a technique that performs high-throughput, multi-parameter analysis of particles in suspension and has the capability of sorting sort out specific subsets. There have been intensive attempts to apply this sorting ability for the quantification and multi-parameter analysis of fluorescently labeled submicron-sized EVs (Groot Kormelink et al. 2016). However, most commercially available flow cytometry systems are optimally designed to accommodate particles in the size range of cells. As a result, a significant limitation is its sensitivity to EVs typically smaller than 100 nm or 200 nm. Moreover, coincidence and swarm effects can occur during flow cytometric measurement of submicron-sized particles at high concentrations.

3.3 Conclusions and Perspectives

The evolution of extracellular vesicle (EV) separation is characterized by new developments in microfluidic-based technologies from traditional established methods such as ultracentrifugation and ultrafiltration. These advances address the limitations of traditional methods such as exosome damage and low yields. Microfluidic approaches utilizing size-based separations, immunoaffinity, dielectrophoresis, and acoustophoresis improve purity and efficiency, but challenges in scalability and analytical integration remain. In addition, techniques to isolate targeted EVs based on multiparameter analysis, i.e., fluorescence-activated vesicle fast sorting, remain challenging.

The future trajectory of EV research and its ability to revolutionize the understanding and use of EVs in clinical and biological contexts will depend on targeted EV isolation technologies that combine high sensitivity for small particles with high throughput.

4 Cell Analysis Methods

This section presents a series of techniques that have been used or developed to quantitatively study the behavior and function of extracellular vesicles within or between cells. Topics include the development of reporter cells producing fluorescently tagged EVs, droplet microfluidic systems for studying EV secretion from single cells, high-dimensional imaging flow cytometry techniques, and multimodal cell analysis.

4.1 Reporter Cell Development

Generally, expression of the fusion protein of an EV marker protein (e.g. CD63, CD9) and fluorescent protein (FP) (e.g. EGFP, mScarlet) in the cells will lead to efficient labeling of EVs released from the cells (Hattori et al. 2022). EVs are heterogeneous, so it is important to choose an appropriate EV marker protein depending on the population of EVs that need to be analyzed (Mathieu et al. 2021; Kunitake et al. 2023). Also, it is reported that a small difference in linker sequences changes the intracellular localization of the reporter protein, which potentially leads to the change of the characteristic of labeled EVs (Mathieu et al. 2021), so careful consideration of the linker might be necessary depending on the purpose.

Choice of cell lines is critical as the nature of the EVs changes depending on the cell origin of the EVs. If you want to focus on some biology of specific cell types, it is of course quite important to choose the appropriate cells to produce EVs.

At the same time, for practical observation of the labeled EVs, the amount of reporter protein expressed on EVs and cell stability is also important. In terms of the ease of the reporter expression, HEK293T cells are one of the best model cell lines to use for adherent cells. For the suspension-adapted cells, we confirmed that the expression of the fusion of CD9 and Superfolder GFP in K562 cells enables bright labeling of EVs released from the cells (Hattori et al. 2022).

For the reporter transduction, lentivirus transduction and transposon-based DNA integration are much more efficient than the random integration of the plasmid DNA into the genome of the cells. Especially, for the laboratory without a BSL-2 setting, the transposon-based method (which can be done in the BSL-1 setting) is a powerful choice. The Piggybac transposase system and Sleeping Beauty transposase system are among the most famous transposon/transposase-based systems. For the latter, users can easily prepare all the required materials at a low cost from Addgene, a non-profit plasmid repository (Backbone; Eric Kowartz's lab's collection: https://www.addgene.org/Eric_Kowarz/, Transposase: Addgene #34,879). We note that we achieved efficient reporter expression in K562 cells with the Sleeping Beauty Transposase system (Hattori et al. 2022), which enabled observation of EVs released from the cells with single-cell resolution by combination with an optimized droplet array system. For the detailed protocol of the establishment of stable cell lines with the transposon-based system, you can find several relevant protocols in another book recently edited by us (Kojima).

5 EV Secretion from Single Cells

Analyzing the dynamics of EV secretion at the single cell level is critical to understanding the mechanisms governing EV secretion. This requires techniques to accurately measure the minute amounts of EVs secreted by individual cells. This technique requires collection of EVs from a cell population and long periods of cell culture to accumulate sufficient amounts of EVs. This technology also needs to be able to monitor multiple identical single cells longitudinally. More specifically, cells must be cultured in isolated compartments to prevent contamination by EVs secreted from neighboring cells and to eliminate variations in EV secretion due to intercellular communication. Furthermore, a thorough analysis of secretion kinetics requires parallel measurements across many cells followed by rigorous statistical analysis. In short, a system that can measure minute amounts of EVs from individual cells and track secretion over many single cells over time is essential for accurate analysis of secretion kinetics.

5.1 Existing Technologies

Various techniques have been developed to measure EV secretion from single cells, and the heterogeneity of EV secretion has been revealed by profiling surface markers of EVs from individual cells.

A method developed by Lu and colleagues allows simultaneous profiling of five different EV secretions (CD63, CD9, CD81, EpCAM, and HSP70) and three protein secretions (IL-6, IL-8, and MCP-1) from over 1000 cells using antibody barcodes spatially patterned on a microchip (Ji et al. 2019). EVs are captured by antibody-coated microchips placed on top of cell culture chambers, and the amount of EVs is quantified in a manner similar to a sandwich ELISA. In a follow-up effort, the team introduced a method to quantify EV secretion in single cells using a home scanner (Zhu et al. 2021). This method minimizes costs and avoids laborious steps by skillfully utilizing gold nanoparticle-enhanced silver staining.

Dittrich and colleagues have not only quantified the abundance of EVs from a single cell, but also enabled profiling of individual EVs from a single cell (Nikoloff et al. 2021). Their method classifies each EV based on the expression levels of four different EV markers. By fixing EVs at the bottom of the chamber with specific antibodies and using total internal reflection fluorescence microscopy, they were able to achieve high-resolution imaging and classify individual EVs from single cells into 15 groups.

Pegtel and colleagues shifted their focus from EVs themselves to the intracellular processes that govern EV secretion (Verweij et al. 2018). Specifically, they exploited the unique properties of a pH-sensitive fluorescent reporter to visualize the fusion of multivesicular body and plasma membrane as a surrogate marker for EV secretion. This approach is based on the principle that the fluorescent signal of the reporter is only activated at neutral pH in the extracellular environment, while it remains off at the acidic pH found within the MVB.

Although these methods are innovative, they are limited in their ability to perform temporal analysis. Several groups addressed this issue and have developed strategies to assess EV secretion dynamics. One of the efforts in this direction was by Lo group (Chiu et al. 2016). They presented a technique to investigate single-cell secretions at various time points. Their device, fabricated using polydimethylsiloxane and designed as a mesh structure for efficient cell loading and precise positioning, facilitates single cell culture and minimizes cross-contamination of secreted factors; EVs are collected on glass slides placed over a culture dish and then subjected to multiple staining steps to visualize the EV signal at each time point. Using this method, EV secretion rates over 24–96 h were evaluated in as little as 3 h. In addition, the technical repertoire was expanded with a single-cell translocation and secretion assay (TransSeA), allowing time-lapse analysis, molecular cargo analysis, and tracking phenotypic shifts between parental and progeny cells (Cai et al. 2018).

Revzin and colleagues introduced a microcompartment array in which single cells are confined within a 20-picoliter space together with antibody-modified sensing beads to monitor the secretory activity of the cells; EVs are captured by the antibody-coated beads, which form a fluorescent antibody complex (Son et al. 2016). Fluorescent signals accumulate on the beads in conjunction with the progress of secretion. They demonstrated that EVs secreted from a single cell can be quantified by the fluorescent signal on the beads over 12 h and analyzed dozens of cells.

Similarly, Varadarajan's group also employed a bead-based strategy and measured EV secretion levels using beads functionalized with capture antibodies (Fathi et al. 2021). Beads and cells were co-loaded onto nanowell arrays, and 45 min after the addition of the detection antibody, the secreted EVs were quantified by fluorescence imaging. This technique allowed them to track EV secretion from dozens of cells at the single-cell level every 2 h for a total of 6 h. The group further extended this technique by sorting single cells with micromanipulators and performing RNA-seq on clonally expanded single cells to correlate EV secretion with the transcriptome (Fathi et al. 2023).

Thus, the field of single-cell EV secretion analysis is advancing rapidly, but continuous tracking of EV secretion against individual and large numbers of cells over long periods remains a challenge. For example, immunoassays are a common method, but the multiple wash steps required are labor-intensive and impractical for frequent measurements. However, omitting these washing steps decreases the signal-to-noise ratio. In addition, bead-based assays limit the total number of target cells that can be analyzed because of the limited cell-bead co-occlusion rate.

5.2 *Droplet Array for Studying EV Secretion from Single Cells*

To address the limitations associated with single-cell measurements of EV secretion dynamics, we have developed a droplet array-based measurement platform (Fig. 1a) (Hattori et al. 2022). This platform enables frequent and long-term measurements of large numbers of cells in a simple manner.

The first key technology is a method that allows thousands of droplets to be held in the micropillar array for extended periods, enabling time-lapse measurements of the same droplet for up to 36 h. The array consists of approximately 5,000 pillars with 80 μm between each pillar (Fig. 1b, c). The droplet arrays were formed by injecting droplets prepared in advance (Fig. 1d). The array was immersed in FC-40, a low-volatile oil, to ensure that droplet size remained almost unchanged for at least 36 h. The oil-soaked micropillar array was placed in a stage-top incubator on an automated stage to allow automated imaging while maintaining an environment of 37 °C, 5% CO_2, and high humidity.

The second key technology is a platform for confining fluorescently labeled EVs in droplets and their computational image analysis. To visualize the amount of EVs released from single cells, sfGFP was fused to CD9, one of the EV markers, and introduced into K562 cells (hereafter, K562 CD9-sfGFP cells). These cells were encapsulated in the 80-μm droplets and incubated for 24 h to see the accumulation of fluorescent EVs inside the droplets (Fig. 1e). Treatment with the EV release promoter increased the fluorescent signal, while that with inhibitors decreased it, confirming that this signal was from EVs. We also quantified that the number of EVs derived from a single cell per 24 h ranged from approximately 1.3×10^3 to 1.9×10^4 by preparing standards of fluorescent EVs. With a Python-based analysis, the fluorescent signal and subsequent secreted EVs were automatically quantified (Fig. 1f). Specifically, each droplet in the phase contrast image was segmented, droplets containing live cells were selected, and the average fluorescence intensity of the extracellular region within the droplet was calculated.

Using this new measurement platform, we achieved measurement of EV secretion dynamics from 332 cells every 2 h for a total of 36 h, clearly identifying heterogeneity in secretion dynamics in the clonal cell population (Fig. 1g, h). Furthermore, by optimizing cell culture conditions, we succeeded in inducing cell division within 24 h in more than 70% of the cells, allowing us to analyze the correlation between cell cycle and EV secretion dynamics. As a result, we found for the first time that the secretion of CD9-positive EVs increases at the timing of cell division (Fig. 1i).

One advantage of our platform is the availability of droplets that can be sorted at high throughput using droplet sorting methods. This capability allows us to enrich specific subsets of the cell population based on EV secretion kinetics. These subsets can then be applied to other analysis platforms such as transcriptome analysis.

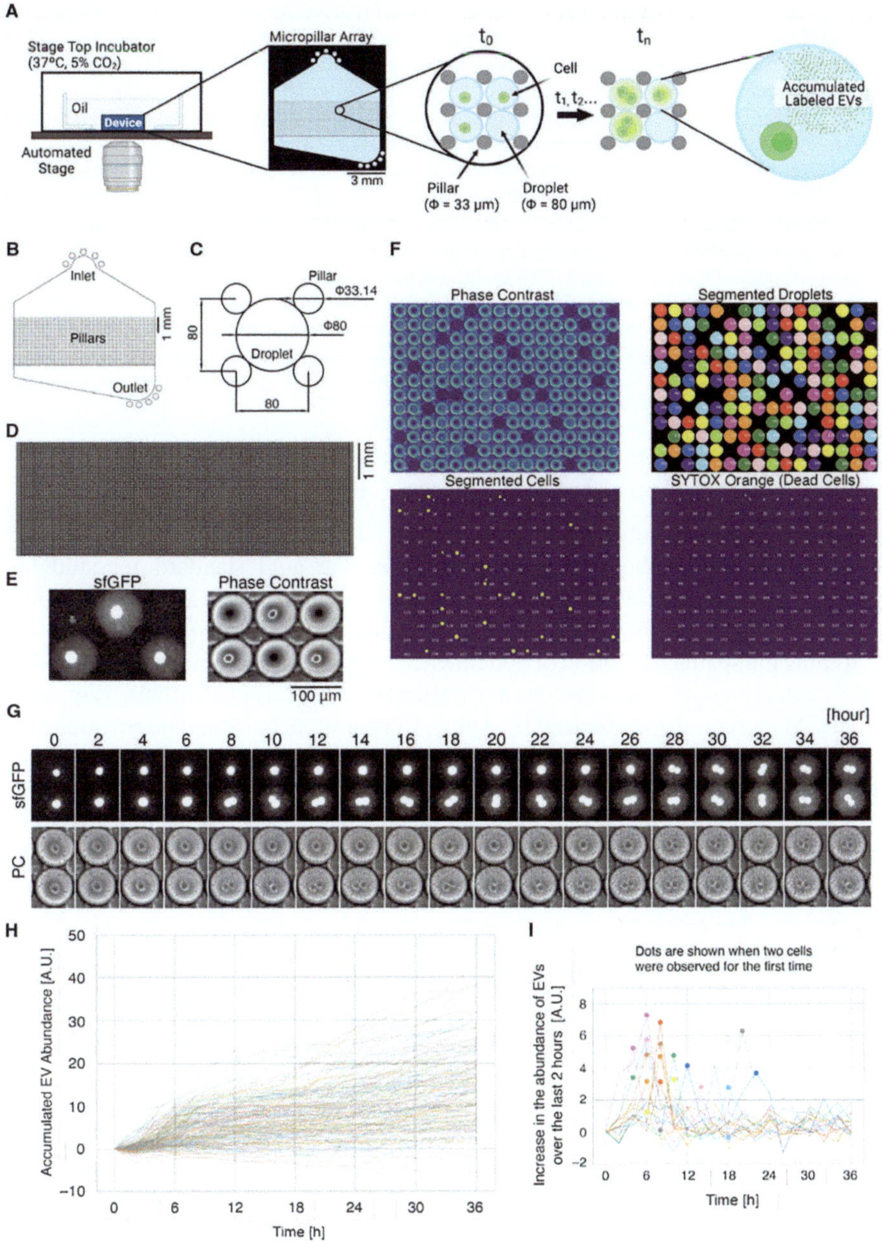
A
Stage Top Incubator
(37°C, 5% CO2)
Oil
Device
Automated Stage
Micropillar Array
3 mm
t0
Cell
Pillar (Φ = 33 μm)
Droplet (Φ = 80 μm)
t1, t2...
tn
Accumulated Labeled EVs
B
Inlet
Pillars
1 mm
Outlet
C
Pillar
Φ33.14
Φ80
Droplet
80
80
D
1 mm
E
sfGFP
Phase Contrast
100 μm
F
Phase Contrast
Segmented Droplets
Segmented Cells
SYTOX Orange (Dead Cells)
G
[hour]
0 2 4 6 8 10 12 14 16 18 20 22 24 26 28 30 32 34 36
sfGFP
PC
H
Accumulated EV Abundance [A.U.]
Time [h]
I
Dots are shown when two cells were observed for the first time
Increase in the abundance of EVs over the last 2 hours [A.U.]
Time [h]

◂**Fig. 1** **a** Schematic illustration of a droplet array-based method for the optical quantification of single-cell EV secretion dynamics by time-lapse imaging. **b, c** Schematic illustration of a micropillar array device. **d** Snapshot of the trapped droplets in the micropillar array (bright-field). **e** Representative fluorescence and phase-contrast images of K562-CD9 cells and their EVs secretion in droplets (24 h after encapsulation). **f** Representative results of the image analysis for the segmentation of droplets from phase-contrast images and that of cells from sfGFP images while dead cells were eliminated based on SYTOX Orange. **g** Representative time-lapse fluorescence and phase-contrast (PC) images of two droplets encapsulating K562-CD9 cells for 36 h. **h** Temporal change of accumulated EV abundance in each droplet, obtained by calculating the mean intensities of sfGFP signals outside the cell region inside droplets (N = 332). For calculating the EV increase, the mean intensities at the initial measurement were substrated from that at each measurement point. In each droplet used in this graph, single cells were initially encapsulated and divided into two cells during the monitoring in general. **i** The difference in the fluorescence intensities between every consecutive image taken every 2 h, representing the EV secretion rate at each measurement point (N = 20). Dots are shown when 2 cells were observed for the first time in the droplet after cell division. Figures are modified from Hattori et al. (2022).

6 High-Dimensional Imaging Flow Cytometry Techniques

While 2D and 3D optical microscopy remains the gold standard for studying the dynamics and spatial distribution of EVs in cells, another rapidly developing approach is high-content flow cytometry (hFCM) and sorters. hFCM is a method combining the spatial resolution of optical microscopy with the rapid sampling ability of flow cytometry. After the first demonstration in 1979 (Kay et al. 1979), a commercial hFCM system was introduced by Amnis Corporation, boosting the research area of the application of IFC to a variety of biological and clinical studies worldwide. Compared with the traditional high-content analysis using 2D or 3D microscopy, IFC cannot capture dynamic cell information or multicellular interactions due to its limited spatial resolution. However, IFC can image many flowing single cells at high throughputs with their positions controlled in space. This simplicity allows users to generate and analyze large image data sets with smaller risks of artifacts, to which machine learning-based analysis is readily applicable.

Another significant advantage of hFCM is its ability to selectively sort cells based on the analysis of the image information. The availability of sorted cells is expected to allow for the use of cells for downstream purposes and analysis, such as therapeutic applications (Koo et al. 2023) and large screening by sequencing (Tsubouchi et al. 2024). Recently, technological developments of cell sorting based on image information analysis have remarkably advanced, with significant increases in throughputs by integrating machine learning methods with the real-time analysis of temporal signals recorded by single pixel imagers both for fluorescence and label-free signals (Adachi et al. 2020; Ota et al. 2018; Schraivogel et al. 2022; Ugawa et al. 2021). Soon, EV dynamics and spatial distribution are expected to become the intracellular imaging phenotype for large-scale screening of drug and gene perturbations, using microscopy and imaging cell sorters.

Furthermore, as the spatial resolution and measurement throughput of hFCMs continue to improve, their capabilities approach those of conventional microscopes.

For example, by combining single objective light sheet microscopy (Ugawa and Ota 2022) with acoustic cell focusing technology, we have recently developed a 3D imaging flow cytometer that operates at a throughput of approximately 2,000 cells per second (Fig. 2a, b). We further applied this platform to 3D image analysis of multicellular cultures of spheroid-like structures on hydrogel beads and achieved a throughput of over 1,000 cells per minute (Fig. 2c, d) (Yamashita et al. 2023). 3D image phenotype is superior to 2D image phenotype in terms of accuracy, especially when observing the intracellular distribution of small objects such as Evs. The 3D image phenotype is therefore more accurate than the 2D image phenotype and motivates the development of a 3D image cell sorter for large-scale high-throughput drug or gene screening of image phenotypes related to the distribution of EVs.

7 Multimodal Cell Analysis

Advances in analytical techniques, such as those described so far, facilitate the acquisition of data of unparalleled quantity and resolution and are promising as a means of investigating the diverse behavior and function of extracellular vesicles (EVs). For example, these techniques have been applied to identify genes closely related to the behavior and functionality of specific EVs. While such target-focused screening methods provide an assessment of one easily interpretable aspect of understanding complex biological mechanisms, they can overlook the complex mechanisms encompassed in large, high-resolution data sets.

A promising approach to deciphering these complex biological mechanisms through high-volume, high-resolution data analysis is multimodal single-cell analysis. This allows for the correlation of different data modalities from individual cells, thereby increasing our understanding of biological processes. Furthermore, by introducing machine learning techniques, it may be possible to build models that simulate complex biological networks in silico. Such simulators may lead to the elucidation of fundamental mechanisms that people have never imagined.

Spatial genomics technology is rapidly evolving in this multimodal single-cell analysis field. These techniques assign single-cell transcriptome data to spatial locations of cells in 2D or 3D tissues. These techniques are extremely powerful tools for analyzing the positioning of cells in tissues without dissociating them and for understanding cell–cell interactions nearby. However, their application to the study of intercellular behavior and function of EVs has been limited.

Alternatively, the strategy of multiplexing the measurement modality of individual cells by assigning large-scale multimodal barcodes within microcompartments containing single cells can be promising. We developed a novel technique using dual image- and DNA-encoded particles to link pooled single-cell microscopic images with transcriptome profiles (Kawasaki et al. 2023). This approach is compatible with dynamic measurements and facilitates high-throughput and large-scale analysis. In the future, such multimodal lab-on-a-particle analysis techniques are expected to play an important role in elucidating the intercellular behavior and function of EVs.

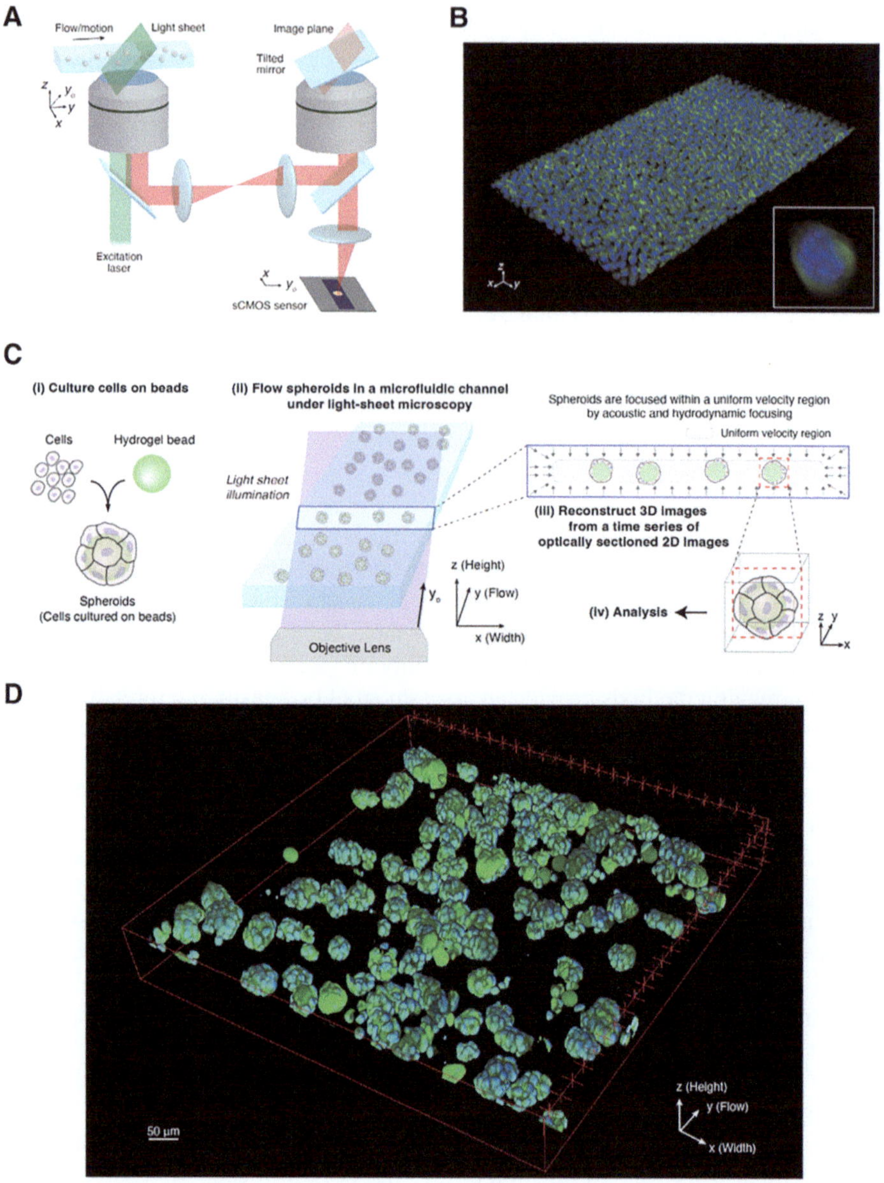

Fig. 2 **a** Optical setup for oblique-plane microscopy. The sample is placed above a sample objective on the left. The remote objective on the right is used along with the tilted mirror to convert the image from the oblique plane to the lateral plane. This converted image is then projected onto a camera. **b** 3D reconstructed image from optical-section images of K562 cells, acquired in a second. Green and blue colors correspond to fluorescence from CFSE (cytoplasm) and DAPI (nucleus), respectively. The inset shows a close-up of a single cell in **b**. **c** A scalable, high-throughput 3D imaging flow cytometry (3D-iFCM) platform for analyzing adherent 3D cell cultures. **d** A 3D view of the merged multicolor images: the green color corresponds to fluorescence from Alexa Fluor 488 (alginate beads) and CellBrite Green (cell membrane), and the blue color corresponds to fluorescence from Hoechest 33,342 (nucleus). Scale bars = 50 μm. A and B are adapted from Ugawa and Ota (2022), and **c** and **d** are from Yamashita et al. (2023).

This will not only lead to a better understanding of EV-mediated communication and processes but will also open new avenues for research in cellular and molecular biology.

8 Concluding Remarks

Advances in analytical methods have enabled the acquisition of unprecedented amounts of data and are beginning to provide insight into the heterogeneity and functionality of extracellular vesicles (EVs) across a variety of biological contexts. From the application of bulk measurement techniques to the high-resolution analysis provided by single-particle assays, each approach uniquely contributes to our understanding of EVs.

However, due to the innate complexity and diversity of EVs, measurement techniques are required to achieve further scalability and increased sensitivity, as well as to gain interpretation into their complex data, Innovative solutions such as multimodal lab-on-a-particle and high-dimensional imaging flow cytometry techniques are being explored. These new approaches, especially when combined with computational advances and machine learning, promise a deeper understanding of the complex biological mechanisms underlying EV behavior and function. Pushing the boundaries of what is technically possible is expected to pave the way for breakthroughs in diagnostic methodologies and therapeutic strategies that will ultimately maximize the potential of EVs to improve human health.

References

Adachi, H., Kawamura, Y., Nakagawa, K., Horisaki, R., Sato, I., Yamaguchi, S., Fujiu, K., Waki, K., Noji, H., and Ota, S. (2020). Use of Ghost Cytometry to Differentiate Cells with Similar Gross Morphologic Characteristics. Cytometry A *97*, 415–422.

Al-Zubeidi, A., McCarthy, L.A., Rafiei-Miandashti, A., Heiderscheit, T.S., and Link, S. (2021). Single-particle scattering spectroscopy: fundamentals and applications. Nanophotonics 10, 1621–1655.

Cai, W., Chiu, Y.-J., Ramakrishnan, V., Tsai, Y., Chen, C., and Lo, Y.-H. (2018). A single-cell translocation and secretion assay (TransSeA). Lab Chip 18, 3154–3162. https://doi.org/10.1039/C8LC00821C.

Chang, W.-H., Cerione, R.A., and Antonyak, M.A. (2021). Extracellular Vesicles and Their Roles in Cancer Progression. Methods Mol. Biol. 2174, 143–170.

Chiu, Y.-J., Cai, W., Shih, Y.-R.V., Lian, I., and Lo, Y.-H. (2016). A Single-Cell Assay for Time Lapse Studies of Exosome Secretion and Cell Behaviors. Small 12, 3658–3666. https://doi.org/10.1002/smll.201600725.

Coughlan, C., Bruce, K.D., Burgy, O., Boyd, T.D., Michel, C.R., Garcia-Perez, J.E., Adame, V., Anton, P., Bettcher, B.M., Chial, H.J., et al. (2020). Exosome Isolation by Ultracentrifugation and Precipitation and Techniques for Downstream Analyses. Curr. Protoc. Cell Biol. 88, e110.

Ding, L., Yang, X., Gao, Z., Effah, C.Y., Zhang, X., Wu, Y., and Qu, L. (2021). A Holistic Review of the State-of-the-Art Microfluidics for Exosome Separation: An Overview of the Current Status, Existing Obstacles, and Future Outlook. Small 17, e2007174.

Doyle, L.M., and Wang, M.Z. (2019). Overview of Extracellular Vesicles, Their Origin, Composition, Purpose, and Methods for Exosome Isolation and Analysis. Cells 8. https://doi.org/10.3390/cells8070727.

Dragovic, R.A., Gardiner, C., Brooks, A.S., Tannetta, D.S., Ferguson, D.J.P., Hole, P., Carr, B., Redman, C.W.G., Harris, A.L., Dobson, P.J., et al. (2011). Sizing and phenotyping of cellular vesicles using Nanoparticle Tracking Analysis. Nanomedicine 7, 780–788.

Fathi, M., Joseph, R., Adolacion, J.R.T., Martinez-Paniagua, M., An, X., Gabrusiewicz, K., Mani, S.A., and Varadarajan, N. (2021). Single-Cell Cloning of Breast Cancer Cells Secreting Specific Subsets of Extracellular Vesicles. Cancers 13, 4397. https://doi.org/10.3390/cancers13174397.

Fathi, M., Martinez-Paniagua, M., Rezvan, A., Montalvo, M.J., Mohanty, V., Chen, K., Mani, S.A., and Varadarajan, N. (2023). Identifying signatures of EV secretion in metastatic breast cancer through functional single-cell profiling. iScience 26. https://doi.org/10.1016/j.isci.2023.106482.

Fraikin, J.-L., Teesalu, T., McKenney, C.M., Ruoslahti, E., and Cleland, A.N. (2011). A high-throughput label-free nanoparticle analyser. Nat. Nanotechnol. 6, 308–313.

Groot Kormelink, T., Arkesteijn, G.J.A., Nauwelaers, F.A., van den Engh, G., Nolte-'t Hoen, E.N.M., and Wauben, M.H.M. (2016). Prerequisites for the analysis and sorting of extracellular vesicle subpopulations by high-resolution flow cytometry. Cytometry A 89, 135–147.

Hattori, K., Goda, Y., Yamashita, M., Yoshioka, Y., Kojima, R., and Ota, S. (2022). Droplet Array-Based Platform for Parallel Optical Analysis of Dynamic Extracellular Vesicle Secretion from Single Cells. Anal. Chem. 94, 11209–11215.

Ji, Y., Qi, D., Li, L., Su, H., Li, X., Luo, Y., Sun, B., Zhang, F., Lin, B., Liu, T., et al. (2019). Multiplexed profiling of single-cell extracellular vesicles secretion. PNAS 116, 5979–5984. https://doi.org/10.1073/pnas.1814348116.

Kashkanova, A.D., Blessing, M., Gemeinhardt, A., Soulat, D., and Sandoghdar, V. (2022). Precision size and refractive index analysis of weakly scattering nanoparticles in polydispersions. Nat. Methods 19, 586–593.

Kawasaki, F., Mimori, T., Mori, Y., Aburatani, H., Yachie, N., Sato, I., and Ota, S. (2023). Computational design of synthetic optical barcodes in microdroplets. Adv. Opt. Mater. https://doi.org/10.1002/adom.202302564.

Kay, D.B., Cambier, J.L., and Wheeless, L.L., Jr (1979). Imaging in flow. J. Histochem. Cytochem. 27, 329–334.

Kojima, R. Mammalian Cell Engineering (Springer US). Mammalian Cell Engineering, Methods and Protocols.

Koo, D., Cheng, X., Udani, S., Zhu, D., Li, J., Hall, B., Tsubamoto, N., Hu, S., Ko, J., Cheng, K., et al. (2023). Optimizing Cell Therapy by Sorting Cells with High Extracellular Vesicle Secretion. bioRxiv, 2023.05.29.542772. https://doi.org/10.1101/2023.05.29.542772.

Kunitake, K., Mizuno, T., Hattori, K., Oneyama, C., Kamiya, M., Ota, S., Urano, Y., and Kojima, R. (2023). Barcoding of small extracellular vesicles with CRISPR-gRNA enables comprehensive, subpopulation-specific analysis of their biogenesis/release regulators. bioRxiv, 2023.09.28.559700. https://doi.org/10.1101/2023.09.28.559700.

Mathieu, M., Névo, N., Jouve, M., Valenzuela, J.I., Maurin, M., Verweij, F.J., Palmulli, R., Lankar, D., Dingli, F., Loew, D., et al. (2021). Specificities of exosome versus small ectosome secretion revealed by live intracellular tracking of CD63 and CD9. Nat. Commun. 12, 4389.

Measuring particle concentration of multimodal synthetic reference materials and extracellular vesicles with orthogonal techniques: Who is up to the challenge? - Vogel - 2021 - Journal of Extracellular Vesicles - Wiley Online Library

Nikoloff, J.M., Saucedo-Espinosa, M.A., Kling, A., and Dittrich, P.S. (2021). Identifying extracellular vesicle populations from single cells. Proc. Natl. Acad. Sci. USA 118, e2106630118. https://doi.org/10.1073/pnas.2106630118.

Ota, S., Horisaki, R., Kawamura, Y., Ugawa, M., Sato, I., Hashimoto, K., Kamesawa, R., Setoyama, K., Yamaguchi, S., Fujiu, K., et al. (2018). Ghost cytometry. Science *360*, 1246–1251.

Schraivogel, D., Kuhn, T.M., Rauscher, B., Rodríguez-Martínez, M., Paulsen, M., Owsley, K., Middlebrook, A., Tischer, C., Ramasz, B., Ordoñez-Rueda, D., et al. (2022). High-speed fluorescence image–enabled cell sorting. Science 375, 315–320.

Shu, S.L., Allen, C.L., Benjamin-Davalos, S., Koroleva, M., MacFarland, D., Minderman, H., and Ernstoff, M.S. (2021). A rapid exosome isolation using ultrafiltration and size exclusion chromatography (REIUS) method for exosome isolation from melanoma cell lines. Methods Mol. Biol. 2265, 289–304.

Son, K.J., Rahimian, A., Shin, D.-S., Siltanen, C., Patel, T., and Revzin, A. (2016). Microfluidic compartments with sensing microbeads for dynamic monitoring of cytokine and exosome release from single cells. Analyst 141, 679–688. https://doi.org/10.1039/C5AN01648G.

Sych, T., Schlegel, J., Barriga, H.M.G., Ojansivu, M., Hanke, L., Weber, F., Beklem Bostancioglu, R., Ezzat, K., Stangl, H., Plochberger, B., et al. (2023). High-throughput measurement of the content and properties of nano-sized bioparticles with single-particle profiler. Nat. Biotechnol. https://doi.org/10.1038/s41587-023-01825-5.

Tsubouchi, A., An, Y., Kawamura, Y., Yanagihashi, Y., Nakayama, H., Murata, Y., Teranishi, K., Ishiguro, S., Aburatani, H., Yachie, N., et al. (2024). Pooled CRISPR screening of high-content cellular phenotypes using ghost cytometry. Cell Rep Methods 4, 100737.

Tsuyama, Y., Xu, B., Hattori, K., Baek, S., Yoshioka, Y., Kojima, R., Cho, Y., Laurell, T., Kim, S., Ota, S., et al. (2023). A 50 μm acoustic resonator microchannel enables focusing 100 nm polystyrene beads and sub-micron bioparticles. Sens. Actuators B Chem. 376, 132918.

Ugawa, M., Kawamura, Y., Toda, K., Teranishi, K., Morita, H., Adachi, H., Tamoto, R., Nomaru, H., Nakagawa, K., Sugimoto, K., et al. (2021). In silico-labeled ghost cytometry. Elife *10*. https://doi.org/10.7554/eLife.67660.

Ugawa, M., Lee, H., Baasch, T., Lee, M., Kim, S., Jeong, O., Choi, Y.-H., Sohn, D., Laurell, T., Ota, S., et al. (2022). Reduced acoustic resonator dimensions improve focusing efficiency of bacteria and submicron particles. Analyst 147, 274–281.

Ugawa, M., and Ota, S. (2022). High-throughput parallel optofluidic 3D-imaging flow cytometry. Small Sci., 2100126.

Verweij, F.J., Bebelman, M.P., Jimenez, C.R., Garcia-Vallejo, J.J., Janssen, H., Neefjes, J., Knol, J.C., de Goeij-de Haas, R., Piersma, S.R., Baglio, S.R., et al. (2018). Quantifying exosome secretion from single cells reveals a modulatory role for GPCR signaling. J Cell Biol 217, 1129–1142. https://doi.org/10.1083/jcb.201703206.

Weigel, A., Sebesta, A., and Kukura, P. (2014). Dark field microspectroscopy with single molecule fluorescence sensitivity. ACS Photonics 1, 848–856.

Yamashita, M., Tamamitsu, M., Kirisako, H., Goda, Y., Chen, X., Hattori, K., and Ota, S. (2023). High-throughput 3D imaging flow cytometry of suspended adherent 3D cell cultures. Small Methods, e2301318.

Yoshioka, Y., Kosaka, N., Konishi, Y., Ohta, H., Okamoto, H., Sonoda, H., Nonaka, R., Yamamoto, H., Ishii, H., Mori, M., et al. (2014). Ultra-sensitive liquid biopsy of circulating extracellular vesicles using ExoScreen. Nat. Commun. 5, 3591.

Yu, L., Lei, Y., Ma, Y., Liu, M., Zheng, J., Dan, D., and Gao, P. (2021). A Comprehensive Review of Fluorescence Correlation Spectroscopy. Frontiers in Physics 9. https://doi.org/10.3389/fphy.2021.644450.

Zaborowski, M.P., Balaj, L., Breakefield, X.O., and Lai, C.P. (2015). Extracellular vesicles: composition, biological relevance, and methods of study. Bioscience 65, 783–797.

Zhu, S., Ma, L., Wang, S., Chen, C., Zhang, W., Yang, L., Hang, W., Nolan, J.P., Wu, L., and Yan, X. (2014). Light-scattering detection below the level of single fluorescent molecules for high-resolution characterization of functional nanoparticles. ACS Nano 8, 10998–11006.

Zhu, F., Ji, Y., Li, L., Bai, X., Liu, X., Luo, Y., Liu, T., Lin, B., and Lu, Y. (2021). High-throughput single-cell extracellular vesicle secretion analysis on a desktop scanner without cell counting. Anal. Chem. 93, 13152–13160. https://doi.org/10.1021/acs.analchem.1c01446.

Magnetic Nanoparticles for Diagnostics and Therapy

Yuko Ichiyanagi

Abstract Magnetic materials have long been essential to human life. Nanometer-sized magnetic particles have recently attracted attention for the development of magnetic materials, such as electronic devices, and other fields such as medicine. Here, we introduce the development of nanosized magnetic particles produced using our unique manufacturing method for biomedical applications. We discuss the possibility of cancer thermotherapy using magnetic nanoparticles (MNPs) for therapeutic applications and magnetic resonance imaging for diagnostic applications. We expect a trend toward nano-theranostics in which MNPs can simultaneously be used for treatment and diagnosis.

1 Introduction

Magnets play an indispensable role in contemporary life. In the twenty-first century, nearly all electrical appliances have integrated magnetic materials, in particular ubiquitous handheld devices such as smartphones. Tracing back to antiquity, ancient Greeks were purportedly acquainted with natural magnets. The earliest documented account pertaining to the magnetic compass was found in the Chinese essay "Mengxi Bitan" from the late eleventh century. Subsequently, during the Age of Discovery, the compass gained prominence, regardless of whether it was brought back by Marco Polo. The publication of William Gilbert's treatise on magnets in 1600 laid the groundwork for electromagnetism, which was further developed by H. C. Oersted, A. M. Ampère, and M. Faraday, thereby fostering a lineage of magnetic material research extended by luminaries like J. A. Ewing and K. Honda and numerous scholars thereafter.

Moreover, elucidating magnetic phenomena through quantum mechanics underscores the rapid progression of magnetic material development (O'Shea and Perera

Y. Ichiyanagi (✉)
Yokohama National University, Yokohama, Japan
e-mail: yuko@ynu.ac.jp

Osaka University, Yokohama, Japan

Y. Baba et al. (eds.), *Extracellular Fine Particles*,
https://doi.org/10.1007/978-981-97-7067-0_18

1996; Cordoyiannis et al. 2009; Kodama 1999; Sukumar et al. 2022). It is noteworthy that Néel was one of the contributors to these theories. Magnetic materials have been a hot topic in application-oriented research in medical sciences.

Nanometer-scale magnetic materials exhibiting unique quantum mechanical phenomena, including magnetic phase transitions, are of great academic interest. Because nanosized particles are sufficiently small to be introduced into cellular structures at the micrometer scale, their application scope in the development of healthcare-related systems is widely explored. Moreover, these magnetic materials exhibit properties such as attraction to magnets, contrast enhancement for imaging (Weissleder et al. 1990; LaConte et al. 2007; Westmeyer and Jasanoff 2007; Sun et al. 2009; Lu et al. 2009; Kattel et al. 2012; Sakai et al. 2012; Casula et al. 2011), and heat generation under external magnetic fields (Jordan et al. 1999; Hergt and Dutz 2007; Suto et al. 2009; Laurent et al. 2011; Ichiyanagi et al. 2012; Kondo et al. 2015; Jadhav et al. 2008; Fabris et al. 2019; Ruta et al. 2015; Roca et al. 2023), suggesting promising features for exploitation. However, it is challenging to maintain ferromagnetism at the nanoscale because ferromagnetic materials tend to form magnetic domains that stabilize their magnetostatic energy. As the particle size decreases, the number of magnetic domains gradually diminish, leading to the formation of a single-domain structure. The magnetic anisotropy of particles within a single domain is determined by the anisotropy constant (K) and volume (v), expressed as $Kv = k_{\mathrm{B}}T$ (where K is the anisotropy constant, v is the volume, k_{B} is the Boltzmann constant, and T is the temperature). As the volume decreases, magnetic anisotropy is lost to thermal fluctuations, impeding spontaneous magnetization. Thus, maintaining strong magnetism at the nanoscale level is challenging. Furthermore, magnetic oxide materials, which act as insulators, are difficult to modify using drugs or other chemical molecules.

This manuscript discusses the fabrication methods for magnetic nanoparticles (MNPs) and the evaluation of their physical properties. Subsequently, the methodologies for biomedical applications are introduced. Specifically, the application scope of MNPs in cancer hyperthermia therapy for treatment purposes is analyzed and their suitability as contrast agents for diagnostic purposes is assessed. By facilitating simultaneous therapy and diagnosis, we aimed to contribute to the emerging field of theranostics (Kelkar and Reineke 2011; Mura and Couvreur 2012; Katayanagi et al. 2022).

2 Preparation and Functionalization of Magnetic Nanoparticles

MNPs with sizes ranging from 2 to 40 nm encapsulated within amorphous SiO_2 were synthesized using a wet chemical method. Aqueous solutions of metal chlorides ($M\mathrm{Cl}_2 \cdot n\mathrm{H}_2\mathrm{O}$) and sodium metasilicate ($\mathrm{Na}_2\mathrm{SiO}_3 \cdot m\mathrm{H}_2\mathrm{O}$) were mixed, thoroughly cleaned, dried at $\sim$350 K, and annealed in air or argon atmosphere using an electric furnace. Depending on the material, the annealing temperature ranged from 473 to 1373 K. The magnetic clusters obtained through these methods include ferrites and metal oxides such as $CoFe_2O_4$, $MnFe_2O_4$, $MnZnFe_2O_4$, Fe_2O_3, $MgFe_2O_4$, $CoTiFe_2O_4$, $CoTiZnFeO_4$, $CoZnFe_2O_4$, $CuFe_2O_4$, $CuZnFe_2O_4$, Gd_2O_3, $BiFeO_3$, $DyFeO_3$, Fe_3O_4, and $NiFe_2O_4$. The particle size was controlled by adjusting the annealing temperature and argon flow rate (Moro et al. 2010). Furthermore, compositional adjustments, such as tuning the composition to $Co_xFe_{1-x}O_4$, are feasible, allowing for the optimization of samples according to the desired magnetic parameters.

Although magnetic materials are typically insulating oxides, thereby impeding their modification with agents such as drugs, the particles obtained through this method feature surface Si, rendering them conducive to silane coupling. Functional MNPs generated using silane coupling agents (Ichiyanagi et al. 2007), with all functional groups, including amino, carboxyl, and thiol groups, were confirmed to be modified on 3 nm iron oxide particles or ferrite particles (Katayanagi et al. 2022). Functional group modification was verified by observing the bonding-derived peaks in the Fourier-transform infrared (FT-IR) spectra. The functionalization of nanoparticles enables the binding of various chemical substances, including drugs and proteins, to magnetic oxide insulators. Hydrophilic and hydrophobic properties were also controlled by modification with amino-groups. Toxicity was also decreased by modification.

3 Localization in Living Tissue and Cancer-Cell-Selective Magnetic Nanoparticles

Functionalized MNPs, with amino groups modified on 3 nm iron oxide γ-Fe_2O_3 particles, were dispersed onto cultured PtK2 cells and observed under a microscope after 24 h of incubation and subsequent washing. The MNPs revealed successful cellular uptake without conventional cationic coating, suggesting uptake via phagocytosis. For comparison, 200 nm fluorescent particles were dispersed under the same conditions but were not internalized by the cells (Moritake et al. 2007).

Furthermore, the ability to localize these MNPs within mouse tissues was confirmed using an external magnetic field (Fig. 1) (Moritake et al. 2007). Functionalized rhodamine-labeled particles were applied to the inner ear of mice with a 350 mT magnet attached to the opposite side as an external magnetic field. After 24 h, the magnet was removed, and the ear sections were washed and observed under a confocal microscope. Remarkably strong fluorescence was observed in the region where the magnet was placed, confirming its localization within the tissue owing to the external magnetic field. The magnetic particles were observed to pass through cellular tissues, such as the epidermis and dermis, stopping immediately before the cartilage.

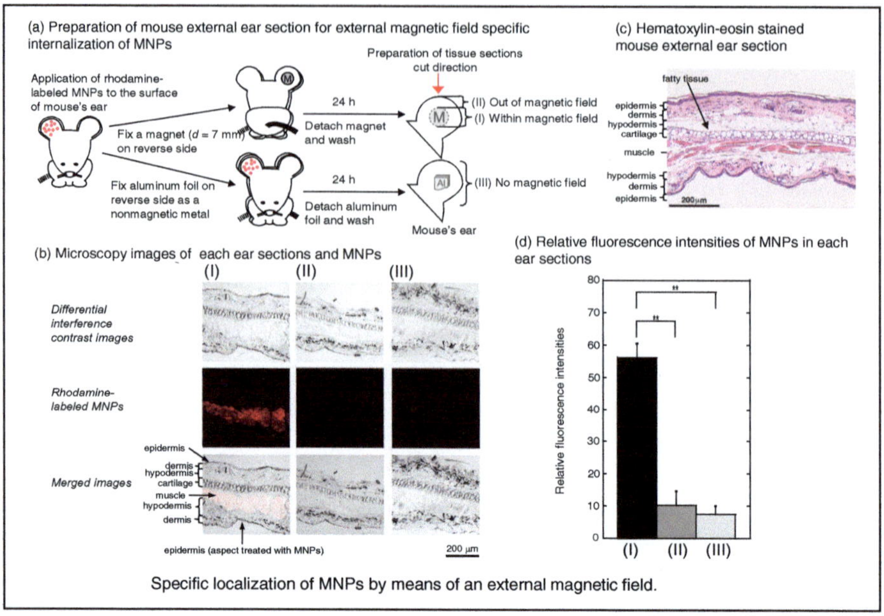

Fig. 1 Preparation of mouse external ear sections for external-magnetic-field-specific internalization of MNPs (a). Confocal laser scanning microscopy images of mouse external ear sections (b). Haematoxylin and eosin-stained external ear sections (c). Differential interference contrast images (upper panel), rhodamine fluorescence images (middle panel), and merged images (lower panel). The localization of the NMPs ($d = 3$ nm) was investigated in the presence (**b-I**) and absence (**b-II**) of an external magnetic field and without a permanent magnet (aluminum foil control) (**b-III**). Relative fluorescence intensity was determined by dividing the sum of all pixel intensities. The values presented are mean ± s.e.m. ($N = 3$). ***P << 0_01 with Student's t-test (**d**) (*Journal of Nanoscience and Nanotechnology,* Vol. 7, 937–944, 2007, American Scientific Publishers)

Considering the overexpression of folate receptors in cancer cells, the selective introduction of functionalized magnetic particles into cancer cells was demonstrated by modifying the particles with folate (Taira et al. 2007). Figure 2 shows the confocal laser scanning microscopy images of the KB cells treated with the folic acid and coumarin fluorophore (FA-CF)-conjugated NPs, KB cells treated with the CF-conjugated NPs, and untreated KB cells used as controls. MNPs were observed inside cancer cells only when the particles were modified with folic acid. Considering these properties, the potential application of MNPs in hyperthermia therapy and the development of drug delivery systems (DDSs) has been considered.

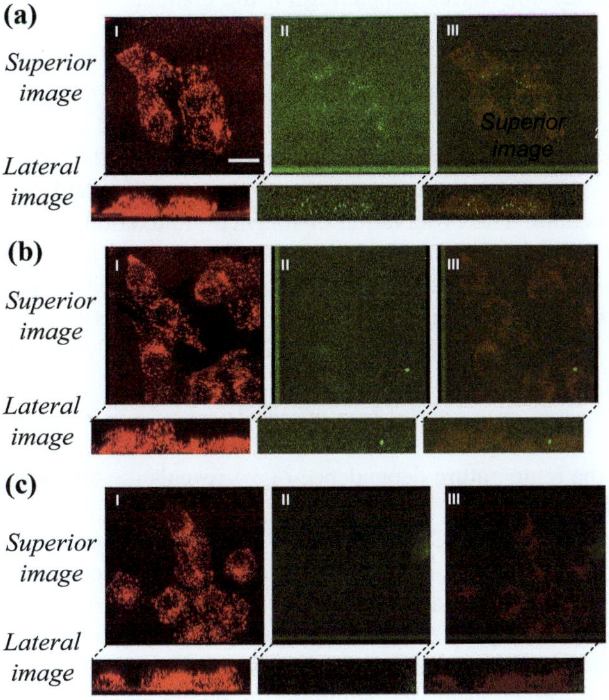

Confocal laser scanning microscopy images of KB cells.

Fig. 2 Confocal laser scanning microscopy images of (**a**) KB cells treated with FA-CF-conjugated NPs, (**b**) KB cells treated with CF-conjugated NPs, and (**c**) untreated KB cells as controls. (**I**) DiI fluorescence images (cells), (**II**) coumarin fluorescence images (NPs), and (**III**) merged images. Scale bar = 20 μm (**e–J** *Surf. Sci. Nanotech.* Vol. 5 (2007) 23–28)

4 Optimization of Magnetic Nanoparticles for Hyperthermia Therapy

4.1 Mechanism of Heat Generation and Temperature Rise Measurement of Magnetic Nanoparticles

Cancer cells are known to be destroyed at temperatures around 43 °C (316 K), making MNP-based hyperthermia therapy a promising approach. Hyperthermia therapy exploits the weakness of cancer cells by heating the affected areas for treatment. Magnetic materials can internally accumulate heat when an external magnetic field is applied. Although there are several mechanisms of heat generation, in these nanoparticles, heat generation predominantly occurs through Néel relaxation, a type of magnetic relaxation. The amount of heat energy (P) is expressed as shown in Eq. (1)

(Rosensweig 2002) and varies depending on the characteristics of the magnetic material.

$$P = f\Delta U = \mu_0 \pi \chi'' f h^2 \tag{1}$$

Here, f is the frequency of the alternating magnetic field, h is the magnetic field, χ'' is the imaginary part of the alternating magnetic susceptibility, and μ_0 is the permeability of the vacuum. The observation of magnetic relaxation phenomena allows the relaxation of magnetic moments within the nanoparticles to be observed based on their relationship with the anisotropy and magnetic field. When measuring alternating magnetization m, the component that lags 90° out of phase with the alternating magnetic field, i.e., imaginary part χ'' in Eq. (2), is related to Néel relaxation.

$$m = (\chi' - i\chi'')h_0 e(i\omega t) \tag{2}$$

Analysis of the temperature dependence of χ'' reveals the temperature at which most heat is released.

For example, we prepared Co-Zn ferrite NPs using the same preparation method mentioned in the previous section by doping with large anisotropic Co and nonmagnetic Zn in ferrites (Shigeoka et al. 2018). It is a very curious and interesting phenomenon that when adding Zn ions to spinel ferrites, their magnetization increases according to the Zn content, even though Zn ions are non-magnetic (Gorter 1950; Ichiyanagi et al. 2004). Aqueous solutions of cobalt chlorides ($CoCl_2 \cdot 6H_2O$), iron chlorides ($FeCl_2 \cdot 4H_2O$), zinc chlorides ($ZnCl_2$) and sodium metasilicate ($Na_2SiO_3 \cdot 9H_2O$) were mixed, thoroughly cleaned, dried at ~350 K, and annealed in an electric furnace. Particle sizes were controlled by annealing temperature. The composition was varied as $Co_{1-x}Zn_xFe_2O_4$ ($x = 0, 0.1, 0.2, 0.3, 0.4, 0.5, 0.6, 0.7, 0.8$) with the AC magnetic susceptibility assumed $Co_{0.8}Zn_{0.2}Fe_2O_4$ ($x = 0.2$), which has the highest DC magnetization value. X-ray diffraction patterns confirmed the single-phase spinel structure of amorphous SiO_2. The particle size was estimated to be 5–17 nm.

AC magnetic field measurement was performed using a SQUID magnetometer.

The temperature dependence of the alternating magnetization susceptibility shows shifts in the peaks depending on the frequency and magnetic field strength (Fig. 3a). Suppose we define the peak temperature as freezing temperature T_f at that temperature. In that case, magnetic spins are frozen, and we can see χ" has a peak at a definite temperature and particle size. Thus, it is clear that this system follows Néel relaxation.

At 300 K, which is room or body temperature, a peak is observed at ~8 nm, in the χ" versus particle size plot, suggesting that effective heat dissipation is expected under this condition.

To demonstrate whether actual heating occurred, an alternating magnetic field generation device was constructed, and the temperature rise of the samples was observed in relation to the initial temperature. A coil was wound, and a function generator generated a sinusoidal wave amplified by a bipolar amplifier. The frequency

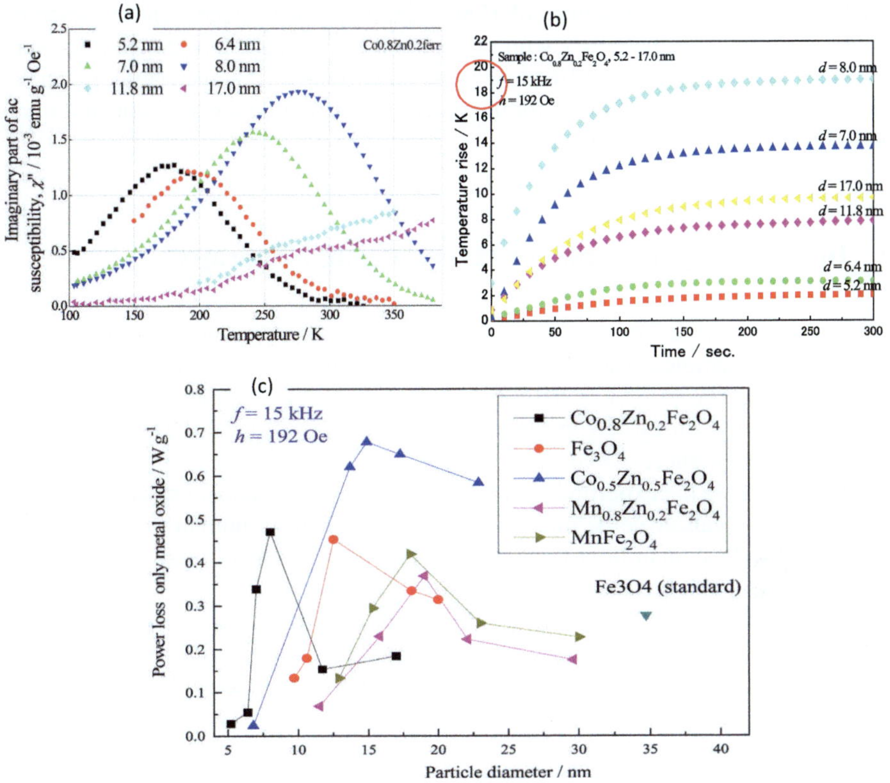

Fig. 3 Imaginary parts of the AC magnetic susceptibility (χ'') of the samples as a function of temperature at 1 Oe, 100 Hz AC magnetic fields for various particle sizes (**a**), increase in the temperature of the samples as a function of time (**b**) and calculated values of heat dissipation depending on the particle size for various particles (**c**)

was adjusted from 100 Hz to 15 kHz, and temperature measurement was performed using radiation thermometers, platinum-resistance thermometers, and optical fiber thermometers unaffected by the magnetic field. Sufficient cooling and insulation of the coil were ensured to distinguish the heat generated from the coil and other sources. Temperature increments were measured for Co-Zn ferrite nanoparticles, Considering the results of AC magnetic susceptibilities in Fig. 3a, the peak of χ'' was found to be near room temperature for the 8 nm sample; accordingly, the temperature rise was the highest (Fig. 3b). The temperature rise curve was fitted with an exponential function. By quantitatively determining the amount of heat generated (specific absorption rate, SAR) based on the slope and heat capacity measurements, a value of 0.5 W g^{-1} was obtained. In the same way, various magnetic and heat dissipation properties of magnetic particles were measured and summarized as Fig. 3c. For comparison, magnetite samples, which are widely used in magnetic hyperthermia, were prepared and measured. Smaller particles (12 nm) generated more heat than commercially available particles with a diameter of 35 nm. Significant heat generation was observed for the $Co_{0.5}Zn_{0.5}$ ferrite sample with a particle size of 15 nm. Thus, the samples were clarified, and by varying the frequency and conditions of the alternating magnetic field, nanoparticles that allowed for the precise control of heat generation at a small particle size scale were prepared.

5 In Vitro Experiments for Hyperthermia Therapy

Finally, we performed an in vitro magnetic hyperthermia therapy experiment using Mn-Zn ferrite samples. The Mn-Zn ferrite samples had the following composition: $Mn_{1-x}Zn_xFe_2O_4$ ($x = 0, 0.1, 0.2, 0.3, 0.4, 0.5, 0.6, 0.7, 0.8$). The magnetic and thermal properties of the Mn-Zn ferrite sample of $Mn_{1-x}Zn_xFe_2O_4$ ($x = 0.2$) with a size of 18 nm were the highest (Kondo et al. 2015).

Figure 4 shows the experimental results for the prostate cancer cells and the three types of breast cancer cells, indicating cell viability of 4a human prostate cancer DU145 cells, 4b human breast cancer KPL4 cells, 4c human breast cancer MCF7 cells, and 4d human breast cancer MDA-MB231 cells.

The cells were cultured in a dish, and 1 mg/ml was distributed in an AC magnetic field for 30 min. They were stained with trypan blue, and the remaining dead and living cells were counted. The magnetic field strength was 31 kHz 90 Oe, which was very weak.

In the cell viability graph in Fig. 4, No. 1 is the control, No. 2 is particles only, No. 3. is magnetic field only, and No. 4 is hyperthermia therapy.

The results indicate a drastic hyperthermia effect, with cell viability reaching 20%. Comparing "1. Control" with "2. Particles only", no difference in cell viability is observed, which also indicates that the nanoparticles do not exhibit toxicity.

6 Magnetic Nanoparticles for Diagnostics

Magnetic resonance (MR) imaging (MRI) is one of the most popular diagnostic techniques, and we aimed to determine if our nanoparticles could function as a contrast agent for MRI.

We prepared 4–34 nm Co-ferrite and measured their MR properties. We observed the magnetization and T_2 relaxation curves (Shigeoka et al. 2018). Figure 5 shows the magnetization curves of the Co-ferrite samples with particle sizes of 4.1, 14, 20, and 34 nm Fig. 5a, T_2 relaxation curves Fig. 5b, and a phantom T_2-weighted image Fig. 5c.

The graph in Fig. 5a shows that as the particle size decreases, the coercivity decreases, transitioning toward superparamagnetism. The Co-ferrite NPs (34 nm) exhibited ferromagnetic behavior and were found to be superparamagnetic at 4 nm.

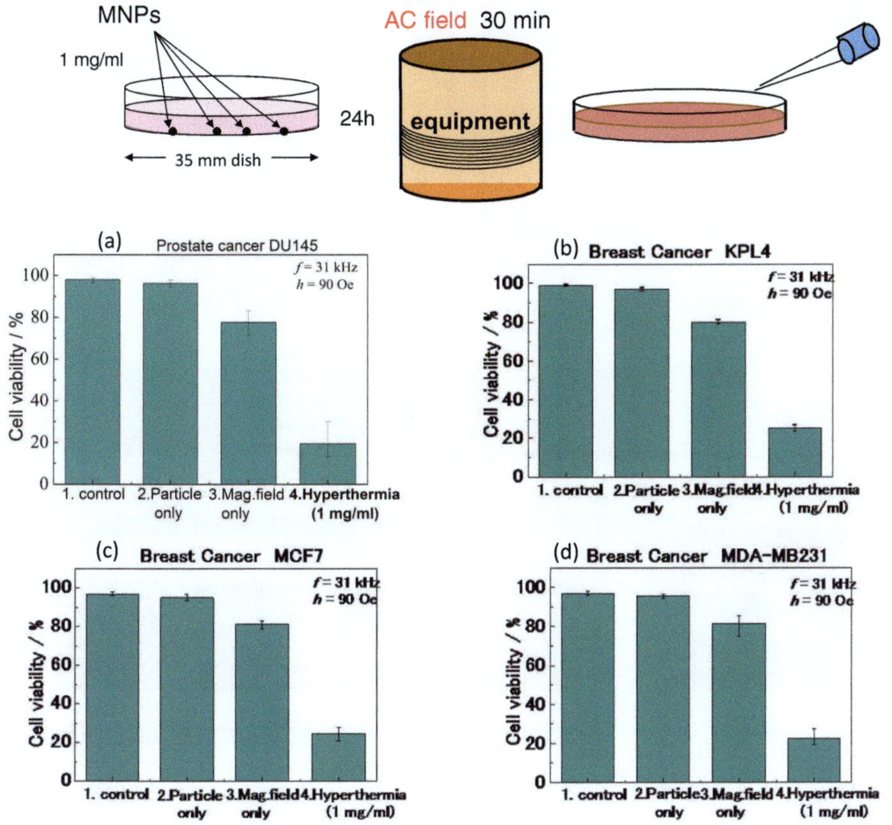

Fig. 4 Cell viability of (**a**) human prostate cancer DU145, (**b**) human breast cancer KPL4, (**c**) human breast cancer MCF7, and (**d**) human breast cancer MDA-MB231 cells (*J. Appl. Phys.* 117 (2015) 17D157)

Figure 5b also depicts the signal intensity of the experimental data, and the solid lines represent the fitting curves according to the T_2 relaxation equation of the spin-echo method. The T_2 relaxation signals were carefully measured thrice while the echo time (*TE*) was increased. A longer echo time results in a lower signal-to-noise (S/N) ratio. The following equation describes the signal intensity of the MR measurement:

$$I_{SE-T_2} = N \cdot K \cdot \exp\left(-\frac{TE}{T2}\right), \tag{3}$$

where N is the repetition constant, and K is the environmental constant.

As seen from the T_2 relaxation curves of the Co–ferrite samples of various particle sizes, the as-prepared particles induced a significant T_2 shortening effect in related to agarose, used as a controlled background for all samples. Relaxivity R_2 can be estimated from the inverse of T_2 of these curves. Excellent R_2 values were obtained.

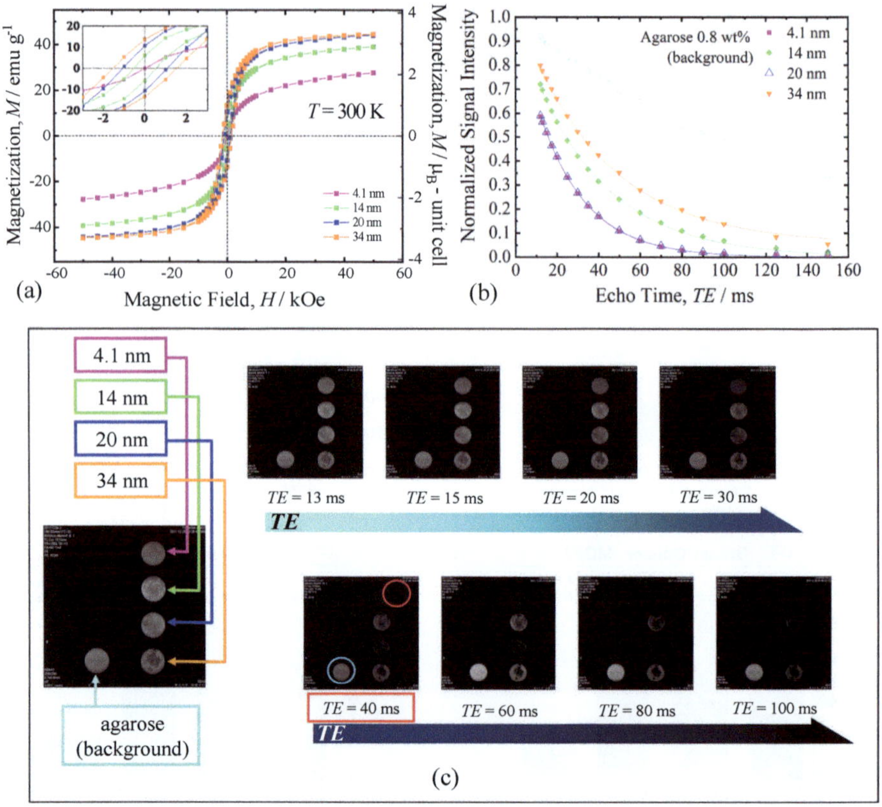

Fig. 5 Magnetization curves of $CoFe_2O_4$ NPs for various particle sizes at 300 K (**a**), T_2 relaxation curves of agarose samples shown for comparison (**b**), and phantom T_2-weighted image (**c**) (*IEEE Trans. Magn.* 54 (2018) 6,100,707)

The R_2 values of the as-prepared MNPs were ~4.0 ms^{-1} mM^{-1}, which are more than two times higher than those of conventional MRI materials.

Figure 5c shows the phantom T_2-weighted images. A clear contrast was observed between the Co ferrite NP and agarose images. In particular, for the 4 nm sample, a clear contrast was observed even with a short echo time of ~40 ms.

The magnetization measurement results confirm that the superparamagnetic behavior and use of smaller particles were applicable for MRI contrast.

7 Other Imaging Techniques

MNPs have been widely used in MRI for diagnostic purposes, and are found suitable for CT imaging. Furthermore, their application scope in other imaging techniques, such as mass spectrometry and magnetic imaging, has been explored.

Mass spectrometry imaging (MSI) exploits the ionization-assisted capabilities of MNPs (Moritake et al. 2008; Taira et al. 2008, 2012; Onuma et al. 2014; Morimoto et al. 2016). The as-prepared particles were utilized as matrices in MALDI-TOF–MS, allowing the observation of the composition and distribution of various components within cellular slices. Owing to their nanoscale size, MNPs possess a distinct advantage in the detection of low-molecular-weight substances, which are difficult to detect by using conventional matrices.

Magnetic particle imaging (MPI) is a novel technique for directly detecting MNPs (Gleich and Weizenecker 2005; Weizenecker 2012). While MRI detects protons, MPI enables the direct detection of particle location if present at the lesion site. This feature holds promise for future diagnostic applications as it directly detects lesion sites where MNPs accumulate.

8 Summary

MNPs were prepared using a unique technique based on proprietary methods and their biomedical application scope is explored. The properties of the as-prepared MNPs, including magnetic and heat dissipation characteristics, were characterized. Considering various compositions and particle sizes, potential applications in the therapeutics field were discussed, particularly focusing on application of MNPs for cancer hyperthermia therapy. In addition, the applicability of MNPs in diagnostics application, particularly MR, was revealed. Looking ahead, there is anticipation for the emergence of nano-theranostics, which could simultaneously facilitate both therapeutic and diagnostic functions using magnetic nanoparticles.

Acknowledgements This study was partially supported by the Precursory Research for Embryonic Science and Technology at the Japan Science and Technology Agency (Y.I.), JST-Mirai Program No.JPMJMI17D7 (Y.I.), Grant-in-Aid for Science Research No. 17H02762, No. 23656013, No.25286041, No.20H00344, (Y.I.), and Collaborative Research Project of the Laboratory for Materials and Structures, Institute of Innovative Research, Tokyo Institute of Technology. The authors thank Mr. M. Kondo of the Instrumental Analysis Center at YNU. We thank Dr. S. Taira and M. Seto's group at Mitsubishi Chemical Life Science Institute and S. Moritake, Y. Kai who were my students at the time for their cooperation in animal experiments. We thank Prof. Gonda's group and Prof. Ouchi's group at Tohoku University for providing us with human breast cancer cells and for teaching us the experimental procedure. We thank Prof. Y. Nakazawa, Prof. M. Nakano, Prof. Y. Miyazaki at the Research Center for Thermal and Entropic Science at Osaka University for their assistance in using the SQUID magnetometer. English language was edited by Editage.

References

M. F. Casula, A. Corrias, P. Arosio, A. Lascialfari, T. Sen, P. Floris, Ian J. Bruce, J. Colloid Interface Sci. 357 (2011) 50–55.
G. Cordoyiannis, L. K. Kurihara, L. J. Martinez-Miranda, C. Glorieux, J. Thoen, Phys. Rev. E 79, (2009)011702.
F. Fabris, E. Lima Jr., E. D. Biasi, H. E. Troiani, M. V. Mansilla, T. E. Torres, R. F. Pacheco, M. R. Ibarra, G. F. Goya, R. D. Zysler, and E. L. Winkler, Nanoscale 11, (2019) 3164–3172.
Gleich and J. Weizenecker, Nature 435, 1214–1217 (2005).
E. W. Gorter, Nature, 165 (1950) 798–800.
R. Hergt and S. Dutz, J. Magn. Magn. Mater. 311 (2007) 187–192.
Y. Ichiyanagi, S. Moritake, S. Taira, and M. Setou, J. Magn. Magn. Mater. 310 (2007) 2877–2879.
Y. Ichiyanagi, D. Shigeoka, T. Hiroki, T. Mashino, S. Kimura, A. Tomitaka, K. Ueda, and Y. Takemura, Thermochim. Acta 532, 123–126 (2012).
Y. Ichiyanagi, T. Uehashi, and S. Yamada, phys. stat. sol. (c) 1 (2004) 3485–3488.
S.V. Jadhav, B.M. Kim, H.Y. Lee, I.C. Im, A.A. Rokade, S.S. Park, M.P. Patil, G.D. Kim, Y.S. Yu, and S.H. Lee, , J. Alloys Comp. 745 (2008) 282–291.
A. Jordan, R. Scholz, P. Wust, H. Fähling, and R. J. Felix, Magn. Magn. Mater. 201 (1999) 413–419.
H. Katayanagi, N. Sakai, S. Hamada, A. Usui, K. Aoki, K. Kodama, K. Nashimoto, Y. Hosokai, and Y.Ichiyanagi, ChemNanoMat, CNMA202200014, (2022) 1–7.
K. Kattel et al., Biomaterials 33(2012) 3254–3261.
S. S. Kelkar, and T. M. Reineke, Bioconjug. Chem. 22 (2011) 1879–1903.
R.H. Kodama, J. Magn. Magn. Mater. 200 (1999) 359–372.
T. Kondo, K. Mori, M. Hachisu, T. Yamazaki, D. Okamoto, M. Watanabe, K. Gonda. H. Tada, Y. Hamada, M. Takano, N. Ohuchi, and Y. Ichiyanagi, J. Appl. Phys. 117 (2015) 17D157.
L. E. W. LaConte *et al.*, *J. Magn. Reson. Imag.* **26**, 1634–1641, 2007.
S. Laurent, S. Dutz, U. O. Häfeli, and M. Mahmoudi, Adv. Colloid Interface Sci. 166 (2011) 8–23.
J. Lu et al., Biomaterials 30, (2009) 2919–2928.
R. Matthew Ferguson, Amit P. Khandhar and Kannan M. Krishnan, J. Appl. Phys. 111 (2012) 07B318.
S. Morimoto, T. Ishikawa, K. Hyodo, T. Yamazaki, S.Taira, K. Tsuneyama, and Y. Ichiyanagi, .Surf. Inter. Anal. 48 (2016) 1127–1131.
S. Moritake, S. Taira, Y. Sugiura, M. Setou and Y. Ichiyanagi, J. Nanoci. Nanotech. 9 (2008) 169–176.
S. Moritake, S. Taira, Y. Ichiyanagi, N. Morone, Si-Y.Song, T.Hatanaka, S. Yuasa, and M. Setou, J. Nanosci. Nanote., 7(2007) 937–944.
Y. Moro, H. Katayanagi, S. Kimura, D. Shigeoka, T. Hiroki, T. Mashino, and Y. Ichiyanagi, Surf. Interface Anal. 42 (2010) 1655–1658.
S. Mura, P. Couvreur, Advanced Drug Delivery Reviews, 64 (2012) 1394–1416.
K. Onuma, T. Hiroki, Y. Ichiyanagi, e-J Surf. Sci. Nanotech. 12 (2014) 269–274.
M.J. O'Shea , P. Perera, J. Magn. Magn. Mater. 156 (1996) 141–142.
A.G. Roca, J.F. Lopez-Barbera, A. Lafuente, F. Oezel, E. Fantechi, J. Muro-Cruces, M. Hemafi, B. Sepulvedo, J. Nogues, Physics Reports, 1043 (2023) 1–35.
R. E. Rosensweig, J. Magn. Magn. Mater. 252 (2002) 370–374.
S. Ruta, R. Chantrell & O. Hovorka, Scientific Reports, 5:9090 (2015) https://doi.org/10.1038/srep09090.
N Sakai, L Zhu, A Kurokawa, H Takeuchi, S Yano, T Yanoh, N Wada, S. Taira, Y Hosokai, A Usui, Y Machida, H Saito and Y Ichiyanagi, J. Phys.: Conf. Ser. 352 (2012) 012008.
D. Shigeoka, T. Yamazaki, T. Ishikawa, K. Miike, K.Fujiwara, T. Ide, A.Oshima, T. Hashimoto, D. Aihara, K. Kanda, A. Usui, Y. Hosokai, H. Saito,Y. Ichiyanagi, IEEE Trans. Magn. 54 (2018) 6100707.
Song, M. Kenney, Y. S. Chen, X. Zheng, Y. Deng, Z. Chen, S. X. Wang, S. S. Gambhir, H. Dai, and J. Rao, Nat. Biomed. Eng. 4 (2020) 325–334.

M. Sukumar, M. Agila, A. Sutha, V. Ravi, A. M. Al-Enizi, M. Ubaidullah, M. S. Samdani, M. Sundararajan, B. Pandit, J Mater Sci: Mater Electron33 (2022)26144–26156.
N. Sun, D. -X. Chen, H. -C. Gu, and X. -L. Wang, J. Magn. Magn. Mater. 321 (2009) 2971–2975.
M. Suto, Y. Hirota, H. Mamiya, A. Fujita, R. Kasuya, K.Tohji, and B. J. Jeyadevan, Magn. Magn. Mater. 321 (2009) 1493–1496.
S. Taira, Y. Sugiura, S. Moritake, Y. Ichiyanagi and M. Setou, Anal. Chem. 80 (2008) 4761–4766.
Shu Taira, Daisaku Kaneko, Kazuki Onuma, Akio Miyazato, Tomoyuki Hiro-ki, Yasuko Konishi-Kawamura, and Yuko Ichiyanagi, Int. Jo. Inorg. Chem., 2012 (2012) 439751.
S. TAIRA, S. MORITAKE, Y. KAI, T. HATANAKA, Y. ICHIYANAGI and M. SETOU, eJ. Surf. Sci. Nanotech., 5 (2007) 23–28.
R. Weissleder, G. Elizondo, J. Wittenberg, C.A. Rabito, H.H. Bengele, L. Josephson, Radiology, 175 (1990), 489–493.
Jürgen Weizenecker, Phys. Med. Biol. 57, (2012) 7317–7327.
G. G. Westmeyer, and A. Jasanoff, J. Magn. Re-son. Imag. 25, (2007)1004–1010.

Engineered and Artificial Exosomes for Non-viral Drug Delivery Nanocarriers

Masatoshi Maeki and Manabu Tokeshi

Abstract This chapter overviews the expanding field of engineered and artificial exosomes as cutting-edge non-viral drug delivery nanoparticles. Exosomes are expected to be natural carriers for drug delivery systems and clinical trials are in progress for several exosome-based nanomedicines. While therapeutic potential of exosomes has been demonstrated, several challenges must be overcome for practical applications including the need for: scalable manufacturing, standardization, and efficient drug loading. To overcome them, various methodologies have been developed for the production of engineered and artificial exosomes through bioengineering and micro/nanobiotechnology, thereby presenting a promising solution to the limitations of natural exosomes. In this chapter, various production methods are discussed, including: top-down, bottom-up, and biohybrid methodologies; drug-loading techniques; and membrane fusion methods for engineered exosome production. Furthermore, this chapter highlights the use of microfluidic devices in the production of artificial exosomes and presents a novel approach for the precise control of particle size and composition. Utilizing nanotechnology, molecular biology, and pharmacology, engineered and artificial exosomes allow the development of novel non-viral drug delivery systems by improving biocompatibility, increasing specificity, and expanding payload capacity.

Keywords Drug delivery system · RNA delivery · Nanomedicine · Engineered exosome · Microfluidic device

---

M. Maeki (✉) · M. Tokeshi
Division of Applied Chemistry, Faculty of Engineering, Hokkaido University, Kita 13 Nishi 8, Kita-ku, Sapporo 060-8628, Japan
e-mail: m.maeki@eng.hokudai.ac.jp

M. Tokeshi
e-mail: tokeshi@eng.hokudai.ac.jp

Y. Baba et al. (eds.), *Extracellular Fine Particles*,
https://doi.org/10.1007/978-981-97-7067-0_19

1 Introduction

The advancement of drug delivery systems (DDSs) has reached a significant milestone in the evolution of nanomedicines. Nanomedicines based on DDS technologies have led to new approaches for the treatment of various diseases. Lipid nanoparticles (LNPs) have attracted significant attention, particularly because of their crucial function in delivering messenger RNA (mRNA), and they were applied to develop vaccines against COVID-19 (Tang et al. 2021; Buschmann et al. 2021). The remarkable efficacy of mRNA vaccines in preventing severe COVID-19 demonstrated clearly that LNPs have potential as nanocarriers that are characterized by their: versatility in encapsulating diverse biomolecules, targeted delivery capabilities, high transfection efficiency, and low cytotoxicity. In parallel with advancements in DDS technologies, exosomes, which are secreted extracellular vesicles ranging in size from 30 to 200 nm, are expected to be natural DDS carriers (Sancho-Albero et al. 2020; Darband et al. 2018; Li et al. 2019; Ortega et al. 2020). These vesicles, which are secreted by cells, present specific membrane proteins on their surface, RNAs, and DNAs encapsulated within a lipid bilayer, making them natural vehicles for molecular communication between cells (Valadi et al. 2007). The surface of exosomes presents specific membrane proteins such as CD9, CD63, CD81, integrins, MHCs, and annexin, and these proteins are used as biomarkers for exosome-based diagnosis. As reviewed by Deng et al. (2018), several studies have demonstrated that exosomes secreted by mesenchymal stem cells (MSCs) have therapeutic effects in various diseases, making them a promising treatment option. In addition, Hoshino et al. (2015) reported that some types of surface integrins play an important role in cancer metastasis and organ targeting. From the viewpoint of LNP-based DDSs, the lipid composition of exosomes affects their biodistribution.

The clinical applications of natural exosomes remain with significant challenges owing to their need for: scalable production, standardized high purity isolation, strict quality control, efficient drug loading, and drug stability (Ortega et al. 2020). Several papers have been published regarding the selective packaging of therapeutic molecules into exosomes (Raguraman et al. 2023), the key factors for targeted delivery to specific organs (He et al. 2022), and biodistributions (Kang et al. 2021); however, exosomes show heterogeneity, and detailed mechanisms to control the characteristics of exosomes as DDS carriers are still unclear. The process of selective packaging within exosomes is essential for determining their potential effectiveness as drug delivery carriers. Understanding the molecular mechanisms that direct the sorting of cargo into exosomes is essential for the advancement of exosomes with increased loading capacity and specificity. Advances in bioengineering and molecular biology are anticipated to elucidate the mechanisms involved in these processes and allow the development of novel exosome-based DDSs.

In addressing these problems, researchers are standing at a new frontier established through the development of engineered and artificial exosomes based on bioengineering and nanobiotechnology (Wang et al. 2022; Tsai et al. 2021; Dai et al. 2020; Jhan et al. 2020; Cheng et al. 2021; Lin et al. 2018; Evers et al. 2022). These approaches aim at combining the benefits of natural exosomes with the versatility of synthetic nanocarriers, resulting in the development of more advanced DDS platforms. Engineered exosomes are produced using a variety of strategies, including top-down, bottom-up, and biohybrid methodologies, each with its own set of advantages and limitations. On the other hand, the ongoing development of engineered and artificial exosomes as potential non-viral drug delivery nanocarriers highlights the gaps in understanding their fundamental mechanisms as DDS carriers, efficacy, and clinical application potential. This chapter provides an overview of the production methods and characteristics of engineered and artificial exosomes and their therapeutic applications. By combining nanotechnology, molecular biology, and pharmacology, engineered and artificial exosomes can overcome the drawbacks of cell-derived exosomes and lead to a new era in non-viral DDSs.

1.1 Methods for Drug Loading into Exosomes

To use exosomes in drug discovery, it is necessary to load drugs into the exosomes. There are two main methods for directly loading drugs into exosomes: pre-secretory loading and post-secretory loading (Zhang et al. 2020; Luan et al. 2017). The gene editing method is one example of the pre-secretory loading approach and it involves modifying the genes of the parent cells that secrete exosomes to load RNAs or proteins (Alvarez-Erviti et al. 2011). This method is suitable for loading RNAs and proteins that cannot directly pass through the lipid bilayer into exosomes. Co-incubation of the parental cells with drugs is another example for pre-secretory loading. The co-incubated drugs are loaded by endocytosis into the parental cells, and then the drug-loaded exosomes can be collected (Kim et al. 2016). This method does not require complicated procedures which is an advantage. However, although hydrophobic drugs exhibit a higher loading efficiency than hydrophilic drugs, overall the drug-loading efficiency is generally relatively low, and the loaded drugs will be cytotoxic to parental cells.

Post-secretion loading is the second approach that directly introduces drugs into the exosomes. Electroporation and sonication are its most commonly used methods. Electroporation uses high-voltage pulses to temporarily form pores on the lipid membranes of exosomes, allowing drugs to permeate into the exosomes (Wahlgren et al. 2012; Lamichhane et al. 2015). Sonication can also efficiently load drugs, and its loading efficiency is higher than that of co-incubation or electroporation (Kim et al. 2016). However, these methods will induce exosome aggregation, deterioration of lipid membranes and membrane proteins, and leakage of loaded drugs.

Both pre-secretory loading and post-secretory methods have their pros and cons, highlighting the need for an efficient technique that minimizes damage when loading drugs into exosomes.

1.2 Production Methods of Engineered Exosomes

As noted in the Introduction, clinical applications of exosomes present problems because of the need for scalable production, purification, quality control, efficient drug loading, and drug stability, and these problems should be overcome for next-generation nanomedicines. To overcome these issues, exosome-engineering strategies have been developed based on nanobiotechnology and genetic approaches (Liu et al. 2022). These approaches aim to provide different and designed exosomes by coupling the features of cell-derived exosomes with DDS nanocarriers. There are three types of methodology: top-down, bottom-up, and biohybrid.

Some examples of drug loading methods into exosomes were briefly described in Sect. 1.1. Genetically manipulated parent cells secrete drug-loaded engineered exosomes with a lipid membrane composition similar to that of natural exosomes. This methodology preserves the biological characteristics and functionality of natural exosomes. In contrast, the bottom-up methodology produces artificial exosomes from synthetic materials to mimic the structural and functional characteristics of natural exosomes. This strategy allows for precise control over the composition and properties of the resulting exosomes, enabling the engineering of tailored functionalities. The biohybrid strategy—the third methodology—produces engineered exosomes by fusing natural exosomes with synthetic liposomes or LNPs. This methodology takes advantage of the biological features of natural exosomes, while incorporating the customisable and scalable properties of synthetic liposomes and LNPs. In this chapter, mainly biohybrid and bottom-up methodologies are overviewed as highly customizable production approaches for engineered exosomes. Contrary to the top-down methodology, the other two can functionalize engineered exosomes using synthesized molecules such as ionizable lipids employed for mRNA vaccines.

1.3 Membrane Fusion Methods for Engineered Exosome Production

Membrane fusion methods have been proposed to produce biohybrid exosomes by combining exosomes with drug-loaded LNPs. This enables the combination of the previously mentioned benefits of exosomes with the characteristics of LNPs as DDS carriers, such as targeting, modifying of surfaces, and controlling of particle size and surface charge.

The freeze–thaw method is used to prepare not only engineered exosomes, but also drug-loaded conventional liposomes. This method promotes the formation of engineered cells by temporarily forming ice crystals that disrupt the lipid membrane on the exosome surface. Although this method can achieve high fusion efficiency, it may reduce drug activity or destroy the exosome membrane (Cheng et al. 2021; Sato et al. 2016). Sato et al. (2016) proposed the preparation method of the engineered exosomes by membrane fusion with liposome using the freeze–thaw method.

Figure 1a shows a schematic illustration of the freeze–thaw method for producing the engineered exosomes. Exosomes have shown promise in mediating intercellular communication and delivering therapeutic molecules; however, their potential is hindered by challenges in customisation and functionality enhancement. The engineered exosomes are designed to combine the favourable characteristics of both constituent parts, such as the specific membrane proteins from the exosomes and the customisable lipid composition of the liposomes. The study by Sato et al. (2016) demonstrated the successful fusion of exosomes derived from Raw 264.7 cells and liposomes, confirmed by fluorescence resonance energy transfer (FRET) assays that quantified the lipid mixing ratio as a measure of fusion efficiency (Fig. 1b). Moreover, exosomes isolated from genetically modified CMS7 cells expressing the human HER2 receptor were fused with liposomes, thereby demonstrating that this method could be applied to genetic modifications to introduce specific proteins into exosomes. Another significant finding was that the lipid composition of the fused liposomes affected the cellular uptake of the engineered exosomes. To conclude, the research indicated that modifying the exosomal membrane through fusion could tailor the interactions between exosomes and target cells, potentially enhancing the delivery efficiency of therapeutic cargoes.

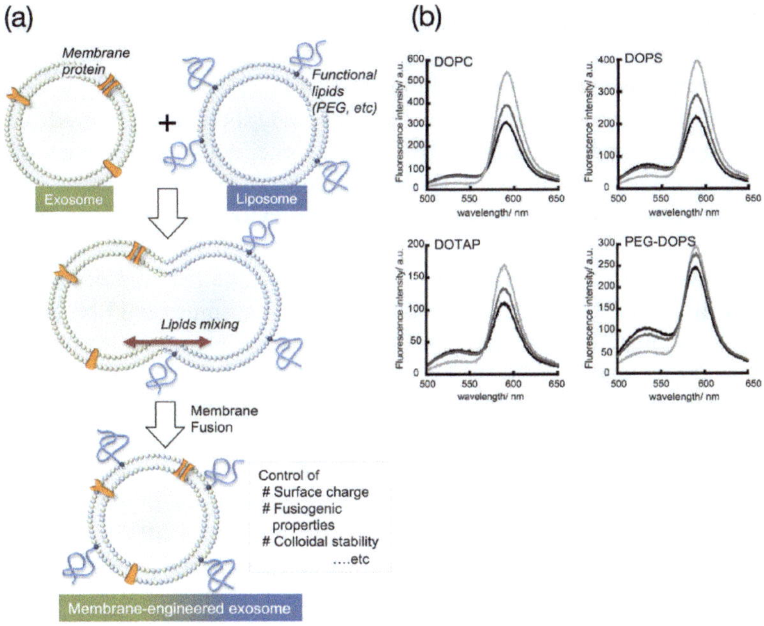

Fig. 1 **a** Schematic illustration of the freeze–thaw method for producing engineered exosomes. **b** The fluorescence spectra were measured at an excitation wavelength of 460 nm. The light grey, dark grey, and black lines depict the spectra before a freeze–thaw cycle, and after five and ten freeze–thaw cycles, respectively. (Reprinted from Ref. Sato et al. (2016) with permission from Springer Nature Publishing Group.)

The incubation method promotes fusion through electrostatic or hydrophobic interactions between exosomes and LNPs as demonstrated by Lin et al. (2018). This method causes minimal damage to exosomes; however, its fusion efficiency is relatively low. Lin et al. developed engineered exosomes to deliver the CRISPR/Cas9 gene editing system into MSCs using the incubation method. The engineered exosomes were produced by incubating a mixture of Lipofectamine 2000 and plasmid for expression of CRISPR/Cas 9 and exosomes containing sgRNA for 12 h at 37 °C. Therefore, the engineered exosomes contained the CRISPR/Cas 9 plasmid and sgRNA. Lin et al. found that these engineered exosomes were able to cleave the target gene, and the cleavage efficiency of the engineered exosomes was higher than that of exosomes and Lipofectamine 2000.

The polyethylene glycol (PEG)-mediated method induces fusion between the exosome membrane and LNPs by mediating the close contact of the lipid bilayer structure through PEG on the LNP surface (Piffoux et al. 2018; Ma et al. 2022; Kannavou et al. 2021). Surface modification by PEG on the engineered exosomes can enhance the circulation time in the blood, although it may reduce cellular uptake. Piffoux et al. (2018) demonstrated a novel approach for enhancing the drug delivery capabilities of exosomes by fusing them with liposomes using PEG-mediated fusion. Employing a FRET-based lipid mixing assay, they found PEG-8000 effectively promoted the merging of exosomes derived from human umbilical vein endothelial cells (HUVECs) and liposomes (Fig. 2a). The fusion efficiency of exosomes and liposomes increased with increasing PEG concentrations (Fig. 2b–d). The engineered exosomes were composed of lipids and proteins from the merged liposomes and exosomes, respectively. In addition, the same study showed engineered exosomes incorporating the antitumor drug mTHPC had a higher delivery efficiency to cancer cells than liposomes containing mTHPC or free mTHPC. In particular, the engineered exosomes enhanced the accumulation and cytotoxic effects of mTHPC in cancer cells.

The membrane extrusion method is also used to tune the size of conventional liposomes. To produce engineered exosomes, natural exosomes and LNPs are passed through size-regulated membrane pores under physical pressure (Jhan et al. 2020; Evers et al. 2022; Sun et al. 2021; Rayamajhi et al. 2019; Zhou et al. 2022; Ji et al. 2022; Hu et al. 2021; Kang et al. 2023). The advantages of the membrane extrusion method are the uniform particle size of the engineered exosomes and the high fusion efficiency. However, the shear stress generated during the extrusion process may damage the exosome membrane and proteins and is labour-intensive compared with other fusion methods. Evers et al. (2022) reported functional siRNA delivery using engineered exosomes prepared by the membrane extrusion method (Fig. 3a). Initially, the lipids were dissolved in a chloroform/methanol mixture and transferred to a round-bottom flask to form a thin lipid film through evaporation. This film was then hydrated with a mixture of siRNA and extracellular vesicles collected from SKOV3 cells at specific protein-to-lipid ratios, such as 1:100 or 1:50 (w/w), in citrate buffer. The hydration step was followed by extrusion, where the hydrated mixture was passed through polycarbonate filters with decreasing pore sizes to form uniformly sized

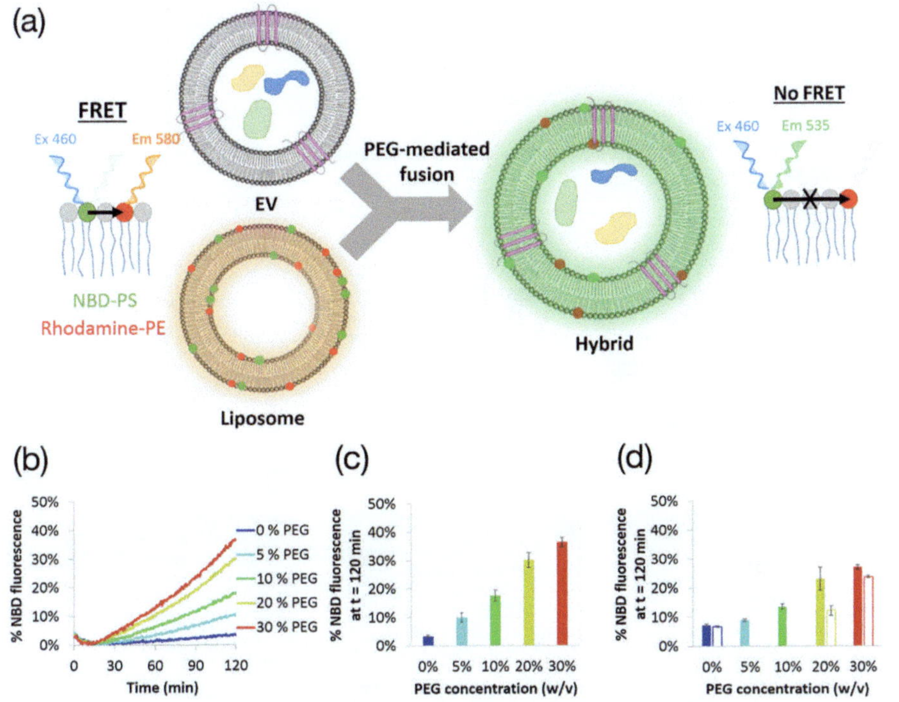

Fig. 2 **a** The concept of FRET-based lipid-mixing assay to evaluate the fusion between exosomes and liposomes. **b** The effect of the PEG concentrations on the membrane fusion efficiency. **c** Average fusion efficiency after 2 h of incubation with liposomes and HUVEC-derived exosomes. **d** Average fusion efficiency after 2 h of incubation with liposomes and MSC-derived exosomes. (Reprinted from Ref. Piffoux et al. (2018) with permission from the American Chemical Society.)

nanoparticles. After extrusion, the nanoparticles were dialysed against phosphate-buffered saline (PBS) to remove unencapsulated siRNA and adjust the pH to physiological conditions. Evers et al. confirmed the particle size, morphology, and inner structure of the engineered exosomes by dynamic light scattering analysis, nanoparticle tracking analysis, and cryo-electron microscopy (Fig. 3b–d). The diameter of the spherically-shaped engineered exosomes was approximately 150 nm, as measured using dynamic light scattering analysis, and 100 nm, as measured using nanoparticle tracking analysis. And they had a unilamellar structure. These engineered exosomes combined the efficient RNA loading capacity of liposomes with the natural cell-targeting and biocompatible properties of exosomes; as well, they had the potential to enhance the delivery efficacy and reduce the toxicity of siRNA therapeutics.

Evers et al. evaluated the cellular uptake, gene silencing activity, and cytotoxicity using the engineered exosomes. The engineered exosomes exhibited reduced toxicity in SKOV3 cells compared to liposomes, which suggests enhanced biocompatibility. On the other hand, there was no significant difference in cell toxicity between liposomes and the engineered exosomes in HEK293T and U87-MG cells. The engineered

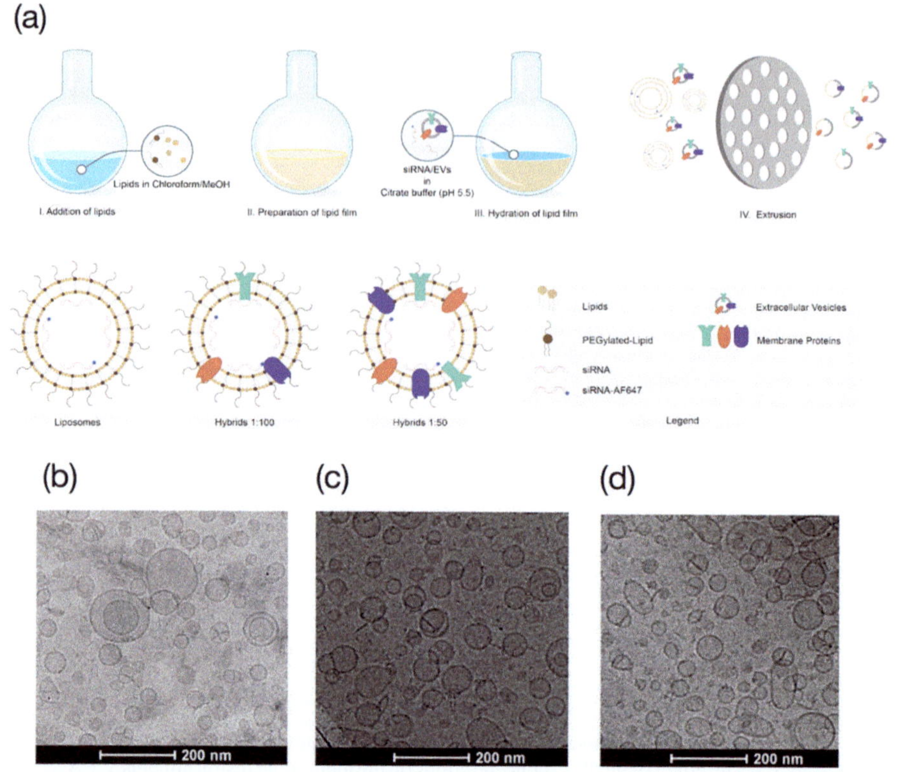

Fig. 3 **a** Schematic illustration of the engineered exosome preparation using thin-film hydration and extrusion methods. Engineered exosomes were produced by mixing liposomes and exosomes with different protein-to-lipid ratios (w/w) of 1:100 and 1:50. **b–d** Cryo-electron microscopic images for **b** liposomes, **c** engineered exosomes (1:100), and **d** engineered exosomes (1:50). (Reprinted from Ref. Evers et al. (2022) with permission from Wiley-VHC.)

exosomes functionally delivered siRNA in a dose-dependent manner in SKOV3, HEK293T, and U87-MG cells and silenced genes. In SKOV3 and HEK293T cells, the gene silencing efficiency of hybrids was lower than that of liposomes; however, in U87-MG cells, the effect was similar to that of liposomes.

Hu et al. (2021) applied engineered exosomes to anabolic therapy for bone loss using a mouse model. The engineered exosomes were produced by merging exosomes that exhibited CXCR4 on their surface with liposomes carrying Antagomir-188. These hybrid NPs specifically accumulated in the bone marrow and released Antagomir-188, promoting osteogenesis and inhibiting adipogenesis in bone marrow mesenchymal stem cells (BMSCs). Age-related trabecular bone loss was reversed, and cortical bone porosity was reduced in the mouse model. The application of engineered exosomes is a promising method for delivering targeted therapeutic agents to the bone marrow, indicating their capacity to enhance bone health by stimulating bone growth.

The delivery of mRNAs by exosomes has also attracted attention. Tsai et al. (2021) reported that exosome-mediated mRNA delivery induces SARS-CoV-2 immunity. Engineered exosome-encapsulated mRNA was produced by mixing and co-incubating mRNA-loaded LNPs with purified exosomes. The use of exosomes for delivering mRNA was found to be more effective than that of LNPs, resulting in higher levels of mRNA expression in human cells. Tsai et al. also developed mRNAs encoding SARS-CoV-2 Spike and nucleocapsid proteins, which were successfully expressed in human cells. Immunization with mRNA-loaded exosomes in mice resulted in the dose-dependent production of antibodies against both proteins and the activation of specific $CD4^+$ and $CD8^+$ T-cell responses, demonstrating the potential of this delivery system to induce immune responses.

In conclusion, engineered exosomes provide versatile and efficient approaches for delivering drugs to specific targets, which combines the biocompatibility of natural exosomes and cell-specific targeting abilities with the adaptable payload capacity of synthetic nanocarriers.

1.4 Bottom-Up Methods for Artificial Exosome Production

Microfluidic devices have been attracting attention in the field of LNP-based DDSs (Maeki et al. 2018, 2022). In comparison with a conventional LNP production method like the lipid film hydration method, the ethanol dilution method using microfluidic devices can produce precisely size-controlled LNPs without adding a size tuning process. Recently, commercially available microfluidic devices or systems for LNP production have been used (Maeki et al. 2023; Webb et al. 2020). Reported applications of microfluidic devices have involved siRNA, mRNA, and plasmid DNA, and ribonucleoprotein and anti-tumor drug deliveries (Maeki et al. 2024; Matsuura-Sawada et al. 2023, 2022; Suzuki et al. 2021; Hassett et al. 2019). Therefore, microfluidic devices represent a promising methodology for the production of a variety of LNPs based on bottom-up techniques and they can be applied to produce artificial exosomes that mimic cell-derived natural exosomes.

As mentioned earlier, exosomes play a role in intercellular communication and they are involved in cancer metastasis. Therefore, exosome drug discovery has attracted worldwide attention. The tetraspanin family (CD9, 63, and 81) and integrins (ITGs) on the exosome surface have been reported to be important in cell signalling and organ targeting. However, exosome heterogeneity is a major problem for understanding the functions of exosomes. In addition, exosome preparation is complicated because it involves multiple purification steps (Ortega et al. 2020). Therefore, the detailed function and role of specific membrane proteins on the exosome surface in DDS applications remain unclear. The development of artificial exosomes that present specific exosomal proteins on the surface at the desired concentration makes it possible to understand the functionality of surface proteins.

Our group has developed a methodology for exosome preparation using a microfluidic device. Figure 4a shows a schematic illustration of the one-step artificial exosome preparation using the microfluidic device. The aqueous solution contained proteins and RNAs in an appropriate buffer solution. The lipid solution contained a mixture of several types of lipids that mimicked cell-derived exosomes. Both the solutions were rapidly mixed in the microfluidic device, and that was followed by the production of artificial exosomes. Figure 4b shows the average sizes of LNP (without proteins) and artificial exosomes. The protein-presented particles were 130–140 nm in size, 10–20 nm larger than those not presented. In addition, the sizes of the tetraspanin-presenting and integrin-presenting particles are almost the same. The presence of CD9, 63, or 81 on artificial exosomes was evaluated by ELISA (Fig. 4c). The amount of the surface proteins was able to control the initial concentration of proteins in the aqueous phase. The presence of ITG $\alpha 6\beta 4$ in the artificial exosomes was measured using flow cytometry (FCM) (Fig. 4d). A shift in fluorescence intensity was observed, suggesting that ITG was present on the surface of the artificial exosomes. The bottom-up method of producing artificial exosomes offers more flexibility in particle design than the top-down approach, and it can also be combined with microfluidic devices to enable precise control over particle size.

1.5 Some Other Microfluidic Approaches for Artificial Exosome Production

The microfluidic technique has also been applied for production of nanovesicles (artificial exosomes) in a top-down manner (Zhu et al. 2018; Jo et al. 2014a, b). Jo et al. (2014a) reported cell deformation-based nanovesicle production using a microfluidic device. The use of hydrophilic microchannels in the device leads to the formation of nanovesicles containing encapsulated RNAs and proteins, along with specific membrane proteins. The process involves cells being stretched as they flow through the microchannels, with the result being the formation of nanovesicles owing to pressure changes and shear stress. Jo et al. fabricated a variety of microchannels by changing the microchannel length as 100, 200, or 400 μm with a constant width of 5 μm or changing the width as 3, 5, or 7 μm with a constant length of 200 μm. The size and amount of proteins and RNA in the nanovesicles can be controlled by the geometry of the microchannels. Longer microchannels and optimal flow rates contribute to the generation of nanovesicles with a uniform size and a higher yield of proteins and RNAs (Fig. 5).

The same research group also reported the large-scale generation of nanovesicles using a centrifuging microfluidic device (Jo et al. 2014b). The device design consisted of simple parts, including a 10 mm pore polycarbonate track-etched filter, a syringe, a PDMS sheet, and a filter holder. The centrifuging function was provided by an ordinary centrifuge machine. The production efficiency of nanovesicles was approximately 250 times higher than that of naturally secreted exosomes. Engineered nanovesicles demonstrate remarkable capacity, with their payload being twice that of

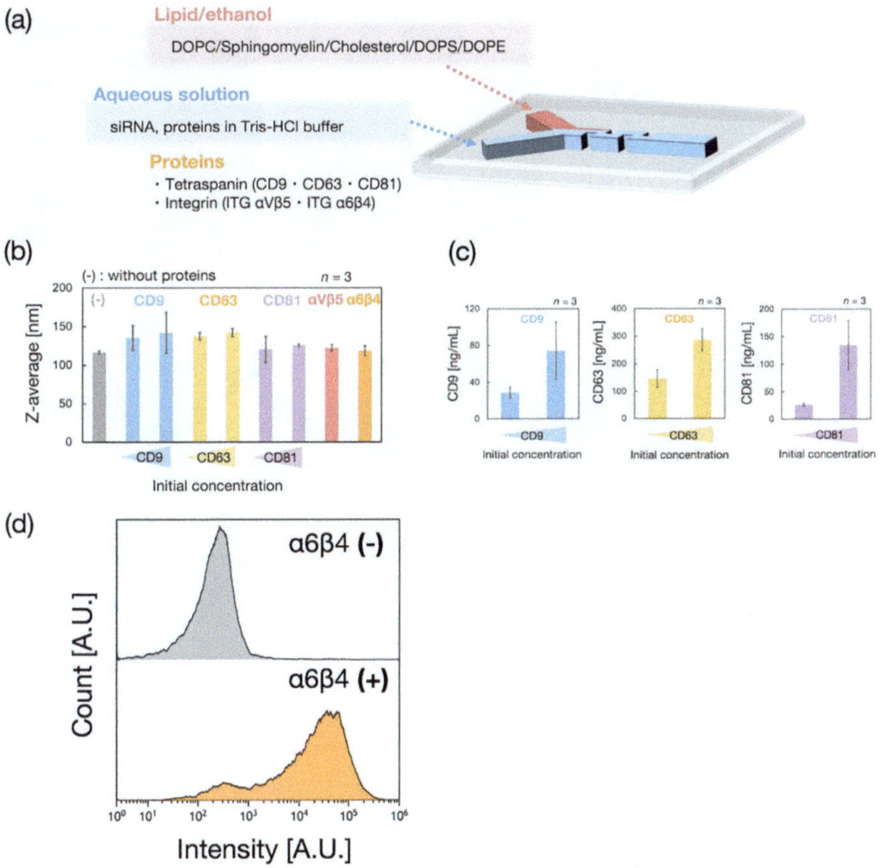

Fig. 4 **a** Schematic illustration of the artificial exosome preparation using a microfluidic device. **b** Average sizes of artificial exosomes. **c** Quantitative analysis by ELISA for CD9, 63, and 81 on the surface of the artificial exosomes. **d** Characterization by flow cytometry for fluorescence-labeled ITG-presented artificial exosomes.

naturally secreted exosomes. This improved loading efficiency highlights their potential for drug delivery and therapeutic use. Consequently, the centrifuging microfluidic device effectively addresses a significant challenge in exosome research, which has been hindered by the scarcity of exosomes. Nanovesicle production using a microfluidic device is based on a top-down approach and yields a higher output than conventional exosome production methods. This method involves applying pressure or centrifugal force to cultured cells to produce nanovesicles, which may have potential therapeutic effects, as in regenerative medicine.

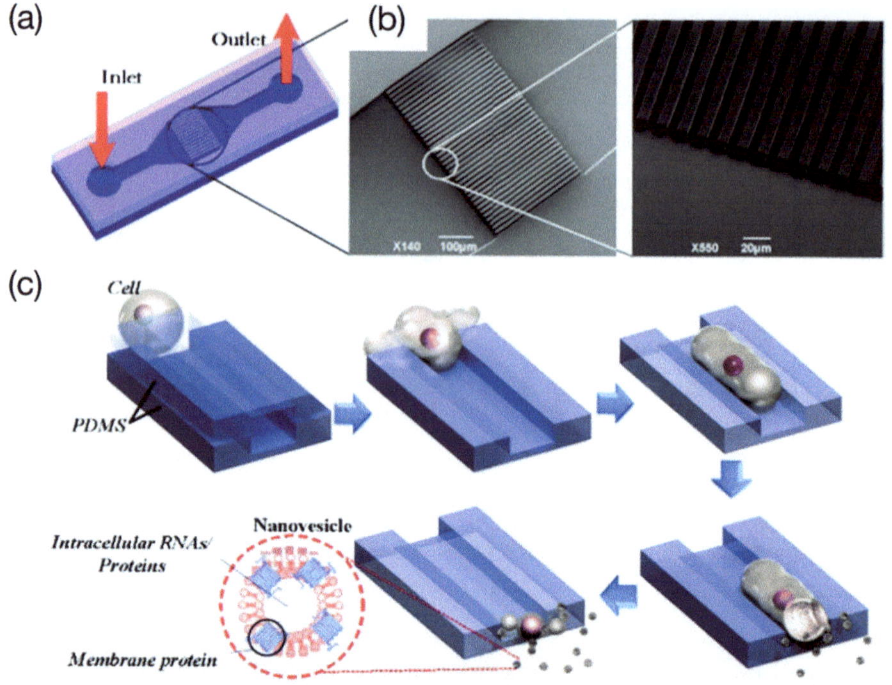

Fig. 5 **a** Schematic illustration and **b** SEM images of the microchannels. **c** The nanovesicle formation process involves introducing the cells into the microchannels. (Reprinted from Ref. Jo et al. (2014) with permission from the Royal Society of Chemistry.)

2 Summary

This chapter reviewed the potential of engineered and artificial exosomes as novel non-viral drug delivery nanocarriers and addressed the limitations of natural exosomes in clinical applications. In particular, diverse drug-loading techniques and production methods encompassing top-down, bottom-up, and biohybrid approaches were discussed. These approaches aim to combine the natural benefits of exosomes with the adaptability of synthetic carriers like LNPs to maximise their potential. In addition to the conventional liposome preparation method, various engineered or artificial exosome preparation methods have been developed using microfluidic technologies.

Advancements in bioengineering and nanotechnology are expected to improve the functionality, specificity, loading capacity, and biocompatibility of engineered exosomes. Understanding the biology of exosomes in detail will allow the development of next generation engineered exosomes to provide customised treatments for various diseases, thereby surpassing the constraints of current drug delivery systems.

Acknowledgements This work was supported by JST CREST (Grant No. JPMJCR17H1, Japan), JST PRESTO (Grant No. JPMJPR19K8, Japan), AMED (Grant No. 22zf0127004h0002), JSPS KAKENHI (Grant Nos. JP19KK0140, JP23H04049, JP23H03719, JP23H00541, and JP24H00038), the Special Education and Research Expenses from the Ministry of Education, Culture, Sports, Science and Technology, and grants provided by the Hokkaido University Support Program for Frontier Research.

References

Alvarez-Erviti L, Seow Y, Yin H, Betts C, Lakhal S and Wood MJ (2011) Delivery of siRNA to the mouse brain by systemic injection of targeted exosomes. Nat. Biotechnol. 29:341–345

Buschmann MD, Carrasco MJ, Alishetty S, Paige M, Alameh MG and Weissman D (2021) Nanomaterial Delivery Systems for mRNA Vaccines. Vaccines (Basel) 9:

Cheng L, Zhang X, Tang J, Lv Q and Liu J (2021) Gene-engineered exosomes-thermosensitive liposomes hybrid nanovesicles by the blockade of CD47 signal for combined photothermal therapy and cancer immunotherapy. Biomaterials 275:120964

Dai J, Su Y, Zhong S, Cong L, Liu B, Yang J, Tao Y, He Z, Chen C and Jiang Y (2020) Exosomes: key players in cancer and potential therapeutic strategy. Signal Transduct Target Ther 5:145

Darband SG, Mirza-Aghazadeh-Attari M, Kaviani M, Mihanfar A, Sadighparvar S, Yousefi B and Majidinia M (2018) Exosomes: natural nanoparticles as bio shuttles for RNAi delivery. J. Control. Release 289:158–170

Deng H, Sun C, Sun Y, Li H, Yang L, Wu D, Gao Q and Jiang X (2018) Lipid, Protein, and MicroRNA Composition Within Mesenchymal Stem Cell-Derived Exosomes. Cell Reprogram 20:178–186

Evers MJW, van de Wakker SI, de Groot EM, de Jong OG, Gitz-Francois JJJ, Seinen CS, Sluijter JPG, Schiffelers RM and Vader P (2022) Functional siRNA Delivery by Extracellular Vesicle-Liposome Hybrid Nanoparticles. Adv Healthc Mater 11:e2101202

Hassett KJ, Benenato KE, Jacquinet E, Lee A, Woods A, Yuzhakov O, Himansu S, Deterling J, Geilich BM, Ketova T, Mihai C, Lynn A, McFadyen I, Moore MJ, Senn JJ, Stanton MG, Almarsson O, Ciaramella G and Brito LA (2019) Optimization of Lipid Nanoparticles for Intramuscular Administration of mRNA Vaccines. Mol Ther Nucleic Acids 15:1–11

He J, Ren W, Wang W, Han W, Jiang L, Zhang D and Guo M (2022) Exosomal targeting and its potential clinical application. Drug Deliv Transl Res 12:2385–2402

Hoshino A, Costa-Silva B, Shen TL, Rodrigues G, Hashimoto A, Tesic Mark M, Molina H, Kohsaka S, Di Giannatale A, Ceder S, Singh S, Williams C, Soplop N, Uryu K, Pharmer L, King T, Bojmar L, Davies AE, Ararso Y, Zhang T, Zhang H, Hernandez J, Weiss JM, Dumont-Cole VD, Kramer K, Wexler LH, Narendran A, Schwartz GK, Healey JH, Sandstrom P, Labori KJ, Kure EH, Grandgenett PM, Hollingsworth MA, de Sousa M, Kaur S, Jain M, Mallya K, Batra SK, Jarnagin WR, Brady MS, Fodstad O, Muller V, Pantel K, Minn AJ, Bissell MJ, Garcia BA, Kang Y, Rajasekhar VK, Ghajar CM, Matei I, Peinado H, Bromberg J and Lyden D (2015) Tumour exosome integrins determine organotropic metastasis. Nature 527:329–335

Hu Y, Li X, Zhang Q, Gu Z, Luo Y, Guo J, Wang X, Jing Y, Chen X and Su J (2021) Exosome-guided bone targeted delivery of Antagomir-188 as an anabolic therapy for bone loss. Bioact Mater 6:2905–2913

Jhan YY, Prasca-Chamorro D, Palou Zuniga G, Moore DM, Arun Kumar S, Gaharwar AK and Bishop CJ (2020) Engineered extracellular vesicles with synthetic lipids via membrane fusion to establish efficient gene delivery. Int. J. Pharm. 573:118802

Ji K, Fan M, Huang D, Sun L, Li B, Xu R, Zhang J, Shao X and Chen Y (2022) Clodronate-nintedanib-loaded exosome-liposome hybridization enhances the liver fibrosis therapy by inhibiting Kupffer cell activity. Biomater Sci 10:702–713

Jo W, Jeong D, Kim J, Cho S, Jang SC, Han C, Kang JY, Gho YS and Park J (2014) Microfluidic fabrication of cell-derived nanovesicles as endogenous RNA carriers. Lab Chip 14:1261–1269

Jo W, Kim J, Yoon J, Jeong D, Cho S, Jeong H, Yoon YJ, Kim SC, Gho YS and Park J (2014) Large-scale generation of cell-derived nanovesicles. Nanoscale 6:12056–12064

Kang M, Jordan V, Blenkiron C and Chamley LW (2021) Biodistribution of extracellular vesicles following administration into animals: A systematic review. J Extracell Vesicles 10:e12085

Kang K, Zhang Y, Zhou X, Yu Y, Zhu N, Cheng J, Yi Q and Wu Y (2023) Hybrid Extracellular Vesicles-Liposomes Camouflaged Magnetic Vesicles Cooperating with Bioorthogonal Click Chemistry for High-Efficient Melanoma Circulating Tumor Cells Enrichment. Adv Healthc Mater 12:e2202825

Kannavou M, Marazioti A, Stathopoulos GT and Antimisiaris SG (2021) Engineered versus hybrid cellular vesicles as efficient drug delivery systems: a comparative study with brain targeted vesicles. Drug Delivery and Translational Research 11:547–565

Kim MS, Haney MJ, Zhao Y, Mahajan V, Deygen I, Klyachko NL, Inskoe E, Piroyan A, Sokolsky M, Okolie O, Hingtgen SD, Kabanov AV and Batrakova EV (2016) Development of exosome-encapsulated paclitaxel to overcome MDR in cancer cells. Nanomedicine 12:655–664

Lamichhane TN, Raiker RS and Jay SM (2015) Exogenous DNA Loading into Extracellular Vesicles via Electroporation is Size-Dependent and Enables Limited Gene Delivery. Mol. Pharm. 12:3650–3657

Li Y, Zhang Y, Li Z, Zhou K and Feng N (2019) Exosomes as Carriers for Antitumor Therapy. ACS Biomaterials Science & Engineering 5:4870–4881

Lin Y, Wu J, Gu W, Huang Y, Tong Z, Huang L and Tan J (2018) Exosome-Liposome Hybrid Nanoparticles Deliver CRISPR/Cas9 System in MSCs. Adv Sci (Weinh) 5:1700611

Liu A, Yang G, Liu Y and Liu T (2022) Research progress in membrane fusion-based hybrid exosomes for drug delivery systems. Front Bioeng Biotechnol 10:939441

Luan X, Sansanaphongpricha K, Myers I, Chen H, Yuan H and Sun D (2017) Engineering exosomes as refined biological nanoplatforms for drug delivery. Acta Pharmacol. Sin. 38:754–763

Ma Y, Zhang Y, Han R, Li Y, Zhai Y, Qian Z, Gu Y and Li S (2022) A cascade synergetic strategy induced by photothermal effect based on platelet exosome nanoparticles for tumor therapy. Biomaterials 282:121384

Maeki M, Kimura N, Sato Y, Harashima H and Tokeshi M (2018) Advances in microfluidics for lipid nanoparticles and extracellular vesicles and applications in drug delivery systems. Adv. Drug Del. Rev. 128:84–100

Maeki M, Uno S, Niwa A, Okada Y and Tokeshi M (2022) Microfluidic technologies and devices for lipid nanoparticle-based RNA delivery. J. Control. Release 344:80–96

Maeki M, Okada Y, Uno S, Sugiura K, Suzuki Y, Okuda K, Sato Y, Ando M, Yamazaki H, Takeuchi M, Ishida A, Tani H, Harashima H and Tokeshi M (2023) Mass production system for RNA-loaded lipid nanoparticles using piling up microfluidic devices. Applied Materials Today 31:101754

Maeki M, Uno S, Sugiura K, Sato Y, Fujioka Y, Ishida A, Ohba Y, Harashima H and Tokeshi M (2024) Development of Polymer-Lipid Hybrid Nanoparticles for Large-Sized Plasmid DNA Transfection. ACS Appl. Mater. Interfaces 16:2110–2119

Matsuura-Sawada Y, Uno S, Maeki M, Wada K and Tokeshi M (2022) Microfluidic Platform Enabling Efficient On-Device Preparation of Lipid Nanoparticles for Formulation Screening. ACS Applied Engineering Materials 1:278–286

Matsuura-Sawada Y, Maeki M, Uno S, Wada K and Tokeshi M (2023) Controlling lamellarity and physicochemical properties of liposomes prepared using a microfluidic device. Biomater Sci 11:2419–2426

Ortega A, Martinez-Arroyo O, Forner MJ and Cortes R (2020) Exosomes as Drug Delivery Systems: Endogenous Nanovehicles for Treatment of Systemic Lupus Erythematosus. Pharmaceutics 13:

Piffoux M, Silva AKA, Wilhelm C, Gazeau F and Tareste D (2018) Modification of Extracellular Vesicles by Fusion with Liposomes for the Design of Personalized Biogenic Drug Delivery Systems. ACS Nano 12:6830–6842

Raguraman R, Bhavsar D, Kim D, Ren X, Sikavitsas V, Munshi A and Ramesh R (2023) Tumor-targeted exosomes for delivery of anticancer drugs. Cancer Lett. 558:216093

Rayamajhi S, Nguyen TDT, Marasini R and Aryal S (2019) Macrophage-derived exosome-mimetic hybrid vesicles for tumor targeted drug delivery. Acta Biomater. 94:482–494

Sancho-Albero M, Medel-Martínez A and Martín-Duque P (2020) Use of exosomes as vectors to carry advanced therapies. RSC Adv 10:23975–23987

Sato YT, Umezaki K, Sawada S, Mukai SA, Sasaki Y, Harada N, Shiku H and Akiyoshi K (2016) Engineering hybrid exosomes by membrane fusion with liposomes. Sci. Rep. 6:21933

Sun L, Fan M, Huang D, Li B, Xu R, Gao F and Chen Y (2021) Clodronate-loaded liposomal and fibroblast-derived exosomal hybrid system for enhanced drug delivery to pulmonary fibrosis. Biomaterials 271:120761

Suzuki Y, Onuma H, Sato R, Sato Y, Hashiba A, Maeki M, Tokeshi M, Kayesh MEH, Kohara M, Tsukiyama-Kohara K and Harashima H (2021) Lipid nanoparticles loaded with ribonucleoprotein-oligonucleotide complexes synthesized using a microfluidic device exhibit robust genome editing and hepatitis B virus inhibition. J. Control. Release 330:61–71

Tang P, Hasan MR, Chemaitelly H, Yassine HM, Benslimane FM, Al Khatib HA, AlMukdad S, Coyle P, Ayoub HH, Al Kanaani Z, Al Kuwari E, Jeremijenko A, Kaleeckal AH, Latif AN, Shaik RM, Abdul Rahim HF, Nasrallah GK, Al Kuwari MG, Al Romaihi HE, Butt AA, Al-Thani MH, Al Khal A, Bertollini R and Abu-Raddad LJ (2021) BNT162b2 and mRNA-1273 COVID-19 vaccine effectiveness against the SARS-CoV-2 Delta variant in Qatar. Nat. Med. 27:2136–2143

Tsai SJ, Atai NA, Cacciottolo M, Nice J, Salehi A, Guo C, Sedgwick A, Kanagavelu S and Gould SJ (2021) Exosome-mediated mRNA delivery in vivo is safe and can be used to induce SARS-CoV-2 immunity. J. Biol. Chem. 297:101266

Valadi H, Ekstrom K, Bossios A, Sjostrand M, Lee JJ and Lotvall JO (2007) Exosome-mediated transfer of mRNAs and microRNAs is a novel mechanism of genetic exchange between cells. Nat. Cell Biol. 9:654–659

Wahlgren J, De LKT, Brisslert M, Vaziri Sani F, Telemo E, Sunnerhagen P and Valadi H (2012) Plasma exosomes can deliver exogenous short interfering RNA to monocytes and lymphocytes. Nucleic Acids Res. 40:e130

Wang Z, Popowski KD, Zhu D, de Juan Abad BL, Wang X, Liu M, Lutz H, De Naeyer N, DeMarco CT, Denny TN, Dinh PC, Li Z and Cheng K (2022) Exosomes decorated with a recombinant SARS-CoV-2 receptor-binding domain as an inhalable COVID-19 vaccine. Nat Biomed Eng 6:791–805

Webb C, Forbes N, Roces CB, Anderluzzi G, Lou G, Abraham S, Ingalls L, Marshall K, Leaver TJ, Watts JA, Aylott JW and Perrie Y (2020) Using microfluidics for scalable manufacturing of nanomedicines from bench to GMP: A case study using protein-loaded liposomes. Int. J. Pharm. 582:119266

Zhang Y, Bi J, Huang J, Tang Y, Du S and Li P (2020) Exosome: A Review of Its Classification, Isolation Techniques, Storage, Diagnostic and Targeted Therapy Applications. Int J Nanomedicine 15:6917–6934

Zhou X, Miao Y, Wang Y, He S, Guo L, Mao J, Chen M, Yang Y, Zhang X and Gan Y (2022) Tumour-derived extracellular vesicle membrane hybrid lipid nanovesicles enhance siRNA delivery by tumour-homing and intracellular freeway transportation. J Extracell Vesicles 11:e12198

Zhu Q, Heon M, Zhao Z and He M (2018) Microfluidic engineering of exosomes: editing cellular messages for precision therapeutics. Lab Chip 18:1690–1703

MIX
Papier aus verantwortungsvollen Quellen
Paper from responsible sources
FSC® C105338

If you have any concerns about our products,
you can contact us on
ProductSafety@springernature.com

In case Publisher is established outside the EU,
the EU authorized representative is:
Springer Nature Customer Service Center GmbH
Europaplatz 3, 69115 Heidelberg, Germany

Printed by Libri Plureos GmbH
in Hamburg, Germany